HANDBOOK FOR THE LIGHT MICROSCOPE

HANDBOOK FOR THE LIGHT MICROSCOPE

A User's Guide

by

James H. Richardson

The Aerospace Corporation
El Segundo, California

np

NOYES PUBLICATIONS
Park Ridge, New Jersey, U.S.A.

Library of Congress Catalog Card Number: 90-27389
ISBN: 0-8155-1269-4
Printed in the United States

Published in the United States of America by
Noyes Publications
Mill Road, Park Ridge, New Jersey 07656

10 9 8 7 6 5 4 3 2 1

Library of Congress Cataloging-in-Publication Data

Richardson, James H. (James Howard), 1933-
Handbook for the light microscope : a user's guide / by James H. Richardson.
p. cm.
Includes bibliographical references and index.
ISBN 0-8155-1269-4 :
1. Microscope and microscopy--Handbooks, manuals, etc. I. Title.
QH205.2R53 1991
502'.8'2--dc20 90-27389
CIP

Dedicated to Him

Whose very nature is light,
I John 1:5

Who is the source of all light,
Genesis 1:3

Who is the maker of all that light reveals,
Hebrews 1:3

And Who brings life and immortality to light through His gospel
II Timothy 1:10

Preface

After visiting numerous industrial laboratories throughout the United States, I am convinced that the optical microscope rivals that venerable musical instrument, the oboe, which was humorously described as "an ill wind that nobody blows well." In stark contrast to its acknowledged importance in showing us the invisible, the optical microscope is generally used in a casual and often incorrect fashion that can only assure us of incorrect information, increased production losses and, at the very least, eye strain for the microscopist.

One might assume that improper use of the microscope could be minimized through use of the instruction manual provided with each instrument or through appropriate training courses. This does not usually happen. The instruction manuals for a given instrument, although accurate, are usually terse and deal only with the mechanics of set-up. Furthermore, the manuals are almost never retained with the microscope beyond a small fraction of the instrument's useful life. The information contained in these manuals is seldom of any value for use with any other instrument.

With regard to training courses, the outlook is little better. A few universities and organizations, such as the McCrone Institute, are providing good training in optical microscopy, but train only a fraction of the beginning microscopists, leaving without help many who are already struggling daily with the instrument.

It is the intent of the author to combine sufficient information in this handbook so that it can serve as an aid to any microscopist from the student to the seasoned veteran. Furthermore, every attempt is made to make the book useful for a wide range of applications such as quality control, biology, medicine, industry, electronics, metallurgy, and other material sciences. The manual should also serve as textbook for a comprehensive course on optical microscopy.

This Handbook of Microscopy contains relatively complete and concise information on how to use a wide variety of optical microscopes correctly and easily. A complete description of each type of optical microscope and its elements is to be found at the beginning of each section. The applications for that particular microscope are also suggested. Where brief discussions of the theory of operation are required, these are inserted in such a way that they will not detract from the efficient use of the handbook in the laboratory. Each discussion continues with a detailed set-up procedure which, as much as possible, is suitable for almost all manufacturers' instruments. Should a problem develop, a listing of causes and their corrections completes the discussion.

Microscopes for special applications are presented, with such topics as long distance, remote, comparison, double view, projection and shop microscopes discussed.

Many of the accessories for the optical microscope are designed to improve contrast of the image. Each of these accessories is discussed in sufficient detail to permit its intelligent use. Again each section on a contrasting accessory is closed with a comprehensive procedure for its set-up.

Other accessories for the light microscope are designed to extend its capability for special studies. A comprehensive chapter categorizes and discusses these unique and often complex accessories.

The accessories for photomicrography and photomacrography are described in separate chapters. Photographic media are discussed and tables for the selection of suitable film materials, including the instant formats, are presented. The matter of correct illumination is crucial for a professional photomicrograph or photomacrograph, thus considerable space is allotted to the details of the correct set-up of the instrument and the illumination of the sample.

The first part of the final chapter deals with the selection of a microscope for a given application. The remaining portion of this chapter deals with the care and maintenance of the instrument; thus assuring that it will continue to give years of reliable service.

Finally a word of thanks to the manufacturers and their representatives that provided numerous photographs and drawings. Special thanks are offered to Mr. Walter Gumpertz, Mr. Jack Reinhart, Mr. Mark Richardson, and Ms. Linda Wagner who read and commented on selected chapters. Greatest thanks are due to my wife, Ann, who patiently endured the writing of the book.

NOTICE

Contents

1. **VISION AND MICROSCOPICAL AIDS** 1
   - I. **Introduction** 1
   - II. **The Simple Microscope** 3
   - III. **The Compound Microscope** 5
   - IV. **The Stereo Microscope** 10
   - V. **The Light Scanning Microscope** 21
   - **References** 30

2. **BRIGHTFIELD TRANSMITTED LIGHT MICROSCOPY** 31
   - I. **Introduction** 31
   - II. **The Stand** 31
     - A. The Upright Stand 31
     - B. The Inverted Stand 38
     - C. Upright vs Inverted Stands 40
     - D. The Culpeper Stand 40
   - III. **The Optical System** 42
     - A. Illumination 42
     - B. The Condenser System 48
     - C. The Objective Lens 51
       1. Primary Objective Lens Design Parameters 52
          - Immersion Media 54
          - Magnification 56
          - Spherical Aberration 57
          - Cover Glass 59
          - Axial Chromatic Aberration 61
          - Lateral Chromatic Aberration 61
          - Astigmatism 63
          - Curvature of Field 65

Coma . . . 65
Distortion . . . 65
2. Other Objective Lens Design Considerations . . . 67
Parfocality . . . 67
Working Distance . . . 67
Tube Length . . . 67
3. Optical Parameters Controlled by the Microscopist . . 69
Lateral Resolution . . . 69
Depth of Field . . . 73
Depth of Focus . . . 74
Axial Resolution . . . 74
Image Brightness . . . 75
4. Types of Objective Lenses . . . 77
Achromatic Objective Lens . . . 77
Semi-Apochromatic Objective Lens . . . 77
Apochromatic Objective Lens . . . 77
Plano Objective Lens . . . 78
Strain Free Objective Lens . . . 80
D. Intermediate Optics . . . 80
1. Corrector Lens . . . 80
2. Telan Lenses . . . 80
3. Tube Lens . . . 81
4. Zoom System . . . 81
5. Bertrand Lens . . . 81
E. The Eyepiece . . . 83
1. The Huygens Eyepiece . . . 87
2. The Ramsden Eyepiece . . . 90
3. The Compensating Eyepiece . . . 90
4. The Fixed Reticle Eyepiece . . . 90
5. The Widefield Eyepiece . . . 93
IV. Procedures for Brightfield Transmitted Light Microscopy 93
A. Instrumentation Location . . . 93
B. Alignment of an External Illuminator . . . 93
C. Set-Up of a Brightfield Transmitted Light Microscope . . 96
D. Adjustment to Minimize Spherical Aberration . . . 102
E. Problem Correction for Transmitted Light Microscopy . 103
References . . . 109

3. BRIGHTFIELD INCIDENT LIGHT MICROSCOPY . . . 110
I. Introduction . . . 110
II. The Stand . . . 110
A. The Upright Stand . . . 110
B. The Inverted Stand . . . 113
C. Upright vs Inverted Stand . . . 113

D. The Culpeper Stand 117

III. The Optical System 117

A. Illumination 117

B. The Condenser System 119

C. The Objective Lens 122

1. Lack of Cover Glass Correction 122
2. Working Distance 122
3. Plano Corrected Objective Lenses 123
4. Radiation Resistant Objective Lenses 123

D. Viewing Screen 123

IV. Procedures for Brightfield Incident Light Microscopy . . 124

A. Instrument Location 124

B. Alignment of the Incident Light Condenser System . . . 124

C. Setup of a Brightfield Incident Light Microscope 125

D. Problem Correction for Incident Light Microscopy 128

Selected Reading 129

4. ENHANCEMENT OF CONTRAST 130

I. Introduction 130

II. Oblique Illumination Microscopy 134

A. General 134

B. Procedure for Transmitted, Oblique Illumination 135

C. Procedure for Incident, Oblique Illumination 136

D. Problem Correction for Oblique Illumination 138

III. Darkfield Microscopy 140

A. General 140

B. Procedure for Transmitted, Darkfield Illumination 140

C. Procedure for Incident, Darkfield Illumination 146

D. Problem Correction for Darkfield Illumination 147

IV. Phase Contrast Microscopy 147

A. General 147

B. Transmitted Light Applications 153

C. Incident Light Applications 161

D. Problem Correction for Phase Contrast Illumination . . . 166

V. Differential Interference Contrast Microscopy 166

A. General 166

B. Transmitted Light Applications 169

C. Incident Light Applications 174

D. General Comments on the Set-Up of a DIC Microscope 178

E. Procedure for Transmitted, DIC Illumination 178

F. Problem Correction for Transmitted, DIC Microscopy . 179

G. Procedure for Incident, DIC Illumination 179

H. Problem Correction for Incident, DIC Microscopy 181

VI. Polarized Light Contrast Microscopy 181

A. General 181
B. Construction of Simple Polarizing Accessories 183
C. Transmitted, Polarized Illumination 184
D. Incident, Polarized Illumination 194
E. Problem Correction for Polarized Light Microscopy . . . 198
VII. Modulation Contrast Microscopy 198
A. General 198
B. Procedure for Modulation Contrast Microscopy 202
C. Problem Correction for Modulation Contrast Microscopy 204
VIII. Amplitude Contrast Microscopy 204
A. General 204
B. Applications 205
C. Procedure for Amplitude Contrast Illumination 206
IX. Fluorescence Microscopy 206
A. General 206
B. Applications 206
C. Fluorescence Instrumentation 207
D. Alignment of the Mercury Vapor Arc Illuminator 212
E. Procedure for Transmitted, Fluorescence Illumination . 213
F. Procedure for Incident, Fluorescence Illumination 214
G. Problem Correction for Fluorescence Microscopy 214
X. Video Enhanced Contrast 215
References 219

5. QUANTITATIVE MICROSCOPY 221
I. Introduction 221
II. Transverse Measurements 222
A. Length Measurements 222
B. Precision and Accuracy of Length Measurements 228
C. Angular Measurements 230
D. Particle Number 230
III. Axial Measurements 233
A. Fine Focus Axial Measurements 235
B. Light Sectioning Axial Measurements 237
C. Two Beam Interferometry 243
D. Procedures for the Incident Light, Two Beam Interferometric Microscopes 250
E. Multiple Beam Interferometry 254
IV. Derived Linear Measurements 257
A. Grain Size 258
B. Volume Fraction 265
C. Particle Spacings 268
D. Computerized Image Analysis 268
V. Crystal Property Measurements 270

A. Optical Crystallography 270
B. Optical Petrography 274
C. Procedure for Petrographic Microscopy 276
D. Problem Correction for Conoscopic Observation 284
E. Dispersion Staining 285
F. Procedure for Dispersion Staining Microscopy 286
G. The McCrone Particle Atlas 290
H. Procedure for Particle Atlas Examinations 291
VI. Microphotometry 291
A. General 291
B. Procedure for Microphotometry 293
References 293

6. SPECIAL MICROSCOPES 296
I. Introduction 296
II. Portable Microscopes 296
III. Long Distance Microscopes 298
IV. Remote Microscopes 303
V. Comparison Microscopes 307
A. General 307
B. Procedure for Comparison Microscopy 309
C. Mask and Wafer Alignment Microscopes 310
VI. Double View Microscopes 311
VII. Hole Inspection Microscopes 311
VIII. Projection Microscopes 317
IX. Shop Microscopes 317
References 322

7. ACCESSORIES 323
I. Introduction 323
II. Accessories for the Stand 324
A. Vibration Isolation 324
B. Powered Focusing Accessories 324
C. Multiple Viewing Accessories 325
D. Miscellaneous 331
III. Illumination Accessories 331
IV. Stage Accessories 333
A. Sample Translation Stages 334
B. Stages for Quantitative Analysis 336
C. Sample Orientation Stages 338
D. Micromanipulators and Microinjectors 342
E. Hot Stages 343
F. Cold Stage 347
G. Stages for Microdetermination of Material Properties . . 348

H. Special Environmental Stages 351
I. Miscellaneous 353
V. Objective Lens Accessories 353
A. Microhardness Testers 355
B. Slide Marker 357
VI. Filters 358
A. Neutral Density Filters 358
B. Color Filters 359
C. Heat Absorbing Filters 363
D. Ultraviolet Absorbing Filters 363
E. Filters for Fluorescence Microscopy 365
F. Polarizing Filters 365
G. Polarizing Microscope Accessories 367
H. Diffusing Screens 368
VII. Eyepiece Accessories 368
A. Screw Micrometer Eyepiece 369
B. The Image Shearing Eyepiece 374
C. Comparison Eyepiece 374
D. Eyepiece Telescope 375
E. Pointer Eyepiece 375
F. Setting Eyepiece 380
G. Pinhole Cap 380
H. Protractor Eyepiece 381
I. Wright Eyepiece 381
J. Soleil-Babinet Compensator 381
References 381

8. PHOTOMICROGRAPHY 385
I. Introduction 385
II. Microscope Considerations 386
III. The Camera 390
A. The Integral Lens Camera 392
B. The Camera Without an Integral Lens 393
IV. Photographic Media 399
A. Wet Process Media 399
B. Instant Process Media 404
C. Suggested Photomicrographic Media 410
V. Preparation of a Photomicrograph 410
A. Framing and Composition 410
B. Magnification 413
C. Flatness of Field 414
D. Focusing 415
E. Determination of Exposure Time 419
F. Correction of Illumination for Color Film 424

G. Contrast Enhancement for Black and White Film 426
H. Photomicrographic Procedures 426
1. Fixed Bellows Camera Procedure 427
2. Flexible Bellows Camera Procedure 429
3. Cine Photomicrography 431
4. Problems 435
References 437

9. PHOTOMACROGRAPHY 438
I. Introduction 438
II. Photomacrographic Instrumentation 438
III. Preparation of a Photomacrograph 445
A. General 445
B. Framing and Composition 445
C. Magnification 445
D. Depth of Field 449
E. Exposure Determination 450
F. Filters 450
IV. Photomacrographic Procedures 450
A. Transmitted, Brightfield Photomacrography 450
B. Transmitted Darkfield Illumination 456
C. Incident Brightfield Illumination 461
D. Incident Darkfield Illumination 469
E. Stereoscopic Photomacrography 476
F. Problems 481
References 481

10. SELECTION, CARE AND MAINTENANCE OF THE OPTICAL MICROSCOPE 482
I. Introduction 482
II. Selection of a Microscope 482
III. The Microscope's Environment 483
IV. Ergonomics 487
V. Testing of a Microscope 488
VI. Operator Care and Maintenance 497
A. The Microscope Stand 497
B. Care of Objective Lenses and Eyepieces 500
C. Installation of Lamps 502
VII. Professional Maintenance 510
References 512

APPENDIX I–IMMERSION FLUIDS 513

APPENDIX II–LONG WORKING DISTANCE LENSES (>10X) 514

INDEX 515

1

Vision and Microscopical Aids

I. INTRODUCTION

Sight is, without a doubt, the grandest of our personal senses. With it we can rapidly perceive the number, size, shape, texture, color and position of those objects around us. In the last few hundred years we have further extended the capability of our eyes to the very remote with telescopes and to the very small with microscopes. It is the methodology of this latter excursion into the "micro" world, specifically with the light microscope, that is the topic of this handbook.

To fully appreciate the microscope, it is important to first consider the eye in its role as the recording device in the viewing process. In this respect, the eye may be considered as analogous to a simple camera with the lens in front and the light sensitive retina covering the interior rear portion of the eye.

The lens system of the eye consists of the outer, transparent cornea, the iris which changes the pupil diameter and the lens proper. The unique aspect of the eye's lens in the world of light optics is that its focal length can be varied, within limits, by the involuntary nervous system. Thus, the eye can focus on a near or distant object without repositioning the lens relative to the retina. When the range of focus includes only distant objects, this condition is known as farsightedness or hyperopia. Conversely, when the range of focus includes only near objects, this is referred to as nearsightedness or myopia. Any deformity of the curvature of the cornea distorts the image; this produces astigmatism.

The retina is the surface in the rear of the eye upon which the lens focuses the image. This surface is composed of two types of light sensors, the cones which give color vision and rods which are very sensitive at low levels of illumination but which provide only black and white images. The most sensitive portion of the retina is the fovea, a small area about 1.5 mm in diameter. The center of the fovea contains densely packed cone sensors; beyond this area the number of cones per unit area decreases and the proportion of rod sensors increases. The total number of both rods and cones gradually decrease as the edge of the retina is approached.

At levels of illumination of 10^{-4} mL (milli-lamberts), vision is accomplished using only the rods; this corresponds roughly to starlight. At this low light level the eye is unable to distinguish color. As the level of illumination increases above 10^{-4} mL, vision is increasingly due to the cones. Since almost all microscopy is accomplished at illumination levels well above 10^{-4} mL, the eye makes use of the fovea with its high density of cones, thus providing fine detail and color vision.

The rods and cones produce a mosaic structure on the light sensitive surface of the retina that determines the resolution of the eye. In general, two small dark spots can be recognized as individuals if their images involve two cones; this corresponds to a spacing of about 70 micrometers at a viewing distance of 25 cm. Another way of expressing this is that the average eye resolves the two dark objects when they subtend about one minute of arc at the eye. The resolution of the eye decreases linearly as the image moves away from the fovea.

The foregoing information would seem to indicate that if we have sufficient illumination and resolution we should be able to distinguish any two objects as individuals. However, an additional parameter, contrast, must be considered. If your sample should be akin to the humorous illustration of "a polar bear eating marshmallows in a snowstorm" you would not expect to see detail even though optimum illumination and resolution were available.

Contrast in the sample's image results from changes in color, opacity, and refractive index for transmitted light observations and from changes in color and reflectivity in the incident light observations. To maximize contrast, the microscopist must preclude the loss of light through misadjustment and introduction of stray light or glare. Where contrast is lacking, a number of techniques, discussed in Chapter 4, are available to dramatically enhance contrast.

Up to this point we have considered each eye as an independent optical system, however working together, they provide us with an additional capability. This capability, stereoscopic vision, is the perception of the three dimensional character of an object or the spatial relationship of two or more objects. Stereoscopic vision arises from the brain's interpretation of the slightly different images seen by the two eyes. The normal unaided eye is capable of noting surface irregularities on the order of 0.1-0.2 mm at a distance of 250 mm.

Now let us consider the need for magnification. We learn from experience that the closer we get an object to the eye the larger it appears. The practical minimum distance between the object and the eye is however controlled by our eye's ability to focus the object clearly. This limitation has been overcome by the use of the microscope. The remainder of this chapter will describe the broad classes of light microscopes that will be covered in detail in the remainder of the book.

One last word is in order before we delve into the technology of the light microscope. We may have eyes that are adequate for normal vision and an excellent microscope, but one additional ingredient is required for successful microscopical examinations, that is, "seeing". Seeing is that learned ability of the observer to obtain useful information from a sample that results from his or her interest in the sample, experience with microscopical examination and his or her facility to make meaningful interpretations from the images in the microscope. One should also realize that numerous factors both internal, such as physiological or psychological conditions, and external, such as illumination and furniture ergonomics can affect the quality of our seeing at any one time.

Optical microscopes can be conveniently divided into several broad categories, the simple microscope, the compound microscope, the stereo microscope, and the confocal microscope.

II. THE SIMPLE MICROSCOPE

As mentioned above, it is necessary when observing a sample more closely, to hold it closer to the eye. In this way we increase the angle subtended by the details of the sample and increase the ability of the eye to resolve the details. This procedure may be used until the sample is brought so close to the eye that the eye can no longer focus sharply on the sample. The only way that further improvement of visual resolution may then be realized is through the use of a microscope.

The origins of the simple microscope probably date back to the ancient Assyrians since crude glass lenses were found in the ruins of Nineveh which was destroyed in 612 B.C. Antoni van Leeuwenhoek developed a number of simple microscopes with magnifications as high as 200X as late as the 17th century; an example of one of his microscopes is shown in Figure 1-1.

The simple microscope consists of a short-focus, positive (magnifying) lens or group of single lenses acting as a single lens. When a small object is placed in the focal plane of a short focus lens held in front of the eye, a magnified virtual[1] image of the

---

[1]This image does not actually exist and cannot be observed on a screen, thus it is referred to as a virtual image.

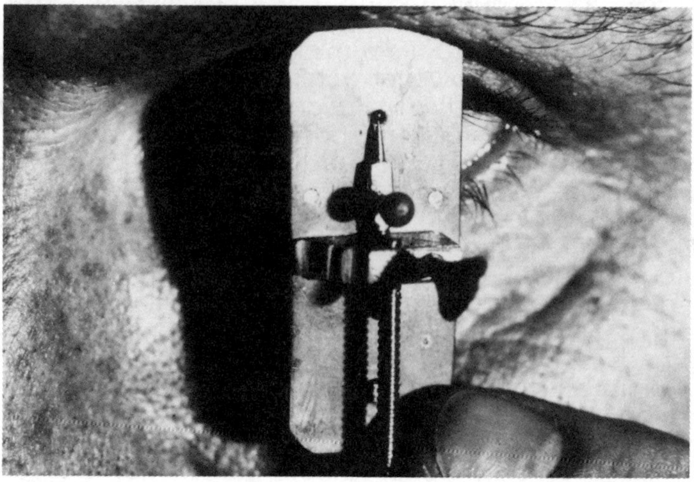

Figure 1-1. Although the compound microscope was available at that time, Antoni Van Leeuwenhoek chose to use simple microscopes in his many observations. This illustration of a genuine Leeuwenhoek microscope shows the proper placement for an observation. The small single lenses of various diameters gave magnifications ranging from 30 to 260X. Courtesy of B. Bracegirdle of The Science Museum London and the University of Utrecht.

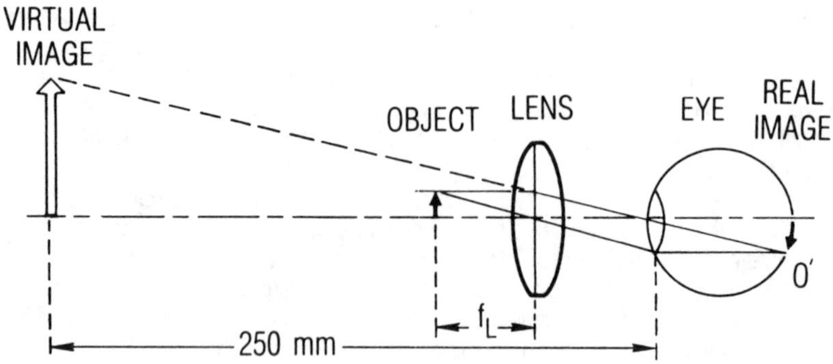

Figure 1-2. Ray diagram of a simple microscope. The object is imaged on the retina of the eye by the lens. The image appears to the eye as if it were from an object at the position of the virtual image.

object is seen. Figure 1-2 shows the effect of this lens on the image of the object. The magnification of the simple microscope, M_{sm} is given by the following expression:

$$M_{sm} = 250\,v\,f \qquad (1)$$

where f is the focal length of the simple lens in millimeters. The magnification ranges of the various simple microscopes are shown in Table 1-1.

The mounting of the modern simple microscope ranges from the support handle of the familiar "reading glass" to tripod support of the "linen tester" to the folding cover of the "hand lens". The bench magnifier has a "no-hands" support that has proven so useful for manufacturing inspection stations. Some versions of these simple magnifiers have built-in illumination sources. Several of these simple microscopes are shown in Figure 1-3.

The proper method for using the hand magnifier or hand lens is to hold it as close as practical to the eye. The object is then focused by adjusting one's position relative to a fixed object or manually moving a smaller object relative to the lens.

The bench magnifier should be adjusted so that it is close to the eyes and with sufficient space behind it so that an object may be held at the correct focus position.

TABLE 1-1. MAGNIFICATION OF SIMPLE MICROSCOPES

TYPE	MAGNIFICATION
Reading Glass	About 3X
Hand Lens	5 - 20X
Bench Magnifier	5 - 10X
Linen Tester	5 - 10X

The linen tester is usually placed directly on the object and spaced by the tripod support. While magnification occurs with the lens closer to the object than to the eye it is not the maximum possible.

III. THE COMPOUND MICROSCOPE

Although many significant studies were performed using the simple microscope, this instrument is severely limited in several ways. At higher magnifications the lens must be very near the sample and even worse, the eye must also be very near the opposite side of the lens to achieve any semblance of an acceptable field of view. Manipulation of the sample is severely hampered by the proximity of the lens and the observer's head. The introduction of the compound optical microscope such as shown in Figure 1-4, has overcome many of these limitations of the simpler instrument.

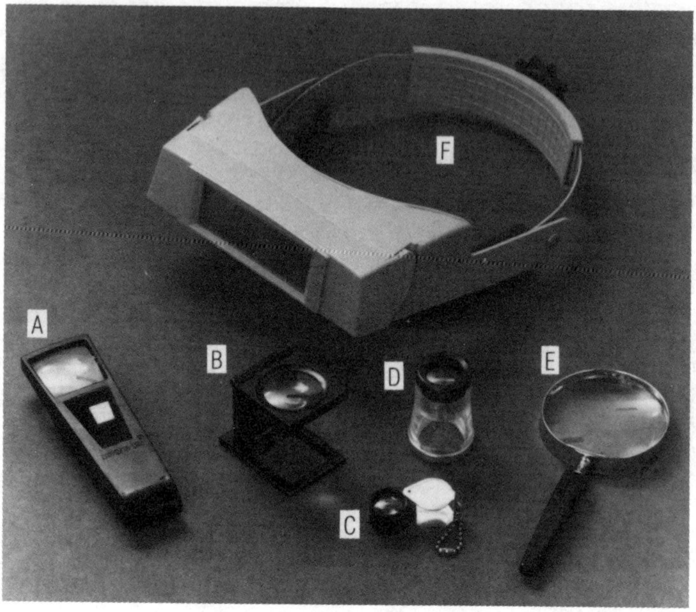

Figure 1-3. Several versions of simple microscopes: A. Self illuminating hand magnifier, B. Linen tester, C. Hastings triplet magnifier, D. Measuring magnifier with interchangeable scales, E. Hand magnifier and F. Visor type stereo magnifier. The two lenses of the latter instrument are so arranged that stereo vision is maintained.

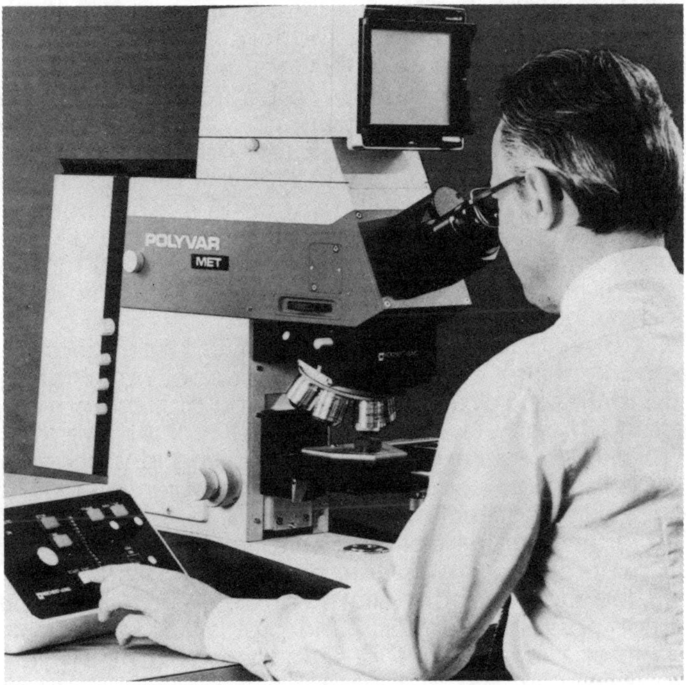

Figure 1-4. A modern, research compound optical microscope. Courtesy of Cambridge Instruments Inc., Research Instruments Division.

The compound microscope in its basic form consists of a lens or group of lenses known as the objective lens and a second group of lenses known as the eyepiece. The function of these lenses is shown in Figure 1-5. The light from the illuminator is focused by the condenser onto the sample, labeled PQ. The objective lens produces an image of the sample normally at P_1Q_1, however because of the interaction of the lower lens of the eyepiece this image is displaced to $P_{1'}Q_{1'}$. The upper or eye lens of the eyepiece, in combination with the lens of the eye focuses the sample image on the retina of the eye at P_2Q_2. This information, however is perceived by the eye to be from an enlarged image of the sample, P_xQ_x at a distance of about 250 mm from the eye. This image is a virtual image and is reversed in appearance from the actual object therefore, any movement of the sample on the stage causes a movement of the image in the opposite direction. A corresponding real image that can be observed on a screen is projected from the eyepiece at a distance of 250 mm from the eyepiece; it is this real image that is used in photomicrography, as described in Chapter 8.

Two developments in optics have been crucial to the improvement of the compound microscope, culminating in the splendid instruments available today. One of these developments is the effort of many to minimize various optical aberrations and distortions in microscope lenses. The resulting lens corrections are so vital to a thorough understanding of the objective lens that they are discussed in Chapter 2, Section III C.

The removal of the optical aberrations in a lens system usually requires a number of glass elements. This stacking of lenses achieves the desired corrections, however, at the expense of image brightness and contrast. The problem is the loss of light at the multiple air to glass interfaces. In the case of a single air-glass interface, for a glass having a refractive index of 1.500, the reflected light amounts to about 4% of the transmitted light. Thus, for a two element, air spaced lens the light loss is 15%. It is not uncommon for a present day objective lens to have 20 air-glass interfaces, therefore, to minimize the resulting reflection losses, all modern microscope lenses have anti-reflection coatings.

The anti-reflection coatings consist of from one to three thin, uniform layers of material(s) with an overall, lower index of refraction than the glass. When the thickness of the coating is one quarter wavelength, the reflection is almost zero. Since the coating can be quarter wavelength for only one color, the thickness is usually adjusted for optimum in the green-yellow region of the spectrum. The small amount of light that is reflected gives rise to the purplish appearance of coated lenses. The magnesium fluoride coatings are the most durable and hence are used on the exterior lens surfaces. The more efficient multilayer coatings are reserved for the inner lens surfaces; these are usually fragile and should only be cleaned by professional service personnel.

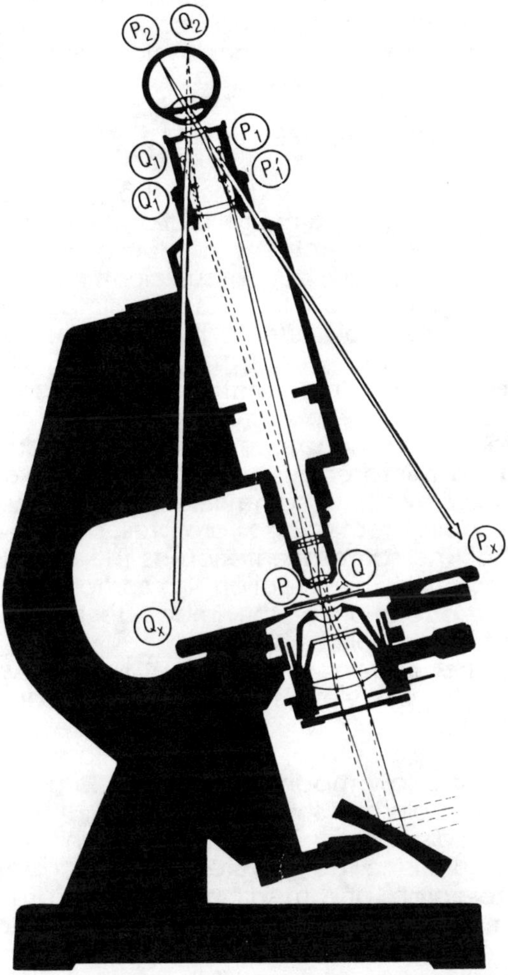

Figure 1-5. Optical arrangement of a transmitted light, compound microscope. The callouts are described in the text. Courtesy of Wild Leitz USA, Inc.

In summary, the compound microscope is an extremely versatile instrument and is vital to the success of the biomedical, physical science and engineering fields. It can be configured for transmitted or incident illumination and can be augmented with a wide range of contrast enhancing accessories. Chapters 2, 3 and 4 deal specifically with the compound microscope in all of its various forms.

IV. THE STEREO MICROSCOPE

Stereo microscopy extends the perception of three-dimensionality to objects at magnifications from about 2X to as high as 250X. To perceive an object in three dimensions with a stereo microscope it is necessary, but not sufficient, that both eyes function properly since it is the brain that fuses the images from the eyes to produce the stereoscopic effect. There is a reasonable number of persons who can see with both eyes that do not perceive stereoscopic images. Unfortunately, these persons cannot derive the full benefit of the stereo microscope.

The stereo microscope, or binocular microscope, consists of two compound microscopes arranged so that each eye has its own individual view of the same field. Through the use of prisms in each microscope, erect images are presented to the eyes, that is the image has the same orientation as the sample. Furthermore, the image moves in the direction the sample moves. This feature permits easy correlation of the image to the sample and easy positioning of the sample.

Since the stereo microscope is a low magnification instrument the field of view is generally large, ranging from about 88 mm diameter at 2.4X to 1.6 mm at 160X. As a result of the low magnifications the depth of field is quite large. Another feature resulting from the low magnifications is a large working distance between the sample and the objective lens. It is not uncommon to have a working distance of 180 mm at a magnification of 40X.

It is not surprising that the stereo microscope is the workhorse of industry, research, and medicine because of the erect image, the broad magnification, the large depth of field and the large working distance. It finds extensive application in the area of microelectronics where assembly cannot be accomplished without such an instrument. Indeed, the stereo microscope is generally an intergral part of each specialized assembly and test instrument for this field of endeavor. The stereo microscope also finds application in the initial stages of most other types of microscopy, especially electron microscopy, where a more detailed examination of a sample may be achieved than with the unaided eye, and small areas may be located and excised for further examination directly under the stereo microscope. Many delicate surgical procedures require the use of the stereo microscope.

The many applications of the stereo microscope has given rise to a wide variety of stands as shown in Figure 1-6. The common fixed stand usually provides for samples that are less than about 100 mm in all dimensions. The coarse position of the stand is usually established with a thumb nut that locks the microscope at a fixed position on the column of the stand. Since these instruments have a relatively low magnification, adequate focusing is possible with a single focus knob. If larger samples are encountered, the table boom stand provides more working area and height capability. For extreme flexibility a wall mount, a ceiling mount or a mobile floor stand are available. The mobile floor stand is important as the basis for surgical microscopes.

Illumination for the stereo microscope may range from the available ambient lighting to one or more external illuminators to built-in illuminators. The easiest and most common form of lighting for the stereo microscopy of bulk and opaque samples is incident, darkfield or oblique illumination. Figure 1-7 illustrates a stereo microscope with incident darkfield illumination; one or more non-axial illuminators is used to illuminate a sample. Figure 1-8 shows several of the more important incident light illuminators. The number and location of the illuminators is dictated by the information required from the sample.

If a sample has structure on a specular surface or has deep crevices in its surface the illumination should be incident, brightfield or normal incident illumination, i.e., illumination that is coincident with the viewing direction. This type of illumination can be achieved with an external mirror which provides a beam of light midway between the two viewing directions or a built-in coaxial illuminator which directs beams of light along both viewing axes. A stereo microscope with the latter type of illuminator is shown in Figure 1-9. Incident, brightfield illumination is commonly used for evaluation of semiconductor and integrated circuits.

Thin sections, whether of biomedical or geological origin are best examined in transmitted, brightfield illumination. For optimum transmitted, brightfield illumination, a special stand with interchangeable condenser lenses that are matched to each magnification range should be used. This type illumination is shown in Figure 1-10. Less critical, transmitted, brightfield observations can be accomplished using a simpler stand with a substage mirror. For thin sections having little or no brightness or color contrast, transmitted, darkfield illumination can be used effectively to delineate the structure; this type of illumination is shown in Figure 1-11.

As a final comment in the discussion of illumination for stereo microscopy, it is often necessary to use more than one of the lighting techniques mentioned above to achieve the optimum illumination of the sample.

The optical arrangement of the stereo microscope takes one of two forms, one with totally independent optics for each of the two individual microscopes, known as the Greenough system,

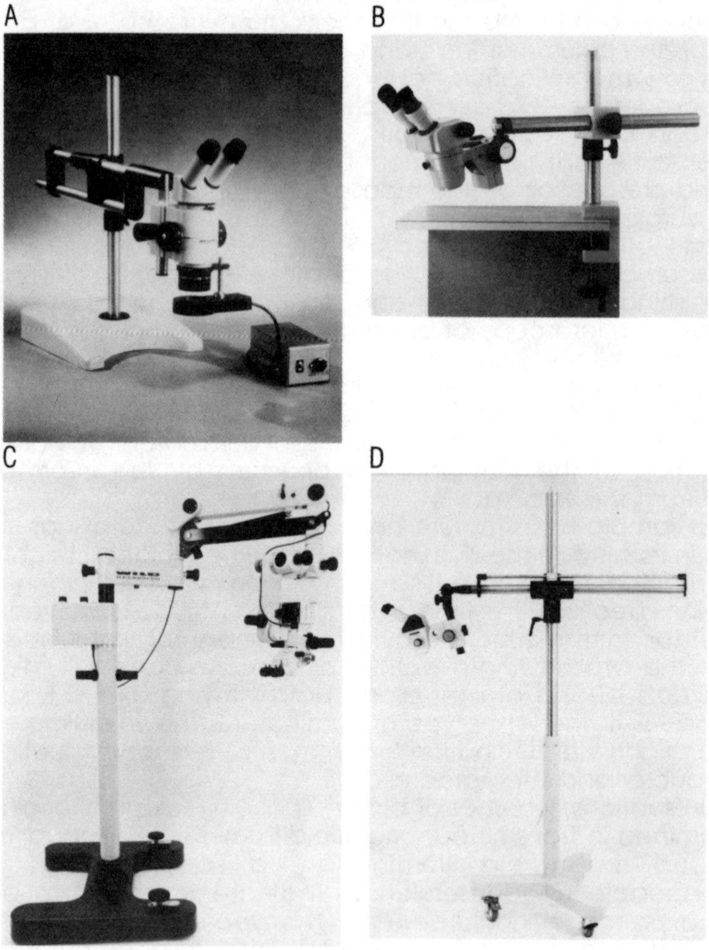

Figure 1-6. A variety of stands for use with stereo microscopes. (A) is a table top, boom stand. The boom allows the microscope body to be moved away from the support so that the central area of large samples may be examined. (B) is a boom stand that is clamped on the edge of the table. (C) is a medical/industrial type floor stand. (D) is a floor stand with boom. (A) and (C) courtesy of Wild Leitz USA, Inc. and (B) and (D) courtesy of Diagnostic Instruments, Inc.

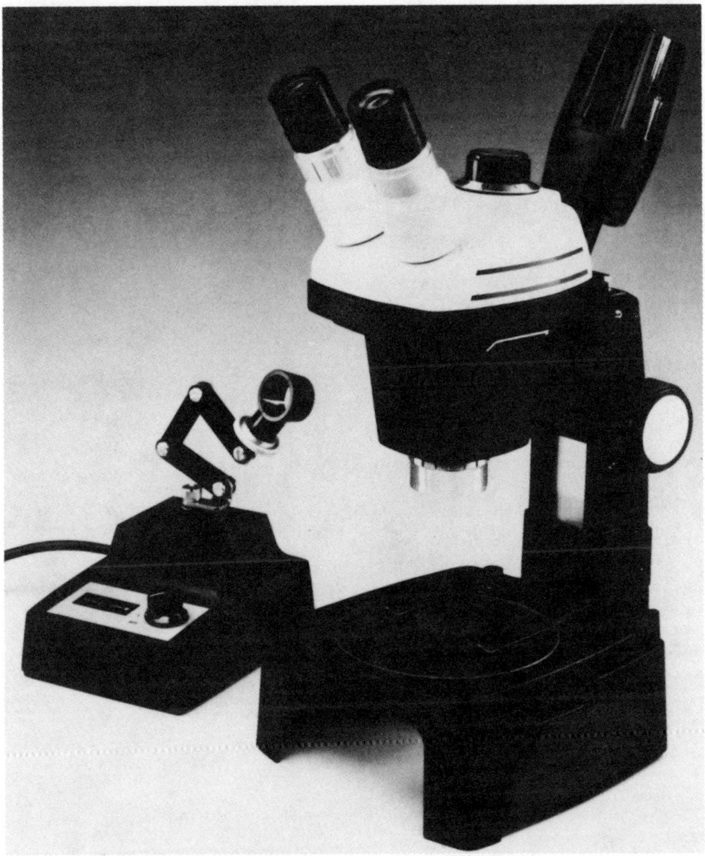

Figure 1-7. Stereo microscope with the most common type of illumination, an external, incident, darkfield illuminator. In this case the illuminator has been removed from its power source and mounted directly on the microscope body. Courtesy of Cambridge Instruments Inc., Optical Systems Division.

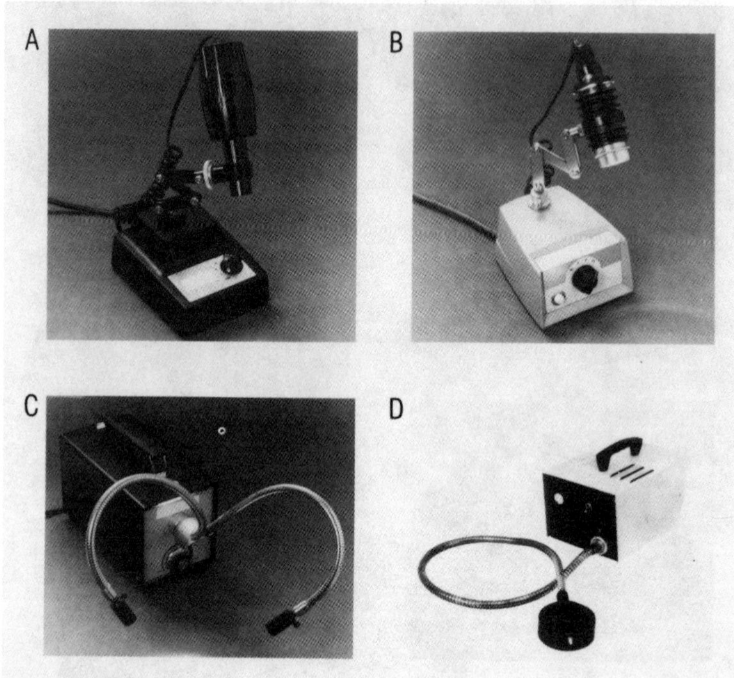

Figure 1-8. Several external microscope illuminators suitable for the incident, darkfield illumination of stereo microscope samples. A. is a fixed focus illuminator, B. is an adjustable focus illuminator, C. is an adjustable focus, dual, fiber optic illuminator and D. is a fiber optic, ring illuminator. Courtesy of Cambridge Instruments Inc., Optical Systems Division and Scientific Instruments Division.

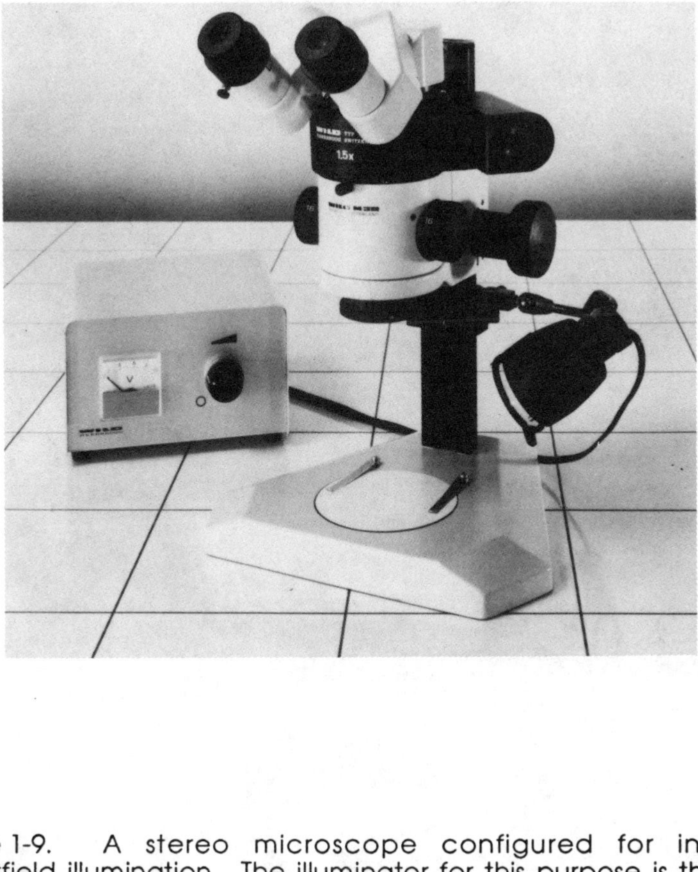

Figure 1-9. A stereo microscope configured for incident brightfield illumination. The illuminator for this purpose is the dark housing at the right of the binocular head. An incident, darkfield illuminator is also attached to the stand at the lower right. Courtesy of Wild Leitz USA, Inc.

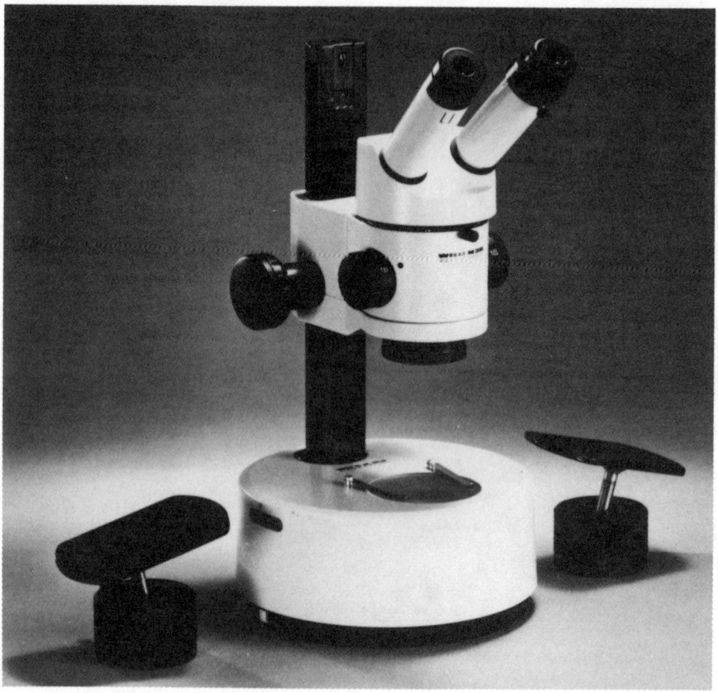

Figure 1-10. A stereo microscope with transmitted brightfield illuminator in the instrument base. Courtesy of Wild Leitz USA, Inc.

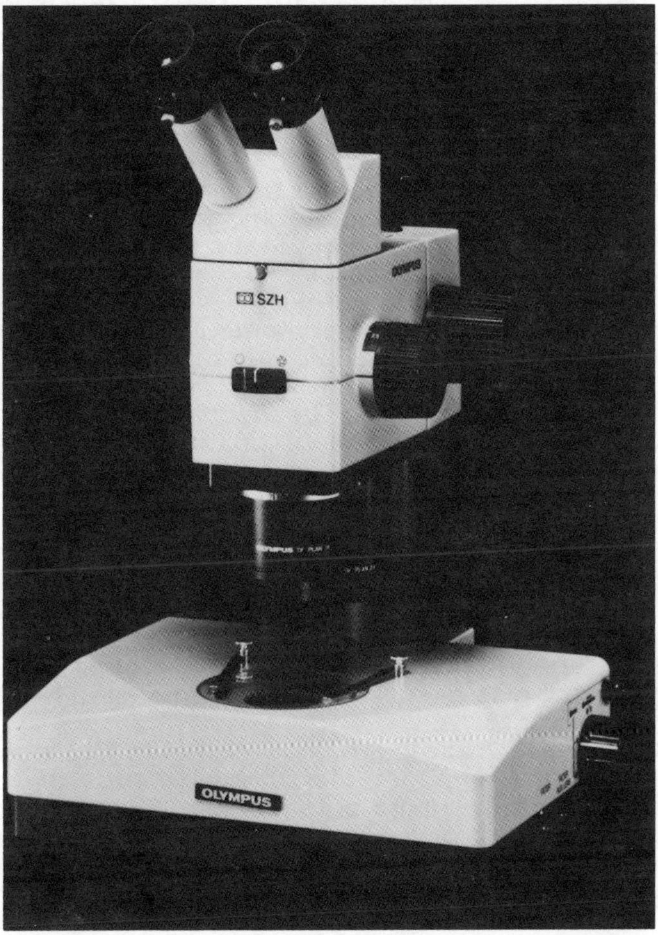

Figure 1-11. A stereo microscope with an illuminator that will provide either brightfield or darkfield transmitted light illumination. Courtesy of Olympus Corporation.

and the common main objective lens (CMO) system in which the CMO is shared by the two individual microscopes. The factors which influence image quality are the same as for the compound microscope, as discussed in Chapter 2, Section III. The magnification is calculated using Equation 4.

In the Greenough system, shown in Figure 1-12, the objective lenses are small so that they may be spaced close enough to each other to target the same viewing area. This system has at least two advantages: (1) the compact size of the nosepiece containing the paired objective lenses makes this an ideal instrument for studies in confined areas, and (2) the optical corrections are less expensive to achieve for the small sized objective lenses. The main disadvantage to the Greenough type instrument is that the intermediate images are inclined to the sample plane. This results in a focus that is sharp only along a line parallel to the axis of tilt between the two optical systems. For this reason the focus is not ideal to the right and left of the optimum focus line for either eye. For photomacrography, the sample should be made normal to the optic axis of the specific objective lens used to give uniform focus across the image.

Total magnification of the Greenough stereo microscope is the product of the magnifications of the objective lens and eyepieces. As with the compound microscope, the higher the magnification of the objective lens, the higher the resolution of the system. The paired objective lenses are generally in a fixed nosepiece on the simpler instruments, therefore, magnification can be changed for these instrument only by exchanging eyepiece pairs. More expensive instruments have objective lenses mounted in nosepieces that can be interchanged on a single dovetail. On the most advanced instruments multiple magnifications are available by changing a slider on which are mounted two or three individual nosepieces.

A CMO stereo microscope, shown in Figure 1-13, has several advantages. The intermediate image planes for the two microscope systems are parallel to the sample plane, therefore, both images are sharply focused over the entire field of view. Care in the design of the common has yielded instruments which have excellent flat fields and minimum distortion. Accessories for drawing, projection and photomacrography are easily adapted to either one of the individual microscope systems. Wild/Leitz further improves the image quality of their M-7 stereo microscope for photomacrography by translating the optical axis of one of the two microscope systems to coincide with the optic axis of the objective lens.

Perhaps, one of the more important features of certain CMO stereo microscopes, such as produced by Olympus, Wild/Leitz and Nikon are the modular infinity systems. The infinity systems, discussed in Chapter 2, Section III C 2, permits the use of various

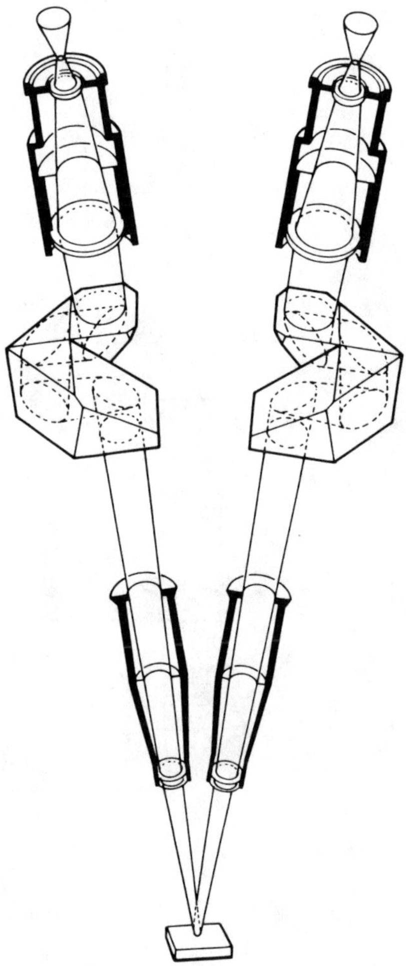

Figure 1-12. Perspective view of the optical elements of a Greenough type stereo microscope. The complex glass bodies in the center of the picture are the erecting prisms that produce an upright image. Note that each eye has a separate objective lens. Courtesy of American Laboratory.

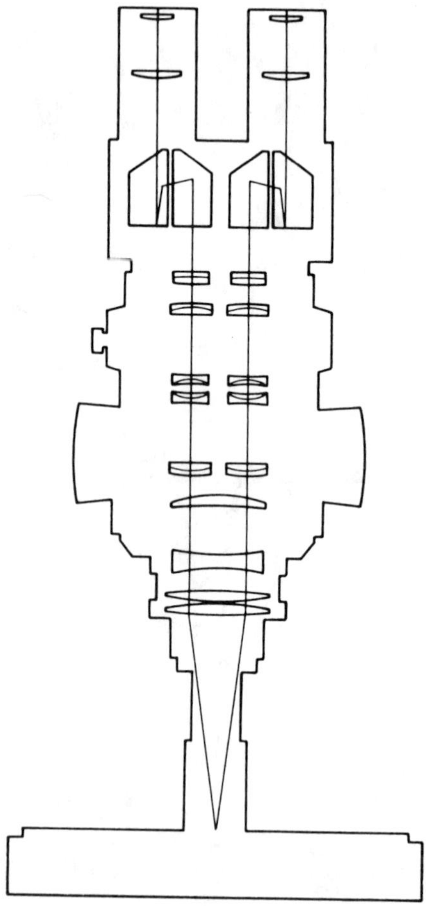

Figure 1-13. Cross section of a common main objective lens (CMO) type stereo microscope. Note that the image information for both eyes is produced by the one main objective lens. Courtesy of American Laboratory.

visual and photo tubes and intermediate accessories such as coaxial illuminators without the introduction of additional optical aberrations.

One disadvantage of the CMO system is that the image information for each eye travels obliquely through the edge of the main objective lens; this introduces chromatic aberration, see Chapter 2, Section III, which is expensive to correct in the large lenses used.

Total magnification of the CMO stereo microscope is also the product of all the magnifications of the system as given by Equation 4. Eyepieces and accessory objective lenses can be used to change the magnification for the simpler instruments. In addition to these means, the magnification can be changed on the more expensive instruments by means of intermediate lenses. These intermediate lenses may be a series of paired, fixed lenses that are mounted in a drum in the two optical paths. Rotation of the drum to various positions gives a number of discrete magnifications. The intermediate lenses may alternatively be part of a mechanical zoom system that gives a continuous range of magnifications.

The set-up of the stereo microscope is relatively simple and if the procedure (Table 1-2) is followed the maximum amount of detail will be observed in the image with a minimum of eye fatigue. The illumination should be chosen based on the type of sample. In general, incident light is used for bulky and opaque samples while transmitted light is used for thin sections or normally thin, translucent or transparent samples. The stand should be chosen based on the sample size and location. If the procedure has been followed but a problem is noted, repeat the procedure carefully. If the problem remains, consult Table 1-3 which contains a general check list of typical problems, their causes, and solutions.

V. THE LIGHT SCANNING MICROSCOPE

One of the main frustrations of the optical microscopist is that of limited depth of field; this results from rigid optical principles and gets progressively worse as the magnification increases. The accommodation of the younger eye can effect some slight improvement in the depth of field and the memory can integrate the various images acquired during sequential focusing to produce a mental image with considerable depth. However, it is in photomicrography that limited depth of field is most bothersome.

Scanning electron microscopy has provided a dramatic improvement in the depth of field, typically 1000 times greater for a given magnification than with the comparable optical microscopy image. This gain is not without cost; for example, the SEM image

TABLE 1-2. SET-UP PROCEDURE FOR THE STEREO MICROSCOPE

1. If a reticle eyepiece is to be used, it first must be focused. Remove the eyepiece from the tube, fully extend the focusing adjustment, gaze through the eyepiece at a broad light source, and focus the eyepiece until the image if the reticle is sharp. If the position of sharp focus is passed, repeat the above procedure. Reinsert the eyepiece.
2. Select the type of illumination to be used, place a sample on the stage, turn on the illuminator to medium brightness and focus the image of the sample using the focus knob. For position of the illuminators see Figures 1-7 to 1-10.
3. Adjust the interpupillary spacing for the binocular body so that both eyes can view the image simultaneously. (If this spacing is calibrated it is desirable to memorize this spacing number since it is the same for any microscope that you use).
4a. Continue the adjustment of the binocular body by performing the interpupillary acuity adjustment. Refocus the microscope critically using the focus knob while observing only with the eye over the eyepiece in the non-adjustable tube. Then focus for the other eye using only the threaded focusing collar beneath that eyepiece. If the zoom stereo microscope has focusing collars on both eyepieces use step 4b.
4b. Interpupillary acuity adjustment for zoom stereo microscopes with focusing collars on both eyepieces is initiated by turning both collars to their zero position. Increase the zoom system to its maximum value and refocus the microscope critically using the focus knob. Then reduce the zoom system to its minimum value and carefully refocus the image for each eye using only the respective focusing collar. The microscope should now be in focus throughout the entire range of the zoom system, if not, repeat this step.
5. The stereo microscope is now ready for sample examination.

---

cannot give a color image and cannot be used for samples that are affected by or affect a vacuum system, e.g., most moist samples.

An optical microscopical approach to increased depth of field for photomicrography was patented by Zampol (1960) and was reduced to practice, first by McLachlan (1964) and later by Schram (1981). In the deep field microscope the sample is illuminated by a thin "sheet" of light that is projected at right angles to the optical axis of the microscope and within the depth of field region of the microscope image. The arrangement of sample, illumination, microscope objective lens and film is shown in Figure 1-14. To acquire a photomicrograph, the room light is reduced, the

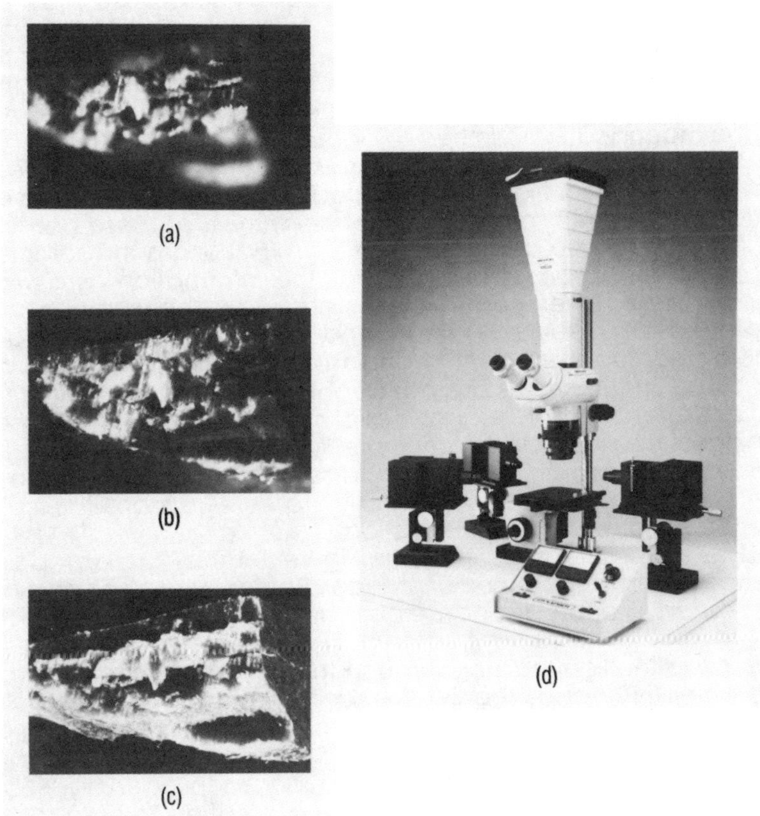

Figure 1-15. The Dynaphot arrangement produces scanning or deep field photomacrographs. (a) A standard photomacrograph of a wood chip (b) A standard photomacrograph of the same wood chip with reduced aperture to increase the depth of field and (c) A Dynaphot scanning photomacrograph of the same wood chip. (d) A Dynaphot arrangement with a Wild M400 photomacroscope to record the image and three illuminators placed around the z-axis scanning stage. The stage and illuminator control is shown in front. Courtesy of Irvine Optical Corporation.

provisions for mounting the microscope of choice. The hydraulically controlled stage provides a slow uniform scan of the sample. The speed of the scan must be adjustable because the effective aperture of the microscope decreases as the magnification increases, requiring longer exposures. At the same time the depth of field is less requiring a thinner illumination slit that results in less light; this too requires a slower scanning rate. Thus, increased magnification requires a reduced scanning rate. As a result the magnification is limited to about 30X.

Confocal scanning laser microscopes have been recently introduced by Tracor Northern, Leitz and Zeiss; these instruments can produce images with great depth of field. Most of these confocal scanning systems use a single laser beam that scans the sample in a relatively slow process and records the information as a video image. These laser scanning systems are the subject of a comprehensive book by Wilson and Sheppard (1987). Tracor Northern's confocal Tandem Scanning Reflected light Microscope (TSM), for example, yields dramatic light section microscopy in real time and even at high magnifications. Scanning in this sense is scanning of the sample by multiple light beams. Thus, the TSM produces for all practical purposes, a real time image that can be viewed and photographed directly; for this reason it can be appropriately discussed in this section.

The TSM instrument, shown in Figure 1-16, is a reflected light instrument which makes use of standard objective lenses and uses incandescent, xenon arc or mercury arc illumination. A simplified diagram of the instrument, shown in Figure 1-17, illustrates the difference in the light passing through the instrument at exact focus of the objective versus the light returning from above and below focus. The illumination first passes through the rotating scanning disk that contains many precisely positioned small circular openings. As the disk rotates, the light passing through each opening "scans" across the sample. The speed of the rotating disk and the multiplicity of holes assures that the entire area of interest is illuminated. The optical system of the TSM is arranged so that light reflected back from the focal plane passes through an identical set of holes on the opposite side of the disk before passing through the eyepiece to the eye. Light from structure above or below the focal plane cannot pass through these latter holes and thus is not seen. The very sharp rejection of light other than that from the exact level of interest in the sample combined with the real time image and a very high resolving power make this instrument useful for a wide range of unusual samples.

The TSM can be used to examine the surface of any kind of sample but it is particularly useful for those materials which are nearly transparent. Studying a single plane of focus or an optical section of relatively translucent materials is one of the most important applications of the TSM. If a sample is stained with a fluorochrome and illuminated with the appropriate wavelength

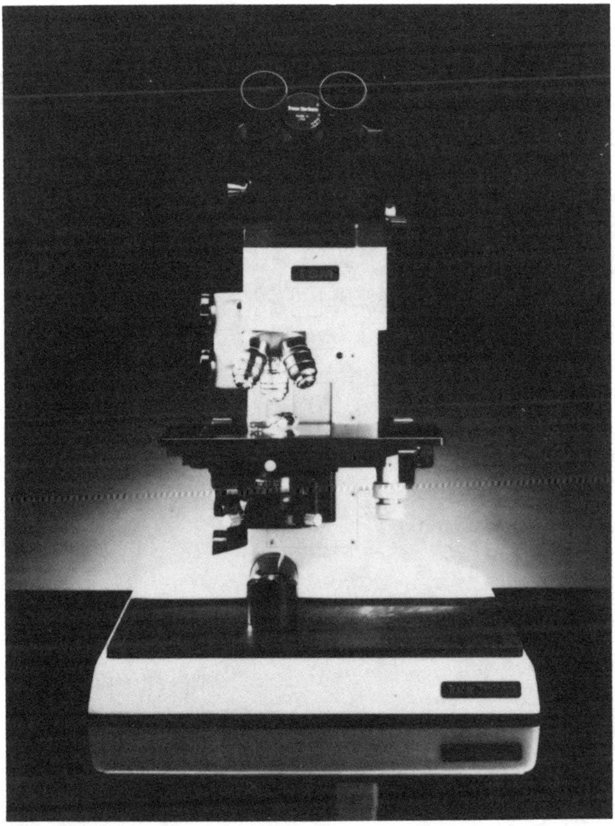

Figure 1-16 A confocal, tandem scanning, reflected light microscope (TSM). Courtesy of Tracor Northern.

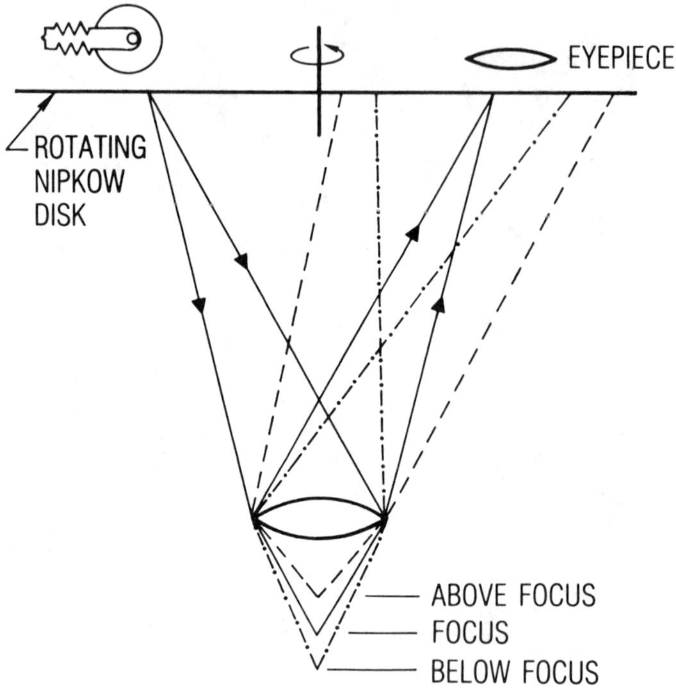

Figure 1-17. Simplified ray diagram of a confocal tandem scanning microscope system. Illumination from the lamp, upper left passes through small apertures in the rotating Nipkow disk and is scanned by the objective lens onto the sample. The light reflected from the sample at the focus is then accepted by apertures on the opposite side of the Nipkow disk and presented to the eye by the eyepiece. Light from above or below focus is not brought to focus on the apertures and is thus blocked by the disk.

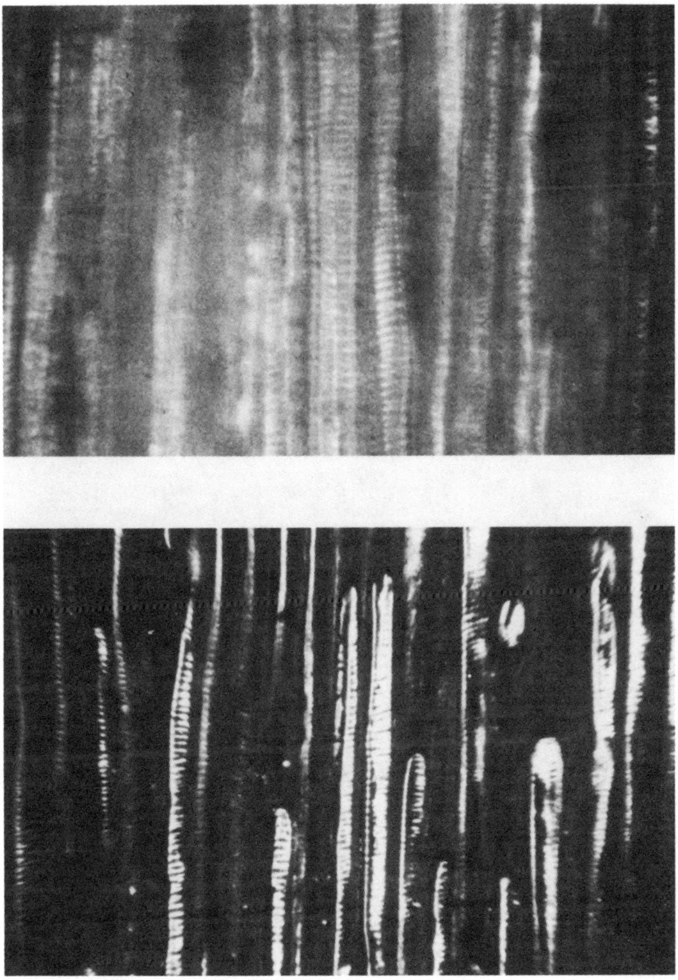

Figure 1-18. A comparison of the photomicrographic images of a wood sample using (A) a conventional microscope and (B) a confocal tandem scanning, reflected light microscope. The conventional microscope shows structure at several depths in the sample, whereas the TSM image shows the structure at a specific level with great detail and contrast. Courtesy of Tracor Northern.

light to produce fluorescent excitation, the effect is impressive. The glare caused by the fluorescence of structures above and below the optical section plane is eliminated and the usual fading of the fluorescence is minimized by the reduced incident illumination as a result of the scanning process. A typical TSM image is shown in Figure 1-18. The TSM can also be used to prepare stereo pair photomicrographs.

REFERENCES

McLachlan Jr., D., Appl. Opt., 3:9, 1009, (1964).

Schram, R.R., Microscope, 29:1, 13, (1981).

Wilson, T. and Sheppard, C.J.R., "Theory and Practice of Scanning Optical Microscopy," Academic Press, London, (1984).

Zampol, P., Patent No. 2,928,734, U.S. Patent Office (1960).

2

Brightfield Transmitted Light Microscopy

I. INTRODUCTION

The brightfield transmitted light microscope is the working instrument of the biological and the medical fields. Since this instrument is quite straightforward we shall use it to describe the detailed features of compound microscopes in general.

II. THE STAND

A. The Upright Stand

The most common style of microscope stand for general use is the upright stand as shown in Figure 2-1. This stand is very popular for the transmitted light medical and biological instruments, although it does find some application for normal incident (or reflected) light microscopy. The base of the stand can take a number of forms; however, each must firmly support the instrument. Many of the the modern bases also incorporate the illuminator and even the power transformer and brightness regulating controls for the illuminator.

The limb is the upright which supports the body of the microscope. In the older style instruments a pivot is provided to permit tilting the microscope to a comfortable viewing angle or to a horizontal position for photomicrography. The limb of most modern instruments is rigidly affixed to the base in a vertical position; with these instruments comfortable viewing is achieved through the use of prisms which redirect the image beam at an angle to the vertical axis of the instrument.

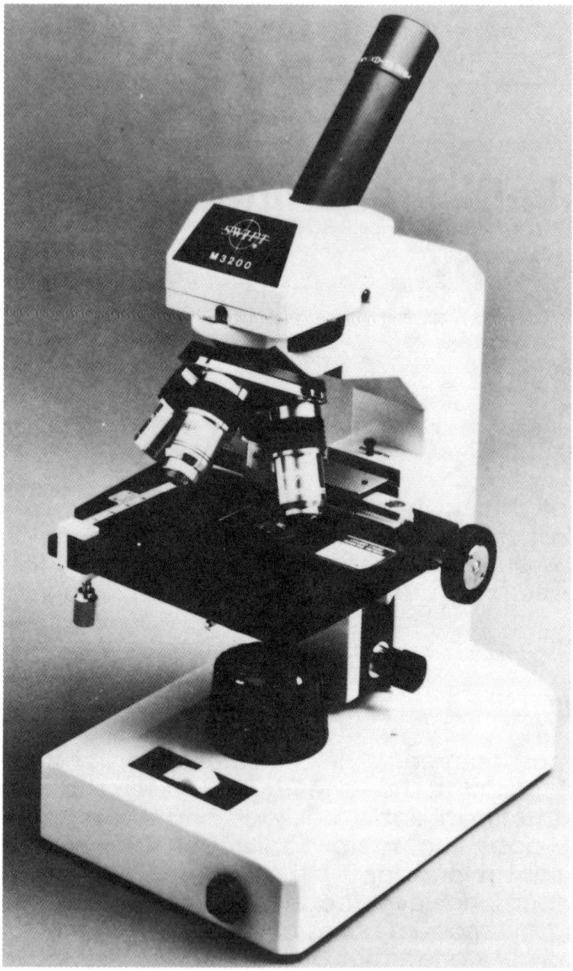

Figure 2-1. An upright, compound microscope with a monocular tube. Courtesy of Swift instruments, Inc.

The coarse and fine focus mechanisms which adjust the spacing between the objective lens are generally contained in the limb portion of the instrument. The coarse focus mechanism is usually a rack and pinion with considerable travel, however some instruments do have a stop which precludes impacting the sample with the objective lens. The coarse focus is usually uncalibrated with respect to travel. In many of the modern instruments the coarse and fine knobs are coaxial.

The fine focus provides the controlled slight adjustment necessary for critical focus at any magnification above about 50X. The mechanisms for achieving this finely controlled adjustment are widely varied and often rather delicate. Furthermore, the range of travel for the fine focus is typically only a few millimeters. Thus, it is imperative that much care be afforded in the use of the fine focus to avoid damaging the mechanism. On the more advanced instruments the fine focus travel is calibrated, usually in one micrometer steps. A few instruments have a combination focus knob which has a fine focus mechanism that operates for about 1/2 turn; if one turns the knob further in either direction the coarse focus then operates. At least one instrument, the Nikon Microphot, has coaxial fine, medium, and coarse focus knobs.

The body tube is that part of the microscope which provides a light tight support and correct spacing for the objective lens and the eyepiece. The body tube is attached rigidly either to the limb or to the focus mechanism which is incorporated in the limb. The body tube is coated internally with a black, non-reflective coating to minimize stray reflections that reduce contrast in the image.

The length of the body tube is the mechanical tube length; this corresponds numerically to the mechanical tube length parameter of the objectives for that specific instrument. Typical mechanical tube lengths range from 140 to 250 mm. A number of research model instruments have an adjustable, graduated draw tube which permits "fine tuning" of the body tube length to match the optical parameters of any specific objective lens. The body tube has various mountings for both eyepieces and objective lenses.

The simplest eyepiece arrangement is the monocular tube. The eyepiece fits into this tube with a slip fit and rests on the upper rim of the body tube. The top surface of the body tube may have one or more notches for critically positioning an eyepiece. Since there is seldom any clamping arrangement for the eyepieces, care must be taken to prevent dropping the eyepiece(s) if the microscope is inverted or knocked over.

The body tube may also be terminated in a binocular body as shown in Figure 2-2. This optical system splits the image into two equal beams at two parallel matched eyepieces and thus permits the use of both eyes. It should be emphasized that each eye sees the same two dimensional image. In addition to splitting the image beam between the two eyepieces the binocular head almost

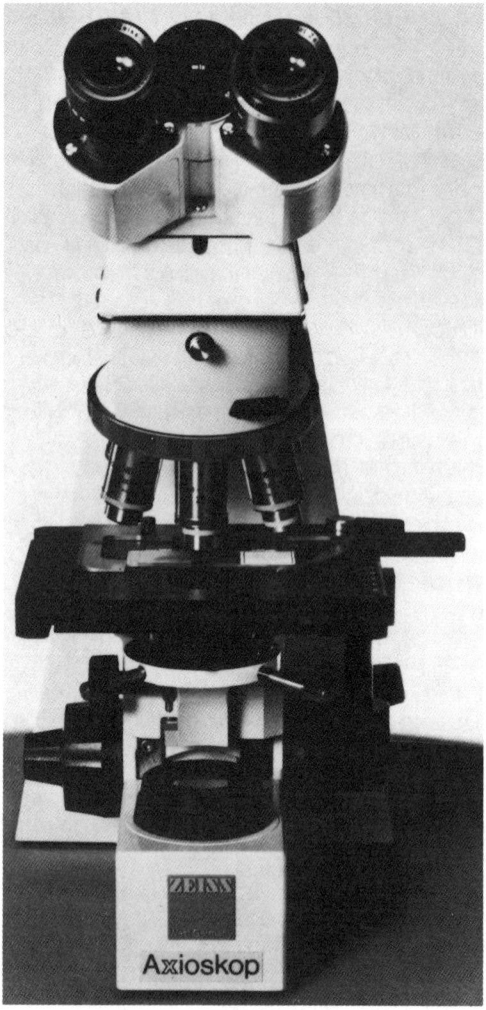

Figure 2-2. An upright, compound microscope with a binocular head. Courtesy of Carl Zeiss, Inc., New York.

always provides for inclining the viewing direction at either 30° or 45° to the body tube axis. This provides for comfortable viewing with the stage in a horizontal position .

The binocular body has two adjustments, one of these, the interpupillary spacing adjustment, varies the spacing between the two eyepieces to accommodate for the eye spacings of the various users. This adjustment is usually calibrated in millimeters of spacing.

One disadvantage of the simpler binocular bodies is that the mechanical tube length is altered by changing the interpupillary spacing and to some extent by changing the interpupillary acuity. One solution to this problem is incorporated in a Zeiss design where the compensation adjustments are made with a focusing collar on both eyepieces; in this way the correct tube length can be maintained. Another approach by Leitz makes use of an ingenious mechanical linkage which simultaneously corrects the tube length for changes in interpupillary spacing.

The second adjustment, known as the interpupillary acuity adjustment compensates for differences in visual acuity of the observer's eyes. This adjustment is usually performed by rotating a threaded focusing collar under one of the eyepieces.

The body tube may also be terminated in a trinocular body as shown in Figure 2-3. This optical system provides for the binocular viewing described above and a monocular tube which is typically used for photomicrography or possibly a video camera. The percentage of light is equal for the two binocular tubes, however the total illumination may be split between the binocular tubes and the monocular tube or may be switched between the binocular tubes or the monocular tube. The latter arrangement is useful when photomicrography of an image with a low level of illumination is required.

In a number of the newer instruments the binocular or trinocular head may be positioned so that when viewing, the limb of the instrument may be adjacent to the microscopist or on the opposite side of the instrument. In the latter arrangement the microscopist has the convenience of easy access to the stage of the instrument.

The termination of the body tube at the objective end also takes a number of forms. The simplest arrangement is merely a threaded opening which will accept the threaded base of an objective lens. Almost all microscopes use the Royal Microscopical Society (1936), RMS, standards for this screw thread.

A second type of objective lens attachment is the so-called quick-change nosepiece, see Chapter 4, Section VI C. In this arrangement a single objective lens is threaded into an interface which mates with the body tube by a dovetail or a twist lock mechanism. For critical work these interfaces are often provided with centering adjustments so that the same area of the sample may be examined with each objective. Speed of changing from

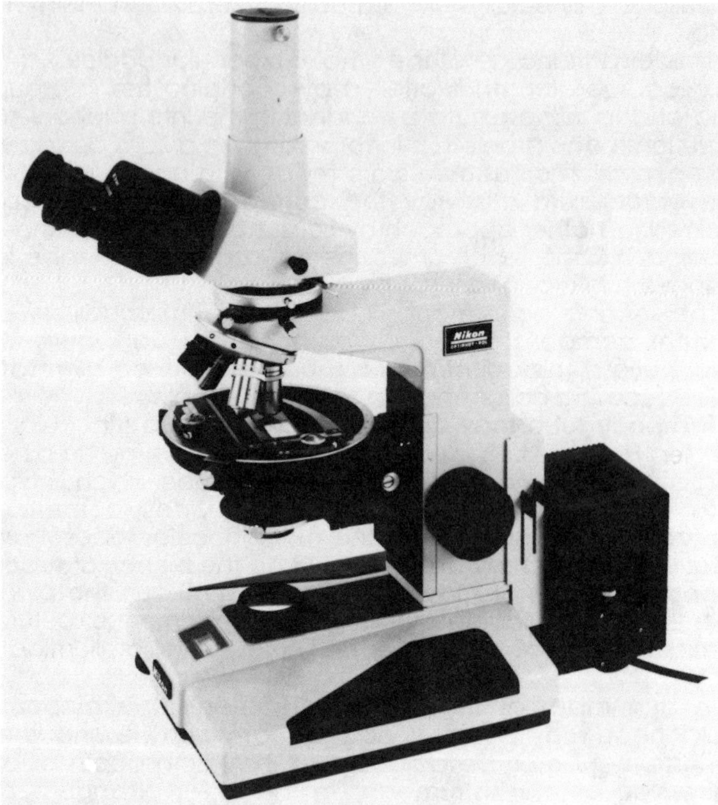

Figure 2-3. An upright, compound microscope with a trinocular head. The vertical tube may be used with a photographic or video camera. Courtesy of Nikon, Inc.

one objective to another is often mentioned as a virtue of the interfaces, however it is doubtful that they can seriously challenge the revolving nosepiece described below for speed.

The most popular means for attachment of the objective lens to the body tube is the revolving nosepiece; this is a disk mounted on the body tube which has from 2 to 5 threaded holes for mounting objective lenses. A detent mechanism causes each hole to line up with the axis of the body tube as that particular objective lens is rotated into position. Although the machining of the revolving nosepiece is generally very good, for critical work some manufacturers provide individual centering for each objective. With the latter type of revolving nosepiece it is imperative to use the knurled rim of the nosepiece disk to change objectives. The temptation to rotate the nosepiece by grasping the objective lenses themselves will assure that they will no longer be aligned.

The advantages of the revolving nosepiece are speed and convenience. Also, the objectives are protected from dust at the threaded end. One disadvantage of this nosepiece is that the mechanism is large enough that it may interfere with accessories to the microscope, e.g., the universal stage and certain mechanical stages. Another disadvantage to the rotating nosepiece is that the objectives must be parfocal (see Section III C 2), otherwise a longer lens may, upon rotation, impact the sample with possible damage to both.

The body tube may also contain other devices between the objective lens and the eyepiece, for example a polarizing analyzer, iris diaphragm for conoscopic observation (see Chapter 4, Section VI), Bertrand Lens, Telan lenses (see Section III C), a zoom system (see Section III C) or an accessory plate slot (see Chapter 4, Section VI).

The stage is the part of the microscope that supports the sample. The stage must be rigidly attached to the limb at right angles to the optical axis of the body tube. Focusing for those instruments with body tube directly attached to the limb is achieved by movement of the stage which in this case is affixed to the fine and coarse focusing mechanisms. For transmitted light instruments the center of the stage must have an opening to conduct the illumination from the condenser system to the sample. The dimensions of the stage can range from about 7.5 x 7.5 cm to 15 x 15 cm.

The revolving stage is imperative for crystallographic and metallurgical studies and are also desirable to orient the sample for proper composition of a photomicrograph. These stages are arranged such that their axis of rotation is coincident with the optical axis of the body tube. "Fine tuning" of this coincidence is done with adjustment screws either on the stage mounting or with the objective centering. This stage is calibrated in degrees around its circumference and is equipped with a vernier that can be read to 0.1 degree.

A mechanical stage provides easy orthogonal translation of the sample in the field of view and thus is a major convenience during the routine use of a microscope. This mechanism takes a number of forms with separate or coaxial knobs that provides for movement in each orthogonal direction. The mechanism may be built into the stage support and thus is an integral part of the instrument or it may be an attachment which mounts on the stage.

The illumination for the sample on a compound microscope is obtained from the condenser system. In the transmitted light case, the condenser is mounted on the side of the stage opposite the objective lens. The transmitted light condenser is ideally provided with a coarse focusing system. Provisions also must be available to make the beam of illumination from the condenser coincident with the axis of the body tube. The condenser may have a number of adjustable elements such as insertable final condenser lens, aperture diaphragm, field diaphragm, shading diaphragms and polarizer. The details concerning these very important adjustments will be discussed in Section IV.

The remaining items of importance to our discussion of the upright stand are the mirror and the built-in illuminator. For those instruments without a built-in illuminator a mirror is essential for redirecting illumination from an external light source into the optical axis of the microscope. The mirror is provided with tilting about two orthogonal axes and thus can direct light into the system from a wide range of directions. The mirror element usually consists of two mirrors back-to-back, a plane mirror used for most applications and a concave surface mirror which may be used without the condenser system for low magnification (<50x) observations. Needham (1977) has described the problem of multiple reflections from the usual second surface mirrors for critical microscopy. He recommends replacing the plane mirror with a first surface plane mirror for this special case. Most modern microscopes have integral illuminators that make the mirror unnecessary; these are usually contained within the base of the instrument.

B. <u>The Inverted Stand</u>

The second major style of compound microscope uses the inverted or Le Chatalier stand. The components of the inverted stand are the same as the upright stand but are arranged in a different manner. Most inverted microscopes are designed for normal incident illumination, as described in Chapter 3. However, transmitted light instruments configured with the inverted stand, as shown in Figure 2-4 are becoming increasingly important for the examination of cultures grown in vessels such as petri dishes or culture flasks. In this latter application, the vessel is placed on the stage of the microscope and the objective is aimed at the sample through the bottom of the vessel.

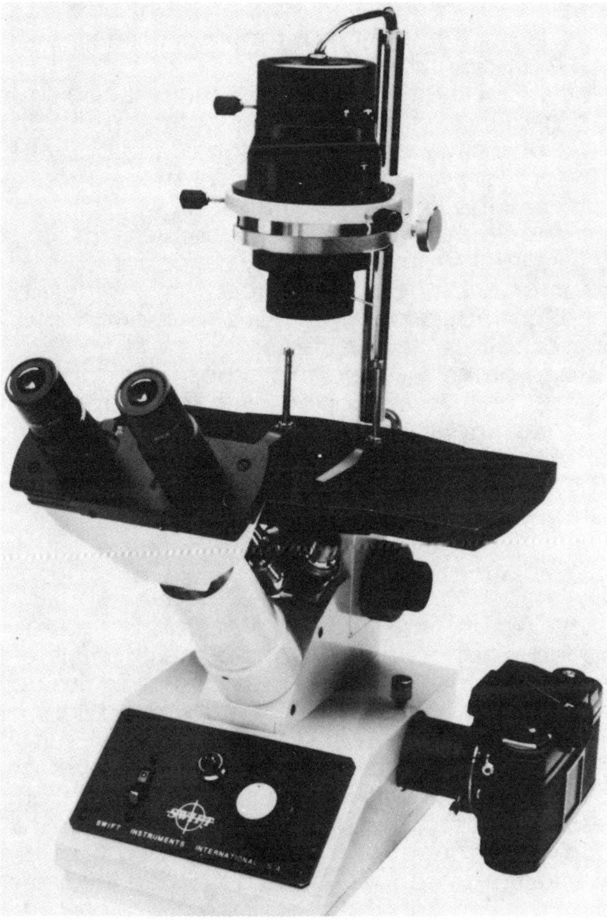

Figure 2-4. An inverted compound microscope with transmitted light illumination. Courtesy of Swift Industries, Inc.

The base may range in size from that used in the upright instrument to a very large base such as used in the metallograph. The latter instrument's base is sufficiently large to support the microscope, the illuminator and a horizontal optical bench for photomicrography.

The limb of the inverted microscope supports the stage. The coarse focus almost always adjusts the height of the stage above the objective lens. The fine focus may adjust the position of either the stage or the body tube.

The body tube of the inverted microscope in identical in function to that of the upright instrument but as one might expect, it is inverted. The objective lens is at the top of the body tube and the image path for the eyepieces is directed below the objective. To comfortably view the image, a mirror or prism arrangement is used to redirect the image beam at a convenient angle for the microscopist. Visual examination may be accomplished with either a monocular or a binocular arrangement. When photomicrography is performed, a prism or mirror is inserted into the image beam thus redirecting it to the camera.

The stage of the inverted microscope is arranged above the objective lens and is equipped with an appropriate central opening for viewing from below. The stage is almost always equipped with an integral mechanical stage. The more advanced instruments may also be equipped with a centerable revolving stage.

C. Upright Vs. Inverted Stands

The naming of these two types of stands is merely a result of history and does not confer any special merits on either instrument. For most applications either stand may be used. The upright stand, the most common configuration for transmitted light application, has few limitations. However, this stand is not suitable for the examination of biological cultures in large diameter containers or even in low containers with a significant layer of liquid over the subject area. As mentioned above, the ideal way to observe a tissue culture, plankton, etc on the bottom of a flask or Petri dish is with an inverted microscope. More details on the selection of the best stand for a given application is given in Chapter 10, Section II.

D. The Culpeper Stand

The Culpeper microscope was developed in England in the 1700s and was characterized by a three-legged stand symmetrically arranged around the body tube. This arrangement is unusually sturdy and provides a great deal of stability between the body tube and the stage. The Carl Zeiss company has extended this concept to a four-posted frame in the Axiomat instrument, as shown in Figure 2-5. This universal microscope has many innovative

Figure 2-5. The Axiomat microscope is arranged with the very rigid corner posts reminiscent of the Culpeper microscope. Courtesy of Carl Zeiss, Inc., New York.

features, but probably the most important of these is the rigidity afforded by the stand.

III. THE OPTICAL SYSTEM

The complete optical system of a microscope starts with the illuminator and concludes with the eyes or the camera. Each element of the optical system will be described in turn, generally following the path of the illumination through a brightfield instrument. The most critical element in the system, the objective lens, will be the subject of a considerable portion of this section.

A. Illumination

Although there a few purists who prefer "good north light" from the brightly lit sky, most modern researchers and certainly those within the deep recesses of a laboratory building must rely on an artificial source of illumination, a lamp. These range from the long outdated oil lamp to lasers. Some of the more notable sources of the past are worthy of a word of description, if for no other reason that some are still in use. However, the common illuminants today, the incandescent lamp, the tungsten halogen lamp, and the high pressure xenon arc lamp will be emphasized. Table 2-1 lists a few of the important characteristics of the various sources of illumination for microscopes.

The low voltage, incandescent lamps are undoubtedly the most widely used and versatile illuminants in use today. They take two distinct forms; the first is the ordinary tungsten filament lamp enclosed in an inert gas atmosphere by a soft glass envelope. These lamps gradually emit less light because the tungsten evaporates from the filament, deposits on the glass envelope, and thus produces a continuously increasing neutral density filter for the emitted light.

The second type of incandescent lamp, the tungsten halogen lamp, appears much like the standard lamp but it differs in having a smaller sized, quartz glass envelope and a small amount of a halogen such as bromine or iodine admixed with the inert gas atmosphere. When the tungsten evaporates from the filament in this lamp it reacts with the halogen to form a tungsten halide gas. This gas decomposes in the hottest part of the tungsten filament to reform tungsten metal and release the halogen for further reaction. This process may be thought of as a self healing process that prolongs the life of the filament while eliminating the tungsten coating which reduces the effective brightness of the lamp.

TABLE 2-1. ILLUMINATION SOURCES

TYPE	INTENSITY	SPECTRUM	COLOR TEMP
Incandescent	Low-Med	Continuous	Variable
Halogen	Low-Med	Continuous	Variable
Xenon Lamp	High	Continuous	Daylight
Carbon Arc	High	Continuous	Daylight
Tungsten Arc	Med-High	Continuous	3100°K.
Mercury Lamp	Med-High	Line	N/A
Sodium Lamp	Low-Med	Line	N/A
Zirconium Lamp	Med	Continuous	3000°K.
Laser	Very high	Line	N/A
Flash lamp	Very high	Continuous	Daylight

The more noteworthy advantages of the incandescent lamps are low cost, ready availability and ease of installation. These lamps are generally operated from a variable low voltage transformer. On the negative side, these lamps have a color temperature that depends on the applied voltage and thus for color photomicrography their operating voltage must be precisely adjusted to achieve a proper color response of the photographic film. Figure 2-6 illustrates this effect for several typical lamps. A further problem with these lamps for photomicrography is the unevenness of the illumination due to the coiled filament. This latter problem may be minimized by careful alignment of the lamp so that the most uniformly heated area of the filament is the illumination source.

Another important illuminant is the high pressure xenon gas discharge lamp. This lamp is characterized by its high intensity and its spectral character that matches daylight quite well. Thus the xenon lamp may be used directly with daylight corrected color film. The light output is very constant after a brief warm-up period.

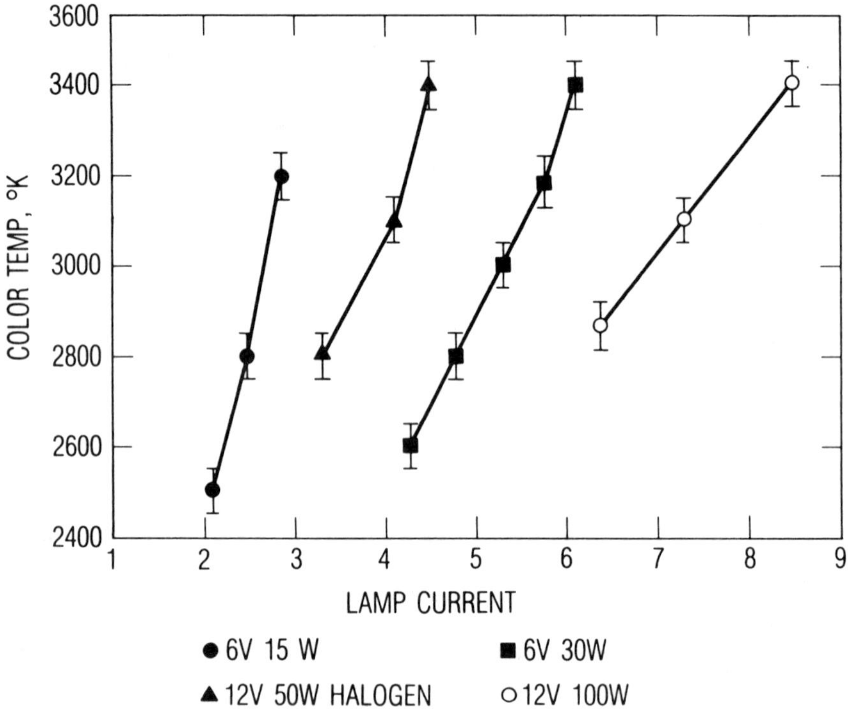

Figure 2-6. Approximate color temperature as a function of lamp current in amps for several common incandescent lamps.

The life of these lamps is very long but is significantly reduced by excessive starting. Therefore, it is best to leave the lamp operating during a number of short investigations rather than start it for each observation.

The xenon lamp consists of two tungsten electrodes sealed in the top and bottom of a thick quartz glass envelope. The lamp is filled with xenon gas under high pressure. Power for the lamp is developed by a high current, low voltage D.C. supply which incorporates a high voltage starting system.

Although the xenon lamp is a superior illuminant, a few words of caution are in order. This lamp, unless specially designed, emits ultraviolet radiation that will generate ozone in the air immediately surrounding the lamp. If this gas is produced in sufficient quantity in a closed room it can pose a health hazard. Adequate air circulation or venting of the air from around the lamp is required if an ozone free lamp is not available.

A second concern with these lamps results from the tendency of the lamp to become brittle with use. Since there are several atmospheres of pressure in the lamp it is necessary to use extreme care in replacing the lamp to avoid serious injury. Details of the replacement procedure are given in Chapter, 10, Section VI.

A final word of caution about the xenon lamp; the lamp also generates considerable infrared radiation. Therefore, a heat absorbing filter is necessary if a temperature sensitive sample is to be observed or if polarizing or color filters are to be used.

Two, relatively new, high intensity illuminators should be mentioned at this point that have limited or special applications, namely, the flash lamp and the laser. The flash lamp is a very intense, pulse source which finds use only in photomicrography. The duration of the flash is typically about one millisecond, thus the problems of vibration and sample movement are effectively eliminated. Since this lamp is a pulse source, a separate lamp that operates alternately with the flash must be used for focusing and framing of the photomicrographic image.

The flash lamp is so intense that glare can be a problem, especially if lenses do not have antireflection coatings. The total amount of illumination can be adjusted by controlling the amount of energy dissipated in the lamp or by using a neutral density filter (see Chapter 7, Section VI A).

The laser has been studied to a limited extent as an illuminator for optical microscopy. A number of lasers such as the He-Ne and the dye lasers have a visible light output of suitable wavelength. This light may be characterized as monochromatic, very intense, coherent, and highly collimated. The monochromatic nature of this light does not pose any special problems for microscopy and can be used to advantage if the emission is in the yellow or green. Due to the extreme intensity of the laser, beam expansion of the beam diameter or filtration must be used to protect the microscopist's eyes. The coherence of the laser beam produces diffraction and an interference pattern in the image of the sample known as speckle or interference noise. Carter and Dougal (1966) compared the images of transmitted light samples illuminated by ordinary incoherent light with those produced by the coherent light of a He-Ne laser. They found a dramatic improvement in contrast and depth of field with the laser illumination, however diffraction fringes were noticeable in the image if the sample is thick.

To complete this section, a brief description of the older and less common illuminants is in order. The carbon arc was used for a number of years as an intense daylight white light source. In this source, two carbon rods in air form the electrodes between which a low voltage, high current arc is formed. Since the arc is formed in air, the hot carbon is slowly consumed or burnt; this requires a motor drive mechanism to maintain the correct spacing for the best and most intense arc. The arc is not self initiating, therefore some provision must be made to establish the arc, once the voltage has been applied to the electrodes, for example, by briefly moving the electrodes together, manually.

The carbon arc may be operated with either AC or DC power; however, the DC arc is more stable. Because the carbon arc has a positive temperature coefficient, current limiting is necessary, usually in the form of a rheostat. The rheostat will provide some adjustment of brightness and color temperature. As in the case of the xenon arc lamp, a heat absorbing filter should be used with carbon arc illumination to protect delicate samples or filters.

The tungsten arc lamp or Pointolite is an enclosed source containing an argon atmosphere, a starter cathode and a tungsten anode. The anode is heated to incandescence by the passage of current through the lamp. The lamp is a medium to high intensity source with a continuous spectrum characteristic of an incandescent body. The color temperature of the tungsten arc is about 3100°K which is sufficient to cause appreciable evaporation of the tungsten. The evaporated tungsten deposits on the envelope thus gradually diminishing the light intensity; otherwise the lamp has a very long life. The lamp is no longer supplied on new equipment but may be found on some of the older instruments.

The mercury vapor lamp consists of a quartz glass envelope containing mercury vapor and tungsten electrodes. The mercury vapor is ionized by the application of voltage across the electrodes. The running voltage is relatively low but a high voltage pulse is required for starting. The mercury lamp is the first lamp discussed that emits light that is significantly different from daylight; it emits largely a line spectrum which to the eye appears bluish-green. These mercury spectrum lines can be separated using a monochromator, absorption filters, or interference filters (see Chapter 7, Section VI B). Table 2-2 lists the more important spectrum lines for the microscopist, their colors and their relative intensities.

The most important of these lines, the intense green line at 546.1 nM, serves to produce the sharpest possible images with achromatic objective lenses since they are corrected for this color (see Section III C 4). The intensity of this line also makes it suitable for use in quantitative interference fringe microscopy (see Chapter 5, Section III C & D). Another important use of the mercury vapor lamp is in fluorescence microscopy (see Chapter 4, Section IX). In this case the blue, violet or ultraviolet lines are isolated and used as the excitation source.

TABLE 2-2. MERCURY ARC SPECTRUM LINES FOR OPTICAL MICROSCOPY

COLOR	WAVELENGTH (nM)	RELATIVE INTENSITY
Yellow	577.0	50
Green	546.1	100
Blue	435.8	20
Blue	434.8	20
Violet	407.8	8
Violet	404.7	10
UV	366.3	70
UV	253.7	30

The sodium vapor lamp is similar in construction and operation to the mercury vapor lamp except that a small quantity of sodium metal is added to the contents of the lamp instead of the mercury. The lamp emits an intense, nearly monochromatic yellow light which is also useful for quantitative interference fringe microscopy.

The zirconium concentrated arc lamp is another enclosed arc source which has an argon atmosphere and a zirconium cathode. The passage of a low voltage current heats a thin layer of zirconium oxide on the cathode producing a largely continuous spectrum with a color temperature of about 3000°K. This lamp operates from a special power supply that provides a high voltage starting pulse in addition to the lower D.C. operating voltage.

It is important to consider the lamp housing as an integral part of the optical system, especially if the lamp is separate from the microscope. The physical structure of the housing should provide adequate cooling for the lamp while minimizing stray light. Safety requires adequate thermal insulation to prevent burns from the hot housing and electrical wiring that has adequate strain relief and sufficient insulation to withstand normal laboratory usage.

The ideal optical arrangement for a lamp housing is shown in Figure 2-7. The mirror nearly doubles the light output if it is provided with centering and focusing adjustments. The lamp mount should also provide adjustments for centering with respect to the lamp condenser lens. The filter slot or slider arrangement permits the modification of the light which exits the lamp housing. The lamp condenser must have a focusing capability. If the lamp housing is separate from the microscope, the lamp condenser should also be equipped with a lamp field diaphragm. Furthermore, the separate lamp housing should have a tilting adjustment to permit aiming the beam toward the microscope mirror. If the lamp housing is to be attached to the microscope, a prime concern is that the attachment is sufficiently rigid to maintain accurate alignment of the illumination.

The illumination mirror is necessary on instruments that do not have built-in illuminators. The purpose of the mirror is to reflect the light from the external illuminator into the condenser system of the microscope. Needham, ibid., makes the point that the mirrors are almost always second surface mirrors, i.e., the major reflecting element is behind glass. He recommends the use of first surface mirrors for critical microscopy. Suitable first surface mirrors for this purpose can be made by vacuum depositing aluminum on the carefully cleaned surface of the existing mirror if the laboratory facilities exist or by purchasing a ready fabricated mirror. An excellent source for the latter is Pancro Mirrors Inc. (6413 San Fernando Rd., Glendale, CA).

B. The Condenser System

The condenser is next in the optical path of the illumination. It is usually arranged on a focusing and centerable mount so that the illumination can be properly aligned to the optical axis of the microscope. Two condensers are shown in Figure 2-8.

A fact that seldom appreciated is that the sharpness and general character of the image is strongly influenced by the illumination of the sample. It is the condenser that provides the necessary conditioning for illumination of the sample. It consists of a number of lenses, some insertable at will, one or sometimes two iris diaphragms and usually an insertable filter carrier. Filters and their application are discussed in Chapter 7 Section VI.

The lenses within the condenser collect the light then produce a cone of light converging at the sample. These are available in three degrees of optical correction. The chromatic condenser uses only simple lenses and is uncorrected for either spherical or chromatic aberration. The next in the series is the aplanatic condenser which is corrected for spherical aberration. The remaining condenser is the achromatic condenser which is

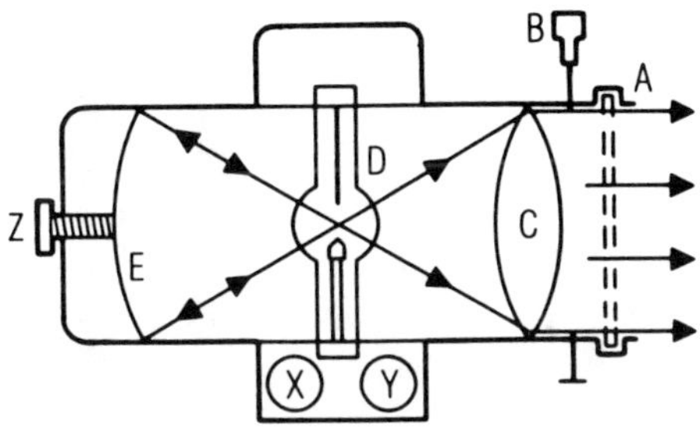

A FILTER HOLDER
B FIELD DIAPHRAGM
C COLLECTOR LENS
D ARC LAMP
E MIRROR
X LAMP X POSITIONING
Y LAMP Y POSITIONING
Z MIRROR FOCUS

Figure 2-7. Cross section of a typical high efficiency illuminator suitable for all types of lamps; shown here with a high pressure xenon arc lamp.

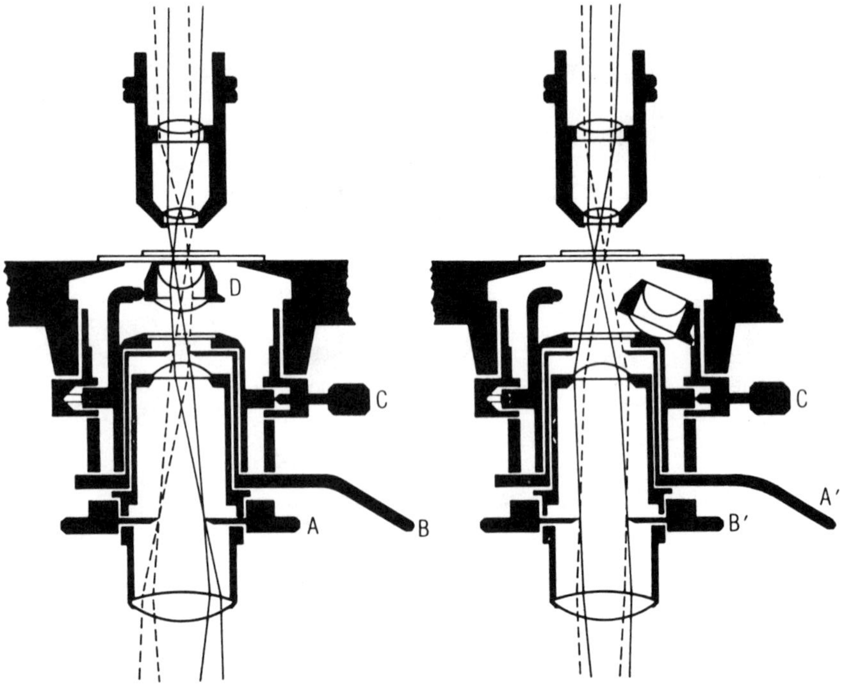

Figure 2-8. Cross section of a condenser system. In the left drawing an upper insertable element, D, is in position for higher N.A. illumination. In this case, the lever B is the aperture diaphragm and the ring, A, is the field diaphragm. The right drawing shows the condenser without the insertable element in position. In this case, the lever, A', is the field diaphragm and the ring, B', is the aperture diaphragm. This drawing emphasizes the fact that the relative location of a diaphragm does not necessarily indicate its function. C is the condenser centering adjustment. Courtesy of Wild Leitz USA, Inc.

corrected for both spherical and chromatic aberrations. The details regarding these corrections are discussed in Section III C; however, it is sufficient to say at this point, that the more the correction, the better the quality of the illumination and coincidently the greater the cost of the condenser. Early microscopists desiring the optimum in image quality often used the much more highly corrected objective lenses as a condenser lens.

The other parts of the condenser which play an important part in obtaining high quality images are the iris diaphragms. The first of these, the field diaphragm adjusts the diameter of the illuminated area of the sample. The second diaphragm, the aperture diaphragm, adjusts the angle of the cone of illuminating light. Some less expensive instruments may not be equipped with the field diaphragm but this omission will not be serious if the external lamp that is used is equipped with a field diaphragm. Some of these inexpensive instruments also employ a rotating aperture disk with a graded series of holes, each of which may serve as an aperture; these are generally inadequate for critical examinations.

The final condenser lens is that lens adjacent to the sample. It is often mounted on a pivoted arm which will allow it to be inserted at will. In this case the final condenser lens in place will permit the use of the higher aperture objectives; with the final lens out of the light path the condenser system is well matched for lower aperture objectives. Some of the final condenser lenses whether fixed or on the pivoting arm are in threaded mounts so that more than one final condenser lens may be used.

The numerical aperture (see Section III C) of the condenser system should be equal to that of the objective lens used in order to achieve the optimum resolution of the instrument. For example, if an oil immersion objective lens is used (see Section III C), a final condenser lens must be used that will allow the use of oil between it and the sample. This final condenser lens will be appropriately marked and under no circumstances should oil be used with an air condenser.

The illumination from the condenser system impinges on the sample and is either reflected from or transmitted through the sample. That light which has been modified in brightness, color and phase by the sample is then accepted by the image forming portion of the microscope.

C. The Objective Lens

If any part of the microscope can be said to be more important or more critical than any other, that part is most certainly the objective lens. The objective lens accepts the information from the sample and produces the magnified primary image of the sample when properly focused.

The objective lens is generally a small cylindrical part, usually tapered on the object side and provided with machined threads on the opposite end. Most objectives use the Royal Microscopical Society (RMS) thread. The threaded portion mates with the body tube of the microscope, a quick change carrier, or a revolving nosepiece. The objective lens system consists of one or more lenses in series to achieve the desired magnification and optical corrections.

1. Primary Objective Lens Design Parameters

Each objective lens is manufactured for a specific instrument and purpose. To achieve the optimum image for a given objective and instrument, a number of lens parameters must be considered; among these are: numerical aperture, magnification, optical corrections, tube length, immersion fluid, cover slip thickness, flatness of field, parfocality, and working distance. These parameters are generally interrelated and therein is the lensmaker's task, to establish certain of these parameters while holding the others at an optimum or at least an acceptable value. Two or more of these parameters are usually engraved on the barrel of the objective lens. Figure 2-9 shows two typical objective lenses with the parameters identified. The various objective lens parameters will be discussed in the following paragraphs as they relate to the microscopist.

Numerical Aperture. The objective lens accepts illumination from an area of the sample to produce an image. A measure of the amount of illumination an objective lens collects is its numerical aperture (N.A.). N.A. is defined as:

$$\text{N.A.} = \eta \sin \alpha \qquad (2)$$

where η is the designated refractive index between the sample and the objective lens and α is the half angle of the most oblique rays entering the front lens of the objective. It is evident from the above expression that the greater the angle of the most oblique rays entering the objective from the sample, the greater the numerical aperture. Since the half angle depends on (1) the diameter of the objective front lens and (2) the distance of the front lens from the sample, it is clear that these two design parameters for the objective strongly influence the N.A. Thus, the larger the front lens can be made and the closer that lens is to the sample, the greater the N.A.

A factor that in not always appreciated is that the effective N.A. of the objective lens is affected by the N.A. of the condenser system. In Section III C 2, it is emphasized that the N.A. of the condenser system should equal that of the objective lens for maximum resolution of the system. The combined effect of the

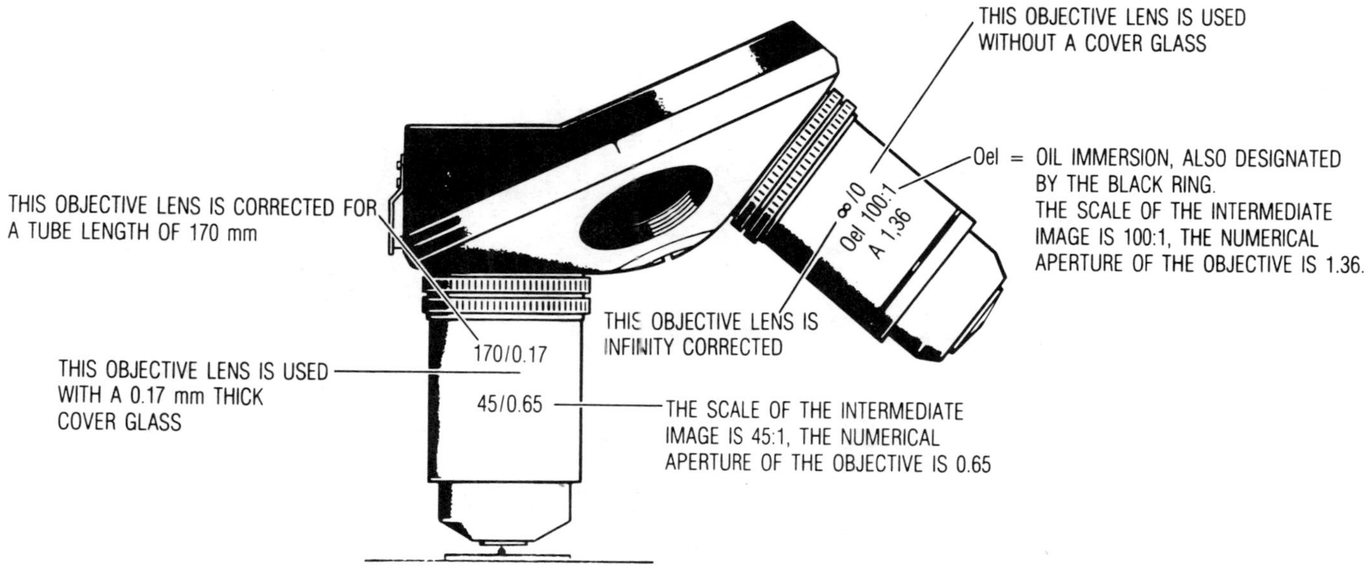

Figure 2-9. Diagram illustrating the significance of the various markings on an objective lens. The scale of the intermediate image is the magnification of the objective lens. Courtesy of Wild Leitz USA, Inc.

condenser N.A. and the objective lens N.A. gives rise to an effective N.A. for total system and may be expressed as:

$$N.A._{effective} = 0.5\,(N.A._{obj} + N.A._{cond}) \qquad (3)$$

This equation is reasonable only for values of $N.A._{cond}$ greater than zero and is a good approximation of $N.A._{effective}$ only for $N.A._{cond} \geq 0.5\, N.A._{obj}$.

The refractive index of the materials in the light path between the objective lens and the sample also has a strong effect on the N.A. In fact, it is the material with the lowest refractive index that is the controlling factor. If air is a part of the light path its refractive index of 1 is the value which must be used in Eq. (1). In this case the N.A. cannot exceed 1.0.

The N.A. is usually marked correctly on the barrel of the objective, however if there is a question about that value, it is possible to determine the N.A. using an aperometer. Aperometers are no longer available commercially but their construction and use are referenced in Chapter 10, Section V.

IMMERSION MEDIA. When suitably constructed objective lenses are used, air in the light path can be replaced with water (η = 1.33), glycerine (η = 1.440), or oil (η = 1.5-1.6); these are known as immersion objective lenses.

The water immersion objective lenses, when fitted with a stainless steel cap, have had considerable popularity in the direct examination of objects in water. The higher magnification objectives have a N.A. up to 1.20; these, however must be used with a cover glass of correct thickness for proper optical correction even though water is present above and below the cover glass. Some of the higher magnification objectives are also equipped with an adjustment collar to permit fine tuning of the objective lens to the actual cover glass thickness.

The oil immersion objective lens is often referred to as a homogeneous system since the refractive index of the oils are designed to match the refractive indices of both the front lens of the objective lens and the cover glass. The oil immersion objective lens has virtually made the water immersion objective lenses obsolete because of its insensitivity to cover glass thickness and its more brilliant image. Typical oil immersion objectives used with the immersion fluid for which they were designed, yield N.A.s as high as 1.4. Figure 2-10 compares the illumination paths for the high dry objective lenses using monobromonaphthalene have been offered with an N.A. of 1.6, however these are uncommon and difficult to use.

Finally two words of warning with regard to the proper use of immersion oils are in order. It must be emphasized that an immersion objective be used only with the immersion medium designated for that particular objective otherwise it will not produce

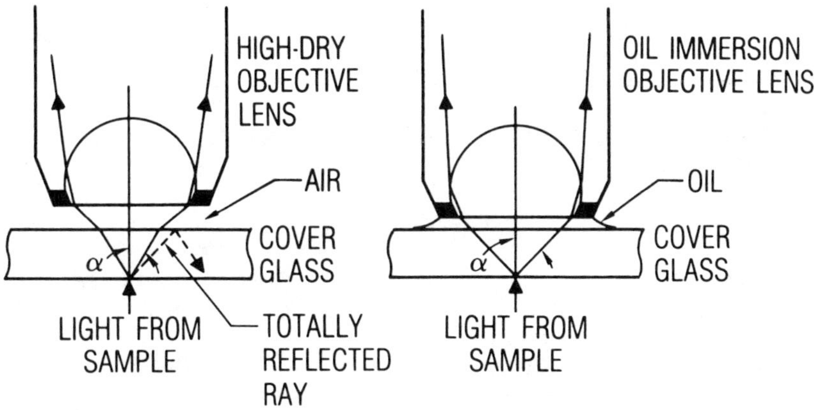

Figure 2-10. Comparison of the maximum aperture rays for a high-dry objective lens and an oil immersion objective lens. The greater the angle, α, the maximum angle rays accepted, the greater the numerical aperture of the lens and the greater its resolving power.

a satisfactory image. A listing of immersion media and their applications are given in Appendix 1. Secondly, under no circumstances should a dry objective lens be used with an immersion medium because it will not only produce unsatisfactory results but may cause permanent damage to the objective.

MAGNIFICATION. The objective lens is a positive lens, that is, it is able to produce a magnified image of an object and in that sense is the equivalent of a simple microscope. The image produced by the objective lens in the compound microscope is further magnified by the eyepiece and any intermediate lenses to produce the image observed by the eye. This image appears to be located at a distance of 25 cm below the eye. Since this image cannot be focused on a screen at this position it is referred to as an virtual image.

The total magnification of the microscope comprised of these lenses is:

$$M_T = M_O \times M_I \times M_E \tag{4}$$

where M_O is the magnification of the objective lens, M_I is the magnification of any intermediate lenses and M_E is the magnification of the eyepiece.

Magnification is described in Chapter 1, in the case of the simple magnifier, as a ratio of the angle subtended by the image of an object in the microscope to the angle subtended by that same object as viewed by the unaided eye at a distance of 25 cm. This distance, 25 cm, is referred to as the "standard distance of comfortable vision."

Another common expression for describing magnification of the compound microscope is linear magnification, that is, the ratio of the image length to the object length. The image that must be used for determining this ratio is that projected, real image found at a distance of 25 cm above the eyepiece. When this same object is viewed in the normal fashion through the eyepiece the virtual image observed is the same size as the projected, real image but located 25 cm below the eyepiece.

Typical magnifications for transmitted light objectives are 3.5X, 10X, 25X, 40X and 100X. There has been no strong rule of standardization and other values appeared from time to time. The main problem with these values is that there is not a systematic range of magnifications. The German microscope manufacturers have adapted a new series of magnifications in which each successive value is a factor of 1.25 or a multiple thereof; this is shown in Table 2-3. The objective magnifications in this scheme are shown emphasized and the appropriate eyepiece values are underlined.

TABLE 2-3. MAGNIFICATION SEQUENCE PER DIN 58 886

1	1.25	**1.6**	2	**2.5**	3.2	**4**	5	**6.3**	8
10	12.5	**16**	20	**25**	32	**40**	50	**63**	80
100	125	**160**	200	250	320	400	500	630	800
1000	1250	1600	2000	2500	3200				

In addition, intermediate lenses are used in the sequence 0.4, 0.5, 0.63, 0.8, 1, 1.25, 1.6 and 2.0 while the camera factors range from 0.32 to 4. The advantage of this system is that a large, systematic range of total magnifications can be obtained, regardless of how the various magnifying elements are combined.

One might ask if there are practical limits, both upper and lower, that should be placed on the total magnification of the microscope for any given objective lens. The answer to this question is affirmative and is based on the resolution of the objective lens and the eye (see Sections III C 3 and III E, respectively).

SPHERICAL ABERRATION. A lens is usually manufactured with one or both surfaces as sections of a spherical surface since these are easiest to generate. These spherical surfaces however, produce one of the most serious problems for critical focusing, namely, spherical aberration.

Spherical aberration arises from the fact that rays of even a single color of light passing through the outer margin of a spherical lens are focused at a different point than the rays passing nearer to the center or axis of the lens; this is shown in Figure 2-11.

Spherical aberration is manifested in the microscope image as a general haze and lack of definition even though the sample appears to be correctly focused. This effect increases with increasing N.A. of the lens. Thus, it would seem reasonable to minimize spherical aberration by simply reducing the effective N.A. of the microscope; this is true, but this simultaneously reduces the resolution of the instrument as will be discussed later.

Spherical aberration is further complicated when white or polychromatic light is used, this effect is referred to as chromatic difference of spherical aberration. John and Peter Dolland, English instrument makers in 1759 combined a convex spherical lens of crown glass with a diverging spherical lens of flint glass in such a way that spherical (and chromatic) aberrations were corrected; the improvement in spherical aberration is shown in Figure 2-11. This combination lens known as an achromatic lens is typically corrected for the yellow green portion of the spectrum. For this reason a green filter is often used in the light path with these lenses to give the sharpest images. The achromatic lens has become the basis for modern lens design.

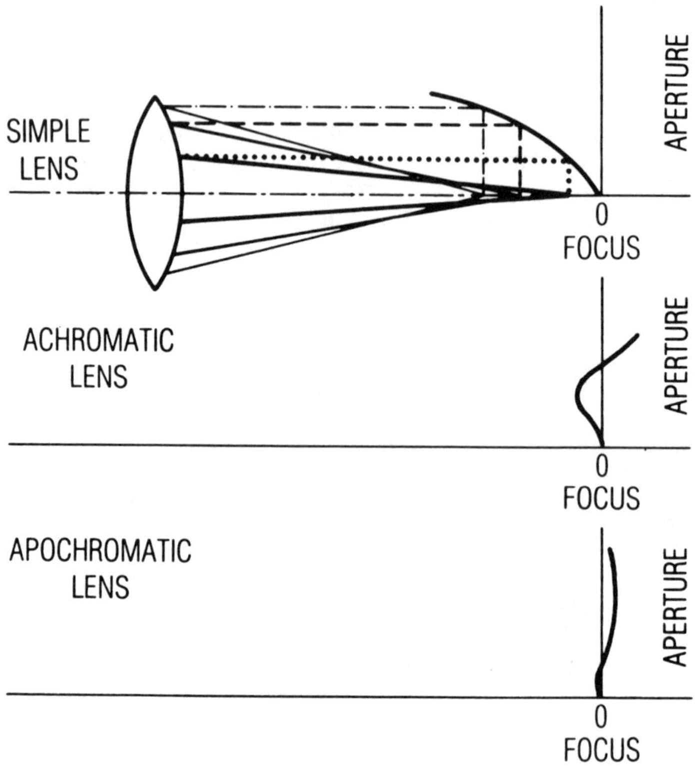

Figure 2-11. The effect of spherical aberration on the focusing of green light for various types of lenses. For the simple lens the central rays focus at zero; the rays are focused progressively closer to the lens as they emerge farther from the center of the lens. The achromatic and apochromatic lens formulas provide successively better focus for non central rays.

<u>COVER GLASS</u>. It is an unfortunate but very real fact that the careful efforts of the manufacturer to minimize spherical aberration can be thwarted through the oversight of the microscopist in the preparation of the sample. Specifically the area of concern is the cover glass. Even though it is physically separated from the objective lens and is not lens like, it has a pronounced effect on the the image quality of the high N.A. objective lenses.

Cover glasses or cover slips are generally made of thin, flat, nonfogging, crown glass (η = 1.515) that is free of bubbles. They are available in round, square, and rectangular shapes and a wide range of sizes. The important factor with respect to the objective lens is the thickness which is divided into three size ranges; #1 - 0.13-0.16 mm, #1 1/2 - 0.16-0.19 mm and #2 - 0.19-0.25 mm. The cover glass thickness designated by most manufacturers for their instruments ranges from 0.17-0.18 mm.

The higher the N.A. of the dry objective lens the more closely the cover glass thickness value must be adhered to for optimum image quality. Cover glass thickness tolerance as a function of N.A. is shown in Figure 2-12. It must be emphasized that the thickness of any mounting medium between the sample and the cover glass must also be considered in the total cover glass thickness.

When preparing the sample, every effort should be made to achieve a nominal thickness of 0.17-0.18 mm for the cover glass plus the thickness of the immersion medium between the sample and the cover glass. If that total thickness differs from the nominal value the microscope must be adjusted to minimize the spherical aberration.

Oil immersion lenses are relatively insensitive to cover glass thickness since the immersion oil has the same refractive index. In this case the thickness of the cover glass need only be less than the working distance of the objective lens so that the sample may be imaged.

If the objective lens has a correction collar for cover glass thickness that may be set to the correct value, if known, or may be adjusted visually to achieve the desired effect. This procedure is discussed in Section IV D. The only other method of correcting for cover glass thickness is by changing the mechanical tube length, however this can be done easily only if the microscope is equipped with a graduated draw tube. This procedure is also discussed in Section IV D.

It is likewise important to remember that if a sample without a cover glass, such as a smear, is to be examined at high magnifications with a dry objective, the highest quality image can be achieved only if the objective is corrected for use without a cover glass, these are available.

Finally some special objective lenses are designed for very thick sections of quartz or crown glass between the sample and the objective lens. These are for special applications such as high

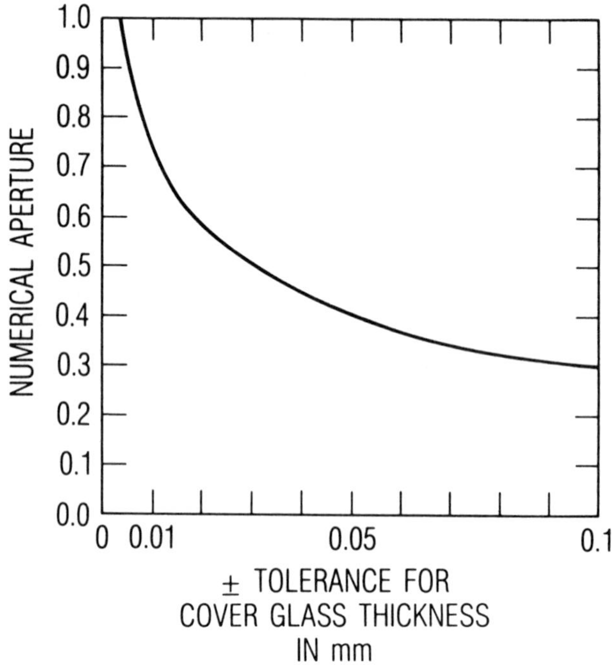

Figure 2-12. The cover glass may be considered an external element of the objective lens. Thus, the correct cover glass thickness is necessary to minimize spherical aberration. This curve gives the maximum tolerance for cover glass thickness as a function of the objective lens numerical aperture. The correct cover glass thickness for a given objective lens is often engraved on the lens body, see Figure 2-9.

temperature microscopy (Chapter 7, Section IV E) but could be used in those cases where samples must be viewed through the wall of a thick container.

AXIAL CHROMATIC ABERRATION. When the effect of spherical aberration was discussed above, the illustration was simplified by considering the effect on only one color. If we extend the illustration to include the visible spectrum, we must consider the effect known as chromatic aberration. A simple lens has a prism-like character and thus, focuses the various colors from a single radial zone of the lens at different distances. If we then include the effect of the spherical aberration, the focusing of white light becomes very complex, indeed. The effect of chromatic aberration in a simple lens is shown in Figure 2-13a.

Axial chromatic aberration is characterized by color fringes at the edges of details in the image. These fringes range from yellowish-brown one one side and violet to red on the other side in the case of minor chromatic aberration to strong blue and red fringes in the more severe cases.

The correction of axial chromatic aberration can be best described by reference to the simple achromatic lens discussed in the section on spherical aberration. This lens, shown in Figure 2-13b, consists of a converging lens made of crown glass and a diverging lens made of flint glass. The combination of these two lenses is known as a doublet lens. When white light passes through the converging lens the green light is refracted more strongly than the red light. The diverging lens has the opposite effect, i.e., it reduces the refraction of the green light relative to the red light. By proper selection of the two lenses in the doublet it is possible to make the focal length the same for red and green light. It will be noted that the blue light still has a significantly different focal length, indeed, it is not possible with the achromatic lens to bring the blue to the same focal point as the red and green. The basic reason for this is that the divergence of the flint glass is most pronounced for the blue light.

In the late 1880s Zeiss produced lenses that substituted the naturally occurring mineral, fluorite, for the crown glass in the simple achromatic objective lens. This mineral refracts the blue light more strongly than the crown glass giving a lens in which the focal length of the blue rays is matched to that of the green and red rays. In the achromatic lenses weak color fringes can still occur around the edges of an image however, the fluorite lenses produce images that are practically free from such effects.

LATERAL CHROMATIC ABERRATION. Even after removing the axial chromatic aberration there remains the lateral chromatic aberration or chromatic difference of magnification, that is, the blue image is more highly magnified than the red image. This aberration is characterized by a blue fringe around the outside of

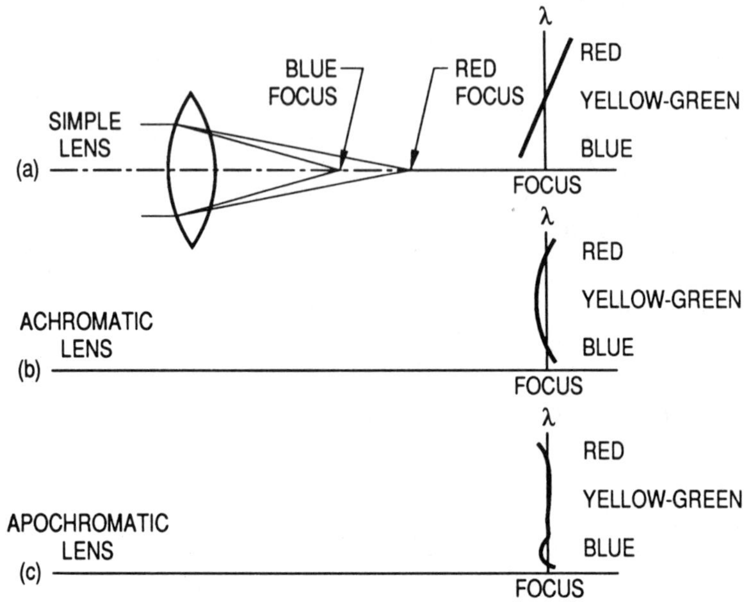

Figure 2-13. The effect of lateral chromatic aberration on the focusing of various colors. The red rays from a single radial zone of the lens focus farther from the lens than the blue rays. The achromatic and apochromatic lens formulas provide successively closer focus of all colors for a given radial zone. (The effects of spherical aberration are disregarded for this simple illustration.)

the image of an object even though the various colors are all focused at the same point. The effect is usually insignificant in the low power objectives because each lens element making up the system is an achromatic lens. However in the higher magnification objectives the front lens is very small and not easily made in an achromatic configuration; this gives rise to the lateral chromatic aberration observed.

Lateral chromatic aberration is usually removed from the final image of the higher power objectives by using a compensating eyepiece in which the red image is more highly magnified than the blue thereby canceling the lateral chromatic aberration of the objective. In this approach standard eyepieces must be used with the low power objectives and compensating eyepieces must be used with the high power objectives.

The modern microscope is equipped with a number of conveniences and one of the more important of these is the rotating nosepiece which permits the easy changing of magnifications. However to realize the appropriate corrections for lateral chromatic aberration described above the eyepieces would have to be changed when changing from low to high power objectives. This problem is solved most easily by deliberately introducing some lateral chromatic aberration in the low power objectives so that all objectives require the compensating eyepieces.

An alternative solution to the problem of lateral chromatic aberration has been introduced by Nikon in their "CF" objectives and eyepieces. In this system they have developed a completely new series of objectives in which the lateral chromatic aberration has been essentially removed. Thus, these objectives can be used with non-compensating eyepieces.

ASTIGMATISM. When an off-axis point is imaged by a lens it might be assumed that it will produce one point in the image plane. However, if the lens has not been corrected for astigmatism there will be two points, one imaged in a plane ahead of and the other imaged in a plane beyond the nominal image plane. Furthermore, these images will not be points but will appear as two mutually perpendicular lines, this is shown in Figure 2-14. Additionally, the further the point is from the axis of the lens the shorter the focal distances the two image planes become. Thus, the off-axis points are are imaged onto two planes that curve toward the lens. These coincide with each other and the nominal image plane on the optic axis.

Astigmatism is corrected when, through lens design, the curvature of the two astigmatic image planes are made to coincide. The single curved image plane that remains after correction of astigmatism is what we refer to as curvature of field. It should be emphasized that astigmatism is well under control in most modern microscopes.

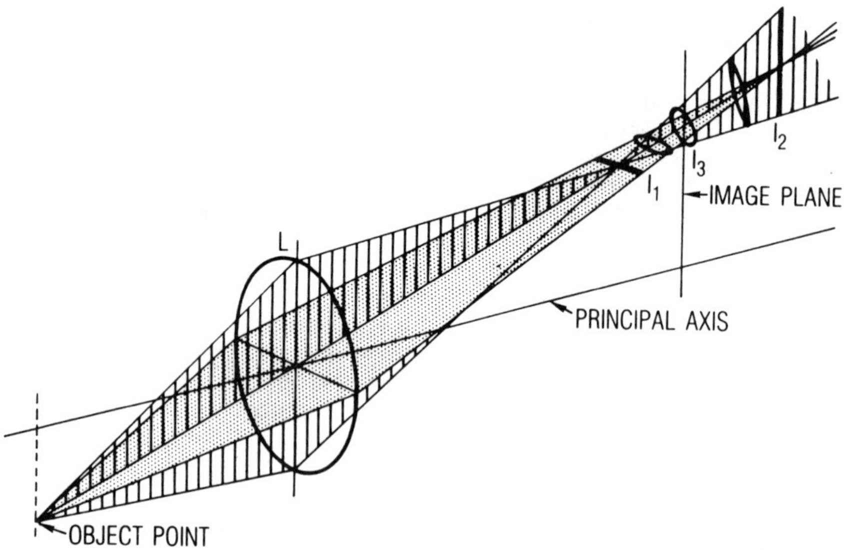

Figure 2-14. Diagram showing the effect of astigmatism on the image of an object point. The point is imaged as a tangential line, I_1, on the lens side of the principal focus, I_3, and as a radial line, I_2, on the opposite side of the principal focus.

CURVATURE OF FIELD. An inherent property of all standard objectives is curvature of field. This curvature of field may be easily observed by focusing carefully toward a thin flat sample. The central portion of the image will first appear in focus followed by progressively larger diameter rings until the outer diameter of the field of view is in focus. The manufacturer minimizes the severity of this effect by limiting the field of view through the selection of a relatively small diameter diaphragm in the eyepiece. As long as the microscopist is performing only visual observations he may wish to accept this curvature of field, since he has the option to refocus to see any desired portion of the image.

Curvature of field is directly proportional to the resolving power of the objective lens and to the aperture of the condenser system and inversely proportional to the magnification of the eyepiece. Thus one approach to minimizing curvature of field is to use a low N.A. objective lens, a narrow condenser aperture setting, and a high power eyepiece. Satisfactory photomicrographs may be achieved with a standard achromatic objective lens if a compensating or a negative camera lenses is used.

Most manufacturers have developed series of plano objective lenses which produce a large flat intermediate image while maintaining corrections for the various other lens aberrations; this type of objective lens is discussed in Section III C 4.

COMA. Coma is the asymmetrical spherical aberration of the images of off axis points. The images take the form of comet tails which become more severe as the the object points are moved away from the optical axis. Figure 2-15 illustrates the effect of coma on the image. Coma is normally well corrected in modern objectives and any appearance of this aberration may be the indication of lens damage.

DISTORTION. Distortion is an aberration in which the reproduction ratio is not a constant along the radius of the field of view. Furthermore this ratio may increase or decrease. Thus, if we image a square object we get with an increasing ratio, a cushion shaped image or with a decreasing ratio, a barrel shaped image, this is shown in Figure 2-16. These effects, known as pincushion and barrel distortion, respectively are usually present in most lenses highly corrected for the more serious aberrations. The complete elimination of this type of aberration is economical only for objectives used in critical lateral measurements or for the imaging of right rectangular objects where the distortion is very obvious and objectionable.

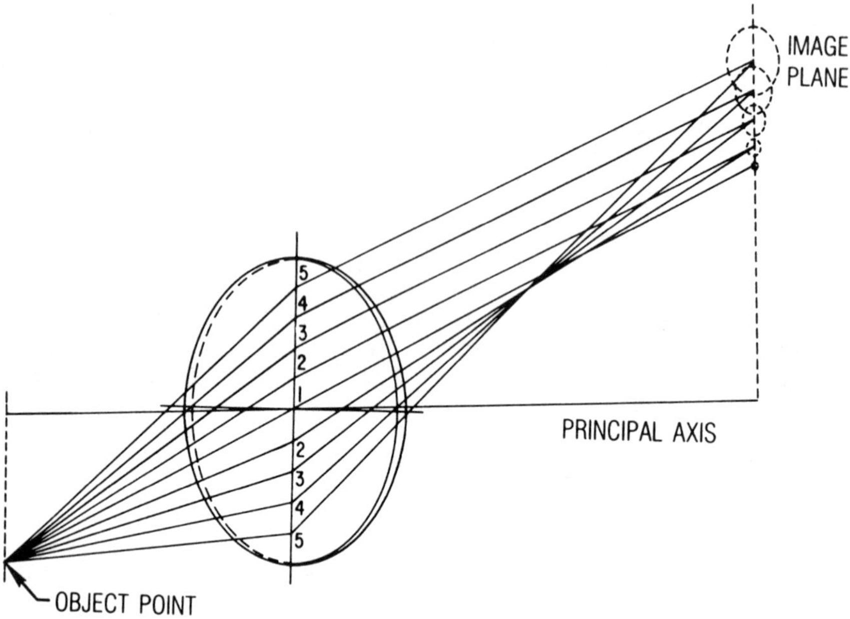

Figure 2-15. Diagram showing the effect of coma on the image of an object point. The object is imaged as a series of progressively more blurred points at the image plane as it emerges at successively greater distances from the center of the lens.

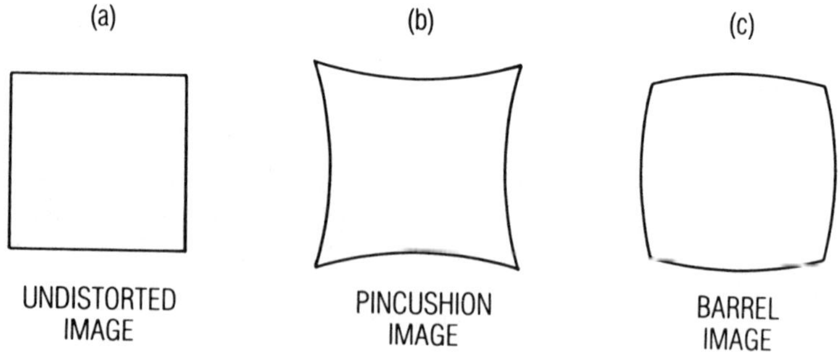

Figure 2-16. The effect of distortion on an image. If a square object, (a), is observed with a lens having pincushion distortion the resulting image is shown in (b); if the the same object is observed with a lens having barrel distortion the resulting image is shown in (c).

2. Other Objective Lens Design Considerations

PARFOCALITY. This design consideration is based on ergonomics rather than any optical considerations, that is, it makes the microscope easier to use. Formerly the changing from one objective lens to another required that the objective lens be moved away from the sample with the coarse focus knob, the objective changed or rotated into position, the approximate focus obtained with the coarse focus and then the final focus achieved with the fine focus. This involved and time consuming process was necessary because the new objective lens usually had a different "adjustment length" as shown in Figure 2-17. Failure to follow these steps could result in movement of the sample away from the area of interest at the very least. A very real possibility of incorrect procedure was also damage to the objective lens or the sample.

Most modern sets of objectives lenses are made parfocal. Strictly speaking, this means that the distance between the sample and the intermediate image is made the same for each member of the objective lens set. In a practical sense the parfocal set of objective lenses has a constant adjustment length as described above. All that is necessary when changing from one parfocalized objective lens to another is that minor refocusing be accomplished with the fine focus knob. This feature is particularly useful in speeding up focusing with the oil immersion lenses because of their small depth of field.

WORKING DISTANCE. Working distance is that distance from the end of the objective lens to the sample or cover glass, if required. This distance, shown in Figure 2-17, is generally sufficient for the transmitted light microscopist as the sample is usually thin and flat. To maximize working distance the cover glass should be no thicker than recommended for the objective and the amount of mounting medium between the sample and the cover glass should be kept to a minimum. Since the oil immersion systems are homogeneous, that is, the refractive index and dispersion of the oil and the cover slip are essentially the same, the thickness of the cover glass is not important with regard to optical correction. However, unless the cover glass is kept sufficiently thin it may be impossible to completely examine the entire thickness of the sample. In the case of specialized long working distance objective lenses the working distance usually includes a window glass of nominal thickness.

TUBE LENGTH. Mechanical tube length is a sufficiently important design parameter that it is usually engraved on the barrel of each objective lens, see Figure 2-9. The mechanical tube length is defined as the distance between the flange of the objective lens and the seating surface on which the eyepiece rests, see Figure 2-17; this value ranges from 140 mm to 240 mm.

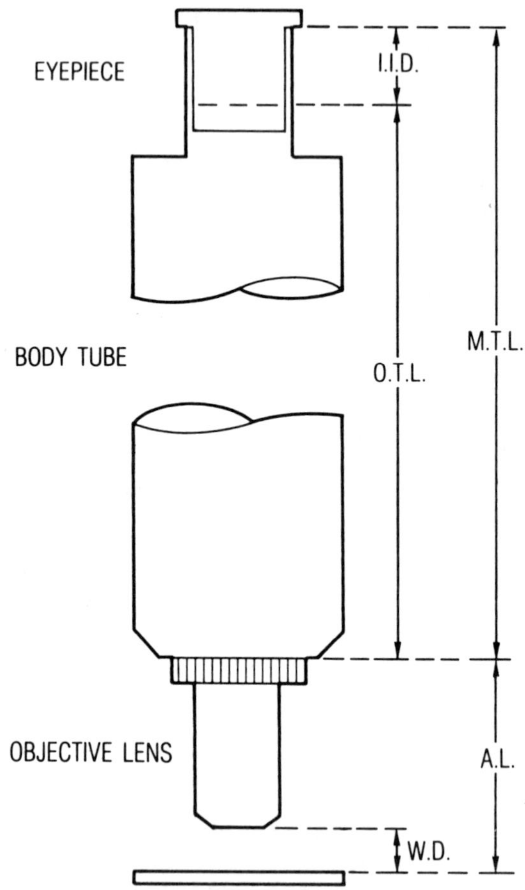

Figure 2-17. Common distance parameters in a compound microscope that must be considered in objective and eye-piece lens design.

W.D. = working distance
A.L. = adjustment length (or parfocal distance)
M.T.L. = mechanical tube length
O.T.L. = optical tube length
I.I.D. = intermediate tube length

The correction for spherical aberration is computed for each lens based on the mechanical tube length, therefore it is important that an objective lens be used with the prescribed tube length. The exception to the above rule is to be found when the spherical aberration correction of an objective lens has been affected by the use of an incorrect thickness of cover glass. This can be corrected by deliberately altering the mechanical tube length of the microscope if it is equipped with an adjustable draw tube. The required change in mechanical tube length, ΔT is given by:

$$\Delta T = 0.4\, d\, m^2 \tag{5}$$

where d is the difference from the correct cover glass thickness and m is the magnification. It must be recognized that a change in tube length does change the magnification of the system, this change is given below as ΔM:

$$\Delta M = \Delta T\ (\text{focal length}_{obj})^{-1} \tag{6}$$

In place of a definite tube length it will be noted on some objective lenses that the infinity symbol, ∞, is used. This means that these objective do not form an intermediate image in the tube but at infinity. Only in combination with a specific lens that is permanently built into the tube do they produce the real intermediate image at the correct point to be accepted by the eyepiece(s). This arrangement allows the manufacturer considerable latitude in the design length of the tube. The value of the objective lens magnification is always computed on the basis of a tube length of 250 mm. If the design of the instrument calls for a different tube length, for example 200 mm or 320 mm, the magnification of the objective is not changed; instead a tube lens is selected that has a magnification factor of 0.8 or 1.25, respectively.

3. <u>Optical Parameters Controlled By The Microscopist</u>

The following are parameters that are strongly influenced by the fixed parameters described above but at the same time are, within limits, under the control of the microscopist. By varying these parameters it is usually possible to achieve the image conditions needed for a specific study.

<u>LATERAL RESOLUTION</u>. The primary function of a microscope is to resolve the structure in the plane of the sample. This is illustrated in the following way. If two closely spaced objects in the sample plane may be distinguished as two separate objects in the microscope image, the objects are said to be resolved. The closer the two objects are to each other, and are still resolved, the greater the resolution of the microscope. Convenient measures of

lateral resolution are "limit of resolution" and "resolving power". The limit of resolution is that minimum distance between objects that may just be resolved. The resolving power is the reciprocal of the limit of resolution.

Resolution is a function of the numerical aperture, wavelength of the illumination and the coherency* of the illumination. Abbe (1873) expressed this for a sample consisting of an infinitely thin periodic grating consisting of completely opaque lines separated by completely transparent spaces as:

$$\text{Limit of resolution} = 0.5\,\lambda(\text{N.A.})^{-1} \qquad (7)$$

where λ is the wavelength of light. The sample in this case is illuminated by a coherent light source. The resolving power is likewise:

$$\text{Resolving Power} = 2\ \text{N.A.}\ \lambda^{-1} \qquad (8)$$

The limit of resolution as a function of N.A. is shown in Figure 2-18.

The limit of resolution is also strongly affected by the N.A. of the condenser system as is shown in the following expression:

$$\text{Limit of resolution} = 4\,\lambda\,(\text{N.A}_{obj} + \text{N.A}_{cond})^{-1} \qquad (9)$$

Recent studies have verified that the Abbe formula is indeed correct but only under the conditions described above. Osterberg (1950) has critically evaluated the resolution obtainable in a number of specific situations; these are summarized in Table 2-4.

Resolution is also a function of the wavelength λ of the illumination; the shorter the wavelength the greater the resolution. Since the ratio of violet and red light wavelengths is approximately 1:2 the limit of resolution can be cut in half by changing from long wavelength, red light to short wavelength, violet light. The usefulness of this effect is less than might be desired because of the decreased sensitivity of the eye in the blue and violet. Furthermore, since most objectives are corrected for green light any color of illuminating light other than green will give less than optimal results.

---

*Illumination is considered to be coherent when the beams of light in the illumination have constant phase differences during the period normally covered by the observations. Non coherent illumination is characterized by numerous and random phase differences during the period of observation.

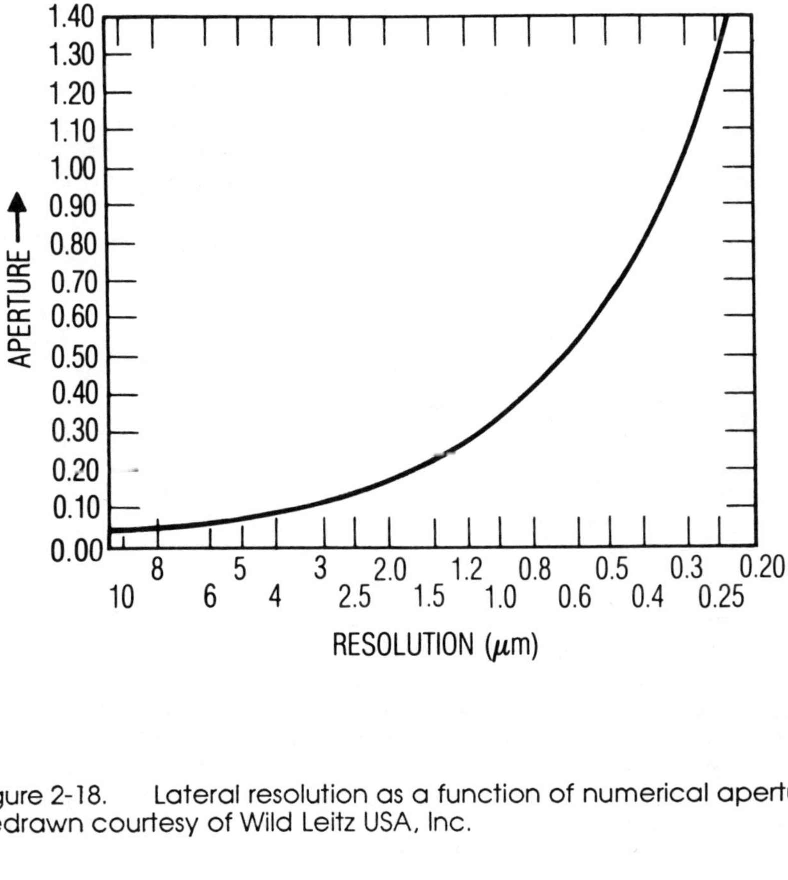

Figure 2-18. Lateral resolution as a function of numerical aperture. Redrawn courtesy of Wild Leitz USA, Inc.

TABLE 2-4.
LIMITS OF RESOLUTION FOR VARIOUS STRUCTURES

TYPE OF STRUCTURE & REFERENCE	EXPRESSION
Infinitely thin periodic structure with alternating opaque and transparent lines Abbe(1873)	$= 0.5\,\lambda\,(N.A.)^{-1}$
Lenticular fine structure Osterberg(1950)	$\leq 2\,\lambda\,(N.A.^{*})^{-1}$ N.A.*is that of the lenticular particles and is 0.05 to 0.1
Two self-luminous particles Osterberg(1950)	$= 0.48\,\lambda\,(N.A.)^{-1}$
Two pinholes in opaque sample Osterberg(1950)	$= 0.85$ to $1.2\,\lambda\,(N.A.)^{-1}$
Two slits in an opaque sample Osterberg(1950)	$= 0.73$ to $1.1\,\lambda\,(N.A.)^{-1}$

With the above information about the resolution of the objective lens we are now in a position to address the question raised earlier with regard to the upper of total magnification. To resolve two closely spaced points on the sample as two distinct points in the image not only requires an objective lens with adequate resolution, but also requires a sufficient total magnification that the observer's eye may resolve the two image points. In other words, the resolution limit of the objective lens and the resolution of the eye must now be related.

Abbe calculated an "Upper Limit" of magnification using Equation 7 and the resolution for a typical eye. It will be recalled from Chapter 1 that the normal unaided eye can resolve points with a spacing of 0.1 to 0.2 mm. Using full aperture illumination, a mean wavelength for white light of 5.5×10^{-4} mm, and a somewhat relaxed value for the resolution of the typical eye of 0.3 mm, he determined the upper limit magnification to be:

$$M_{UL} = \frac{0.3\,mm}{d} = \frac{0.3mm}{0.5\lambda N.A.^{-1}} = \frac{0.6\,mm\,N.A.}{\lambda} = \frac{0.6\,mm\,N.A.}{5.5x10^{-4}\,mm} = M_{UL} = 1090\,N.A. \quad (10)$$

The value of 1000 N.A. has come to be considered as the upper limit for total magnification which should not be exceeded. Indeed, magnifications above this value are referred to as "empty" magnification. However, Van Duijn (1957) has written an excellent series in which he states that the value, 1000 N. A. would be better as a lower limit for total magnification. He further contends that with care, M_{UL} values as high as 3200 N.A. can be used without empty magnification. Van Duijn has also has given an amplified definition of empty magnification as "every magnification value that does not increase the amount of information already contained in an image obtained at some lower magnification, regardless of the amount of trouble that has to be expended in the act of perceiving."

From the above discussion it is clear that N.A. is the important parameter of an objective lens with regard to resolving detail on a sample. It would also seem reasonable that the ideal situation would be to adjust the microscope to obtain the maximum resolution for each objective used; this is, however not the case. When the condenser system is carefully adjusted to completely fill the objective lens aperture the maximum resolution is obtained but the careful observer will note that the image lacks contrast. To markedly improve the contrast it is necessary to reduce the aperture of the condenser until about 0.8 to 0.9 of the objective lens aperture is filled.

Other aspects of a sample such as flatness of field or depth of field may dictate even further reduction of the resolution of the objective. In summary, it is safe to say that we want the maximum resolution obtainable from a given objective lens, commensurate with the other sample considerations, to give the optimum image information or produce the best photomicrograph.

DEPTH OF FIELD. Every transmitted light sample and even the lightly etched incident light sample is a three dimensional object. If one considered the objective lens to be in focus in only an infinitely thin plane we could gain information easily about the structure in the lateral plane but could get information about the structure in the axial direction only by continuously focusing up and down through the structure and mentally developing the third dimension. This would seem to indicate that it would be difficult to obtain an easy understanding of the sample structure without building up a series of photomicrographs of successive layers; to some extent this is true if we limit ourselves to photomicrography. We are spared this problem to some degree, however, by an inherent property of the objective lens, the depth of field.

Depth of field is the measure of the distance on either side of the precise focus that the image is acceptably in focus. Structure outside of the depth of field value will appear indistinct and with low powers it may interfere with the desired structure in the image. Depth of field is a function of the numerical aperture of the objective lens as shown in the following expression.

$$\text{Depth of Field} = \pm 0.5\,\lambda\,(\text{N.A.})^{0.5} \qquad (11)$$

and the total depth of field, TDF is:

$$\text{Total Depth of Field} = \eta\,\lambda\,(\text{N.A.})^{0.5} \qquad (12)$$

Eye accommodation affects the total depth of field as follows:

$$\text{Total Visual Depth of Field} = \text{TDF} + 250\,M^{0.5} \qquad (13)$$

where M is the total magnification of the microscope. In this equation the assumption is that the observer's eye can accommodate on an image that is 250 mm distant.

Depth of field is increased by increasing the wavelength of light and by decreasing N.A. of the system. Typically if we want to increase the depth of field we use an objective with a lower N.A. or decrease the aperture diaphragm setting for a given objective lens. Depth of field can be increased by using a red filter in the illumination but only with a semi-apochromatic or an apochromatic objective lens.

<u>DEPTH OF FOCUS</u>. Depth of focus is often confused with depth of field. The relation of these two may be thought of in this way; depth of field, as discussed above, is in the sample area whereas, the depth of focus is in the sample image. Depth of focus is related to depth of field by the following expression:

$$\text{Depth of Focus} = \text{Depth of Field} \times M_t^2 \qquad (14)$$

where M_t is the total magnification of the microscope. Depth of focus is increased in the same manner as depth of field as well as by the following additional factors; (a) select an objective with the lowest magnification for the desired numerical aperture and (b) use a low power eyepiece.

<u>AXIAL RESOLUTION.</u> If we wish to resolve fine detail in the direction of the microscope axis, i.e., axial resolution, the requirements are exactly the opposite from that of depth of field. These

requirements are: (a) use objectives with the highest N.A. and magnification; (b) use eyepieces of high power; (c) open the condenser aperture stop to use the full aperture of the objective lens; (d) use a blue filter, but only with an apochromatic objective lens; (e) use a long bellows extension in photomicrography. Typical axial resolving powers for various N.A. values are given as a function of total magnification in Figure 2-19.

IMAGE BRIGHTNESS. The brightness of the image depends on the brightness of the illumination, the light loss due to the absorption and reflection of the sample and other elements in the optical path and the numerical aperture on the image side of the objective lens, $N.A._I$. $N.A._I$ is given by the following expression:

$$N.A._I = N.A. \times M^{-1} \qquad (15)$$

where N.A. and M are the numerical aperture and the magnification of the objective lens, respectively. Image brightness, B_I, is then proportional to the square of $N.A._I$:

$$B_I \propto N.A._I^2 \qquad (16)$$

A comparison of image brightness for several objectives is given in Table 2-5.

TABLE 2-5. PROPORTIONAL BRIGHTNESS OF SEVERAL OBJECTIVE LENSES

MAGNIFICATION	N.A.	$N.A._I$	PROP. BRIGHTNESS
6	0.10	0.017	2.8^{-4}
25	0.50	0.020	4.00^{-4}
63	0.85	0.013	1.69^{-4}
100(oil)	1.30	0.013	1.69^{-4}

It must be remembered that the effective N.A. of the objective lens is affected by the N.A. of the condenser system. Therefore, the image brightness can be reduced by reducing the setting of the aperture diaphragm in the condenser system; this however, simultaneously reduces the resolving power of the system. In summary, one can control the brightness of the image to some extent by changing the N.A. of the system but since it affects other more important parameters, brightness should be changed by altering the power applied to the lamp or through filtration.

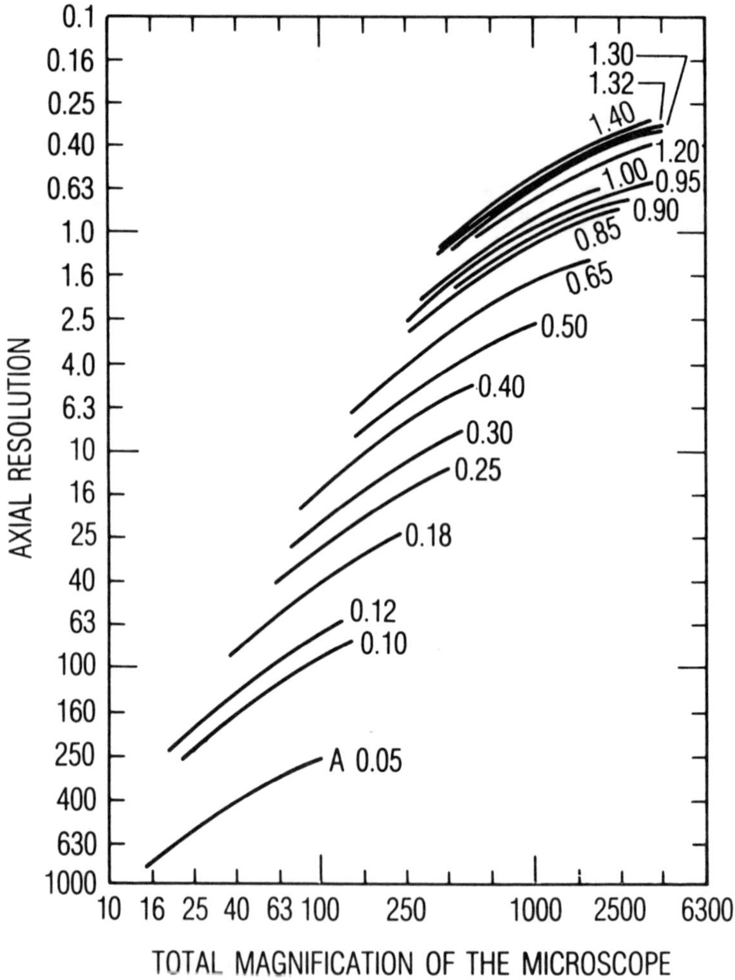

Figure 2-19. Axial resolution as a function of total magnification for objective lenses of various numerical apertures. Redrawn courtesy of Wild Leitz USA, Inc.

4. Types of Objective Lenses.

When all of the above parameters are considered it might seem that there would be a large number of possible objective lens types that could be produced. This is theoretically true but most objective lenses are classified by their correction for chromatic aberration. It is not surprising that as color correction is improved, the means to achieve this becomes more complicated, so that the classification into achromats, semi-apochromats and apochromats correlates directly with price for the objective lens.

ACHROMATIC OBJECTIVE LENS. These objective lens are fully corrected spherically for one color, green and chromatically for red and blue-green; this is shown diagrammatically in Figures 2-11b and 2-13b. The colors intermediate to red and blue-green are also greatly improved over the full aperture of the objective lens. The image quality is particularly good in the green to yellow portion of the spectrum, to which the eye is quite sensitive. The use of a yellow-green filter is recommended for black-and-white photomicrography. The green filter particularly absorbs violet light that does not focus in the plane of the yellow-green image, but will nevertheless expose the photographic emulsion.

SEMI-APOCHROMATIC OBJECTIVE LENS. The semi-apochromatic or fluorite objective is the next in the order of optical correction. The slight residual aberration that remains in the achromatic objective lens for colors whose intercept lengths are not equalized is called secondary spectrum. This secondary spectrum has the effect of reducing the color contrast during observation of the sample. To eliminate this problem, the crown glass in the achromatic lens system is wholly or partially replaced in the semi-apochromatic lens system with cubic, crystalline calcium fluoride known as fluorite or fluorspar. Synthetic fluorite is used almost exclusively in the modern fluorite objective lenses. These objective lenses are spherically and chromatically corrected for two colors.

As a result of the improvements which are brought about by the use of the fluorite in the optical system, the semi-apochromatic objective lenses can be designed with higher numerical apertures that give brighter images and higher resolution. These lenses are suitable for color photomicrography. The newer fluorite objectivelenses are particularly desirable for fluorescence microscopy because of the very low background fluorescence of the synthetic fluorite.

APOCHROMATIC OBJECTIVE LENS. These objective lenses represent the ultimate of the lensmaker's science with regard to color correction. Not only has the secondary spectrum been eliminated but the remaining tertiary spectrum is for all practical purposes insignificant. The apochromatic objective lenses are chromatically corrected for three colors and spherically corrected for two colors; these corrections are shown schematically in Figures 2-11c and 2-13c.

The image produced by the apochromatic objective lens is characterized by its crispness and sharpness; the contrast and resolution are the best that can be obtained. These factors make the apochromatic objective lens ideal for the most discriminating work where the recognition of fine structural features is critical. These objective lenses also are the optimum for color photomicrography.

PLANO OBJECTIVE LENS. In visual observations the microscopist can accept some curvature of field since any area that is slightly out of focus can be observed by simply adjusting the fine focus knob. Photomicrography however, is a different situation, where sharpness is required over the entire field of view. Thus, as the importance of photomicrography has increased so has the need for optics which will yield a flattened field. It is true that field flatness can be improved by reducing the effective N.A. of the system but this also reduces the resolution of the image. Perhaps one of the most important advances in microscope optics in the last 30 years are the plano-corrected optics. New objective lens formulations making use of fluorite or completely new glass compositions in conjunction with appropriate eyepieces have almost eliminated curvature of field. It should be emphasized that these changes could not be realized without the use of low reflection coatings to reduce the glare from the many glass surfaces in these lenses. The results have been so impressive that in some cases larger than normal diameter eyepieces have been developed to fully display the larger fields of view. Figure 2-20 shows the marked improvement in the flatness of a selected field of view.

The plano objective lens clearly produces an excellent, flat image, therefore one might ask why these objective lenses are not used to the exclusion of the standard series of lenses. There several reasons why this is not the trend. First, the large, flat field is achieved at the expense of greater sensitivity to cover slip thickness. The oil immersion objective is usually unaffected by minor differences in cover glass thickness because the immersion oil has about the same refractive index as the cover glass. However, in the case of the plano oil immersion objective the slight differences in both the refractive index and the dispersion of the oil and the cover slip are sufficient to reduce the quality of the image.

Secondly, the plano objective lens is at its best with relatively thin specimens, e.g., a smear.

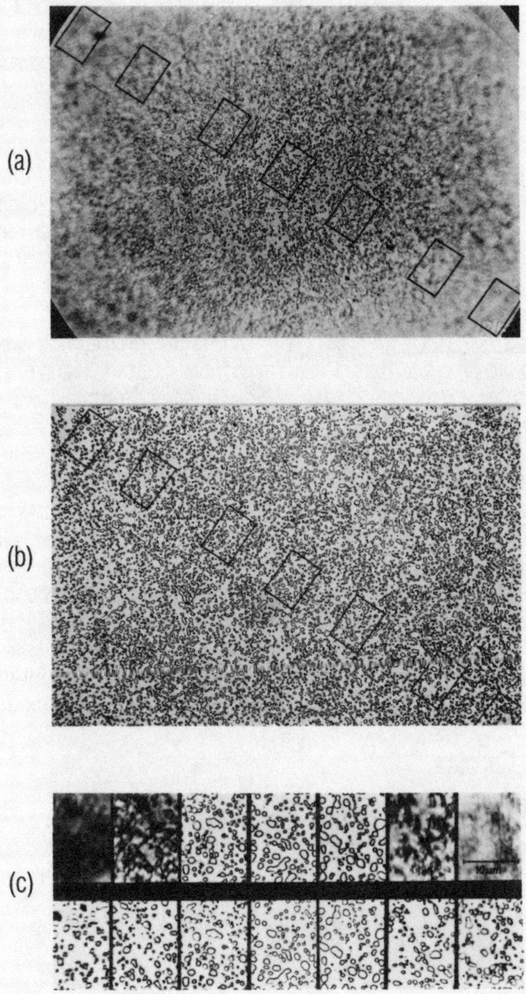

Figure 2-20. A comparison of the images of a sample take with a standard achromatic objective lens, (a), and plano objective lens, (b); (c) shows equivalent areas from (a) and (b). Courtesy of Wild Leitz USA, Inc.

Thirdly, the aperture diaphragm should be as wide open as possible. Since this reduces the contrast significantly samples should be chosen with pronounced absorption detail. One way to improve marginal samples is to over stain or over etch them.

Finally, one must consider the cost since the plano objective lenses are considerably more expensive than the equivalent, standard, achromatic objective lenses.

In summary, the plano objective lenses give best results when flatness of field is the principal concern. They require the use of specific eyepieces which complete the optical correction of the objectives. They also require carefully prepared samples consisting of a thin preparation with good contrast and covered with a cover slip of the correct thickness.

STRAIN FREE OBJECTIVE LENS. When transparent materials are examined in crossed polarized illumination (see Chapter 4, Section VI), any anisotropy due to crystallographic properties or strain is shown as light areas in a dark field. Since such measurements are possible using the optical microscope it is important that none of the observed effects are due to strain in the microscope optics, hence the need for strain free objective lenses and condensers. These lenses are produced from selected blanks of glass which are relatively strain free and mounted in the lens barrels in such a way as to minimize strain on the glass elements. Even with this care the finished objective lenses are subjected to a careful selection process to assure that the strain is minimal. Since these objective lenses are essentially strain free, every effort must be made to avoid mechanical shock which might induce strain in the lenses.

D. Intermediate Optics.

Intermediate optics in this text refers to those lenses or groups of lenses that are an integral part of the instrument and that are positioned in the tube between the objective lens and the eyepiece.

1. Corrector Lens.

Objectives for transmitted and normal incident illumination are usually corrected for different tube lengths. To use these objectives on a single microscope it may be necessary to use a corrector lens for one or both types of objectives to match them to the tube length of that instrument.

2. Telan Lenses.

Telan Lenses are a matched pair of intermediate lenses which allows the insertion of an accessory, such as a prism polarizer; that requires parallel illumination in the normally diverging beam from the objectives. The first of the Telan pair produces a parallel beam which traverses the polarizer; the second Telan lens following the polarizer then reproduces the original diverging beam from the parallel beam.

3. Tube Lens.

The tube lens, as discussed in Section III C 2, is an intermediate lens used in conjunction with infinity corrected lenses to produce the diverging beam that produces the intermediate image at the correct point for the eyepieces. Since the beam from the infinity corrected objectives is parallel this allows considerable latitude in the tube length and simplifies the addition of a binocular head.

The German manufacturers have standardized the tube lens magnifications in the same geometrical progression of 1.25 that they have used for objective lens and eyepiece magnifications. The values of the tube lens magnifications normally used in this series are 0.4, 0.5, 0.83, 1.0, 1.25, 1.6, and 2.0X.

To achieve longer than normal tube lengths on microscopes that do not use infinity corrected objective lenses another intermediate lens system, the relay lens, must be used. The relay lens is positioned such that it accepts the intermediate image of the objective lens and projects it further on in the tube to the eyepiece focal plane. The relay lens system has been used to provide the additional tube length necessary for binocular viewing bodies and the additional space required between the objective lens and the eyepieces for large console research microscopes. Typical magnifications for relay lenses range from 1.25 to 2.0X.

4. Zoom System.

Several manufacturers are producing compound microscopes with continuously variable intermediate lens systems known as zoom systems; a typical zoom systems is shown in Figure 2-21. The zoom system is capable of magnification ranges as high as 4:1 on some instruments. The major advantage of the zoom system is that the optimum compromise can be obtained between magnification and other parameters such as working distance, field of view, depth of focus and image brightness. Total magnification can also be adjusted so that the units of an eyepiece reticle can be made to exactly correspond to a unit dimension on the sample.

5. Bertrand Lens.

The Bertrand lens is a convergent lens which may be introduced at will into the image beam within the tube of the microscope. The purpose of the Bertrand lens is to form a simple

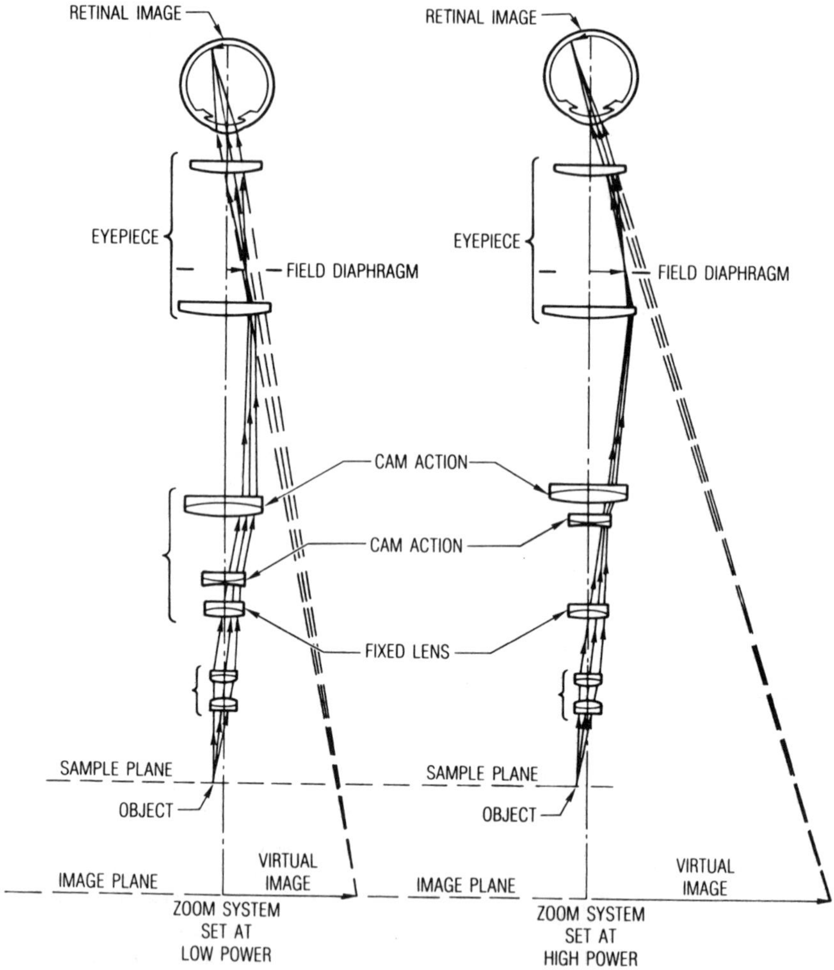

Figure 2-21. Cross sections of a microscope with zoom system set at two different magnifications. Redrawn, courtesy of Cambridge Instruments Inc., Optical Systems Division.

telescope which is focused at the rear focal plane of the objective lens. The Bertrand lens on the research quality instruments is usually equipped with focusing and centering adjustments.

The major use of the Bertrand lens is to produce a "conoscopic image" that is used extensively in mineralogy, ceramics and chemical microscopy to determine crystal orientation and phase identification. This application will be discussed further in Chapter 5, Section V.

Other uses for the Bertrand lens include observing the aperture diaphragm setting and verifying uniform illumination of the sample. In phase microscopy, the Bertrand lens may be used instead of the usual telescope eyepiece to set the illumination annulus on the phase ring, see Chapter 4, Section IV.

E. The Eyepiece.

The eyepiece is a self-contained, cylindrical object that is designed to fit into the upper end of the microscope body tube, binocular body or trinocular body. The nominal diameter of the eyepiece is 23.2 mm but larger sizes such as the 30 mm widefield eyepieces are becoming increasingly more common. The eyepiece is terminated at either end with a lens, the upper end having a wide flange on which the eyepiece rests when in the microscope. The lens at the upper end is the eye lens, which provides the major magnification of the eyepiece. The opposite lens, known as the field lens is designed to efficiently collect the illumination from the entire diameter of the intermediate image and present it to the eye lens. The eyepiece also has a field limiting aperture but the location of this varies with the eyepiece type. Since the eyepiece is frequently handled, it is worthwhile to caution against touching the lens elements since the resulting contamination can degrade the image and sometimes the antireflection coatings.

The eyepiece is designed to form a magnified image of the intermediate image from the objective lens, this can take place in one of two ways. The light from the eyepiece first converges to a minimum at the exit pupil of the eyepiece, often referred to as the "Ramsden disk", then diverges to form a real image of the sample at about 25 cm above the eyepiece. This image can be seen on a ground glass or can expose a photographic emulsion in a suitably darkened camera assembly. The second and the most common way in which this image may be seen is by placing the eye so that the lens of the eye is situated at the exit pupil. The eye then perceives an inverted virtual image of the sample which appears to be about 25 cm below the eyepiece. These two cases are shown in Figure 2-22.

Before we leave the discussion of the exit pupil it is worthwhile to indicate that it is actually the image of the rear focal plane of the objective lens and as such it depends on the N.A. of the objective

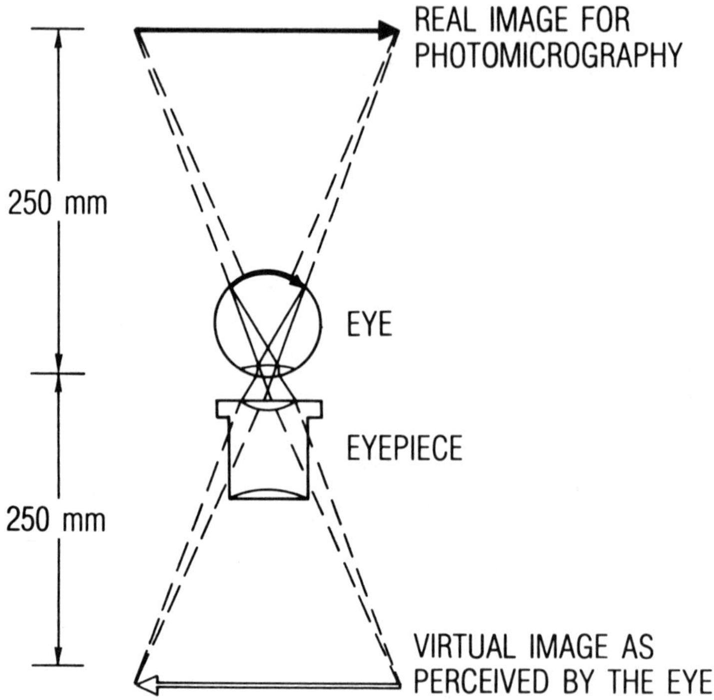

Figure 2-22. Schematic relationship between the virtual image as perceived by the eye and the real image that exists above the eyepiece; this latter image is the one used for photomicrography.

lens and the total magnification of the microscope. Hartley (1979) describes the relationship between the diameter of the exit pupil, D_E, the N.A., and the total magnification as

$$D_E = \frac{2D_V \text{ N.A.}}{M_T} = \frac{500 \text{ N.A.}}{M_T} \quad (17)$$

where D_V is the distance, 250 mm, between the eye and the virtual image of the sample.

It will be recalled from Chapter 1 that the pupil of the eye in bright light is about 3 mm in diameter. If the diameter of the exit pupil is greater than 3 mm, the effective N.A. of the microscope is reduced. If we require the diameter of the exit pupil to be that of the pupil of the eye and rearrange equation 18 we have

$$M_T = 166 \text{ N.A.} \quad (18)$$

Therefore, to take full advantage of the resolution capabilities of the objective lens, the minimum useful magnification M_U of the microscope must be at least 166 times the N.A. Minimum eyepiece magnifications to fulfill this requirement are give in Table 2-6.

TABLE 2-6. MINIMUM USEFUL MAGNIFICATION (M_U)

OBJECTIVE MAG.	N.A.	M_U	MIN. EYEPIECE MAG.*
3.5X	0.10	17	5X
6 X	0.18	30	5X
10 X	0.25	42	5X
25 X	0.50	83	4X
40 X	0.65	108	3X
63 X	0.85	141	3X
100 X	1.30	216	2X

*It should be emphasized that most modern eyepieces for visual observation have a magnification from 6X to 15X.

The optical parameters of the eyepiece that are of concern to the microscopist are; (1) magnification which influences the total magnification of the microscope, (2) diameter of the field limiting aperture, (3) the optical corrections incorporated in the eyepiece, (4) the eye relief, and (5) viewing angle. These will be discussed in detail below.

Magnification of modern eyepieces is clearly marked on the upper surface or around the periphery where it is clearly visible when in the microscope. These magnification values are followed by "X". Older eyepieces have been marked variously with Roman numerals, Arabic numerals or letters; while these are not the magnification of the eyepiece they are generally in order of increasing magnification. Eyepieces for visual observation are manufactured in the range 3X to 30X, however the most useful eyepieces range from 6X to 15X.

A number of the German manufacturers have also applied the geometrical progression factor of 1.25 to the magnifications of eyepieces. In this case the magnifications include 6.3X, 8X, 10X, 12.5X, 16X, and 25X. These magnifications in combination with those of similarly designed objective lenses and intermediate lenses, if used, give a wide range of total magnifications that also are related by the 1.25 progression factor.

The field of view in the microscope is limited in its extent by the diameter of the field limiting aperture. This aperture's purpose is to exclude that outer portion of the image that is inferior in quality due to the various aberrations of the objective lens. In general, the higher the quality of the objective lens, the larger the acceptable area of the intermediate image and the larger the field limiting aperture. Since the external diameter of the eyepiece is 23.2 mm, the maximum diameter of the field limiting aperture is limited to about 18 mm in diameter. If a larger diameter field of view is desired, a larger diameter eyepiece must be used.

The field number of an eyepiece is the diameter of the field limiting diaphragm in mm (or, more correctly, the theoretical diaphragm diameter necessary to contain the same field of view as does the actual diaphragm with the field lens). The field number, if available, can be divided by the objective magnification to determine the actual diameter (in mm) of the sample seen through the eyepiece. For example an eyepiece with a field number of 12.5 used with a 10X objective lens permits an area of the sample 1.25 mm in diameter to be viewed. Typical field numbers for standard eyepieces range from about 6 to 20, whereas the 30 mm diameter, widefield eyepieces have field numbers from 24 to 28.

The optical corrections of an eyepiece are of two types, the simple form and the compensating form. The simple or non-compensating eyepieces generally are constructed with simple lenses and as such are not able to correct any of the residual aberrations in the intermediate image. As the name implies, the compensating eyepieces do correct the remaining aberrations. Since these residual aberrations are very specific for a given manufacturer's objective lenses it is obvious that only that manufacturer's eyepieces could be reasonably be expected to give an optimum image. Each of the specific eyepiece types will be discussed in detail later.

Eye relief is the distance between the eye lens and the exit pupil. The pupil of the observer's eye must reside at the exit pupil to survey the entire field of view. Generally, eye relief decreases as the power of the eyepiece is increased. Eyepieces with a small eye relief (12 mm or less) are generally acceptable for those not requiring eyeglasses and may even be useful for beginning microscopists since the minor contact of the face with the upper surface of the eyepiece does provide some stabilizing influence for easier viewing of the image.

A large eye relief (~ 20 mm) is almost a necessity for those who require eyeglasses and generally desirable for those who do not. Most manufacturers today provide a series of eyepieces with sufficient eye relief that eyeglasses may be used comfortably. Leitz and Zeiss have even identified their "high-point" eyepieces of this design by a small eyeglass emblem on the barrel of the eyepiece. Figure 2-23 illustrates the difference in the normal and high-point eyepieces. Soft rubber guards may be available to protect the eyeglass lenses from contact with the eyepiece and possible scratching.

A parameter of less importance is the intermediate image distance (see Figure 2-17) since it applies only if one manufacturer's eyepieces are being used on another manufacturer's instrument. Since the intermediate image distance may differ for the various manufacturers, this provides yet another reason for not interchanging eyepieces. The simple eyepieces used in microscopy are almost exclusively either the Huygens or the Ramsden eyepieces; these are shown in cross section in Figure 2-24. Both lenses use plano convex, simple lenses but that is where the similarities end.

1. The Huygens Eyepiece.

The Huygens eyepiece, developed by Christian Huygens in 1681, is the most common and least expensive of the eyepieces. The major feature of this eyepiece, aside from its low cost, is that it can be designed to be free from lateral chromatic aberration. Therefore, this eyepiece is suitable for use with most low and medium power (less than 40X) achromatic objective lenses. The Huygens eyepiece may be equipped with a reticle which is mounted at the level of the field limiting aperture.

Unfortunately, this eyepiece does have several shortcomings, among these are a low eye relief, especially for the higher magnifications and a large amount of chromatic aberration and distortion in the image of any reticle used. The latter is minimized by keeping the reticle so small that it is seen only in the center of the image.

The Huygens eyepiece has several identifying characteristics: (1) the convex or curved side of the field lens is on the exterior of the eyepiece, (2) the field limiting aperture is within the body of the eyepiece, (3) the edge of the field limiting aperture has a bluish fringe when viewing an image through the eyepiece and, (4) the eyepiece by itself cannot be used as a hand magnifier.

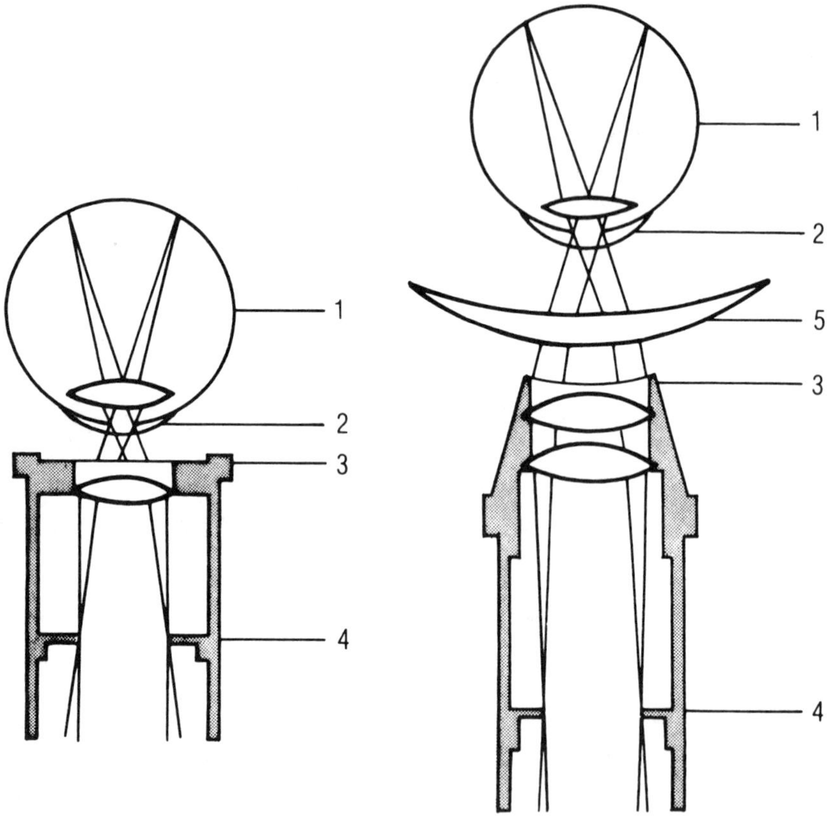

Figure 2-23. A comparison of the eye position for standard and high eyepoint eyepieces. Redrawn courtesy of Wild Leitz USA, Inc.

1. Eye
2. Eye lens
3. Eyepiece
4. Field limiting aperture
5. Eyeglass lens

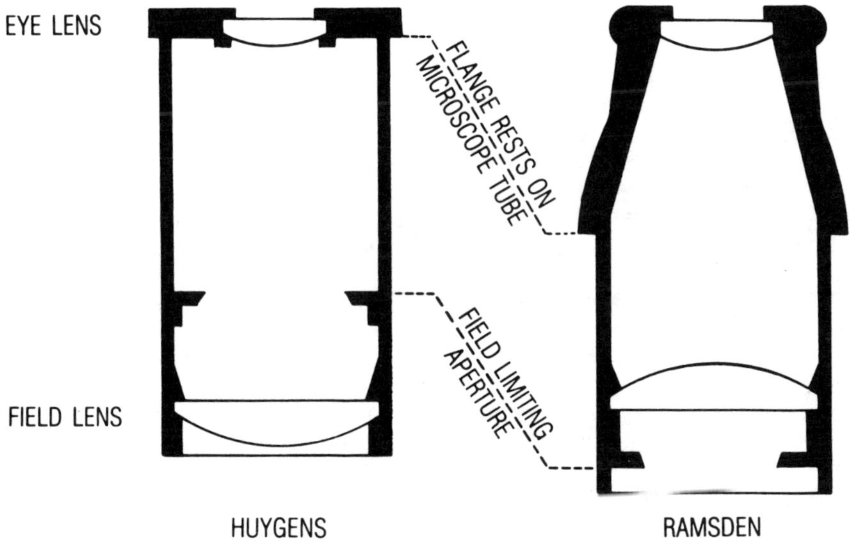

Figure 2-24. Cross sections of the two basic types of microscope eyepieces. The two types may be distinguished in that the Ramsden eyepiece can be used as a magnifier because its field limiting aperture is external to the two lenses.

2. The Ramsden Eyepiece.

The second simple eyepiece, the Ramsden eyepiece, developed by J. Ramsden in 1782 is characterized by a flatter field than the Huygens eyepiece. A major feature of this eyepiece is that the field limiting aperture is located below the field lens; thus any reticle placed there has no more distortions or aberrations than the sample image because both are acted upon by the field lens and the eye lens. As a result of this feature the Ramsden eyepiece is often used in the micrometer eyepiece which requires an undistorted field of view.

The chacteristics of the Ramsden eyepiece are: (1) the plano or flat side of the field lens on the exterior of the eyepiece, (2) the field limiting diaphragm is below or outside of the field lens, (3) the edge of the field limiting aperture has a bluish fringe when viewing an image through the eyepiece, and (4) the eyepiece may be used by itself as a hand magnifier.

3. The Compensating Eyepiece.

These eyepieces are designed to complete the corrections necessary for a specific series of objective lenses of a given manufacturer. As was mentioned earlier these corrections are so unique to each objective lens series that the chances are remote that proper compensation could be obtained with any other manufacturer's eyepieces. These eyepieces are generally based on either the Huygens or the Ramsden eyepieces, but differ from these in that achromatic lenses are used rather than simple lenses.

The compensating eyepieces are over corrected with regard to lateral chromatic aberration they should be used with achromatic objectives having N.A.s > 0.65, all semiapochromatic, all apochromatic and all plano objectives since they are all under corrected for lateral chromatic aberration.

These eyepieces are usually identified on the upper rim by the manufacturer with the trade name for their specific compensation. A few of these trade names are listed in Table 2-7. If unmarked, these eyepieces can usually be identified by the yellow to orange fringe at the edge of the field limiting aperture when viewing the sample.

4. The Fixed Reticle Eyepiece.

The measurement of the size, shape, number, or angular relationship of objects in the image can be accomplished most simply with a fixed reticle eyepiece which superimposes an image of a reticle on the image of the sample. Reticle eyepieces differ from the ordinary versions by virtue of the reticle mounted at the

TABLE 2-7. TRADE NAMES FOR COMPENSATING EYEPIECES

EYEPIECE TYPE	MANUFACTURER
Hyperplan	Bausch & Lomb
Periplan	Leitz
H, HK, & H.K.W.	Nikon
WHK & H.E.P.BiWF	Olympus
WPK	Reichert
Holoscopic	Watson
Compensating	Wild
UW	Unitron
Complan	Vickers
C (Compensating	Zeiss
Kpl (Komplan)	Zeiss
PK	Zeiss (aus Jena)

level of the field limiting aperture and a threaded adjustment to provide focusing of the reticle image. In some eyepieces the reticle may be changed by removing the threaded cell holding the field lens; this exposes the reticle that is held in place with a threaded ring.

Typical reticles are shown in Figure 2-25. The cross-wire reticle is most useful in locating a particular area of interest in the image and defining the polarizer and crystallographic directions in petrographic microscopy (see Chapter 5, Section V). In the latter application the barrel of the eyepiece is also fitted with a precisely located pin which fits in any one of several slots in the body tube to provide a fixed orientation of the cross-wire relative to other elements in the microscope.

The micrometer reticle is very important since it permits the measurement of objects in the sample with speed and modest accuracy. The scale of the micrometer reticle is generally numbered every ten divisions. The actual length of the scale units is unimportant since the corresponding values for the sample image depend on (1) the magnification of the objective lens, (2) magnification of any intermediate lenses, (3) tube length setting and (4) the interpupillary distance setting of the binocular body (unless compensation for tube length difference is provided). Thus this reticle must be calibrated for each total magnification using a stage micrometer (see Chapter 5, Section II A). Other special reticles are discussed in Chapter 5, Section II and Chapter 7, Section VII.

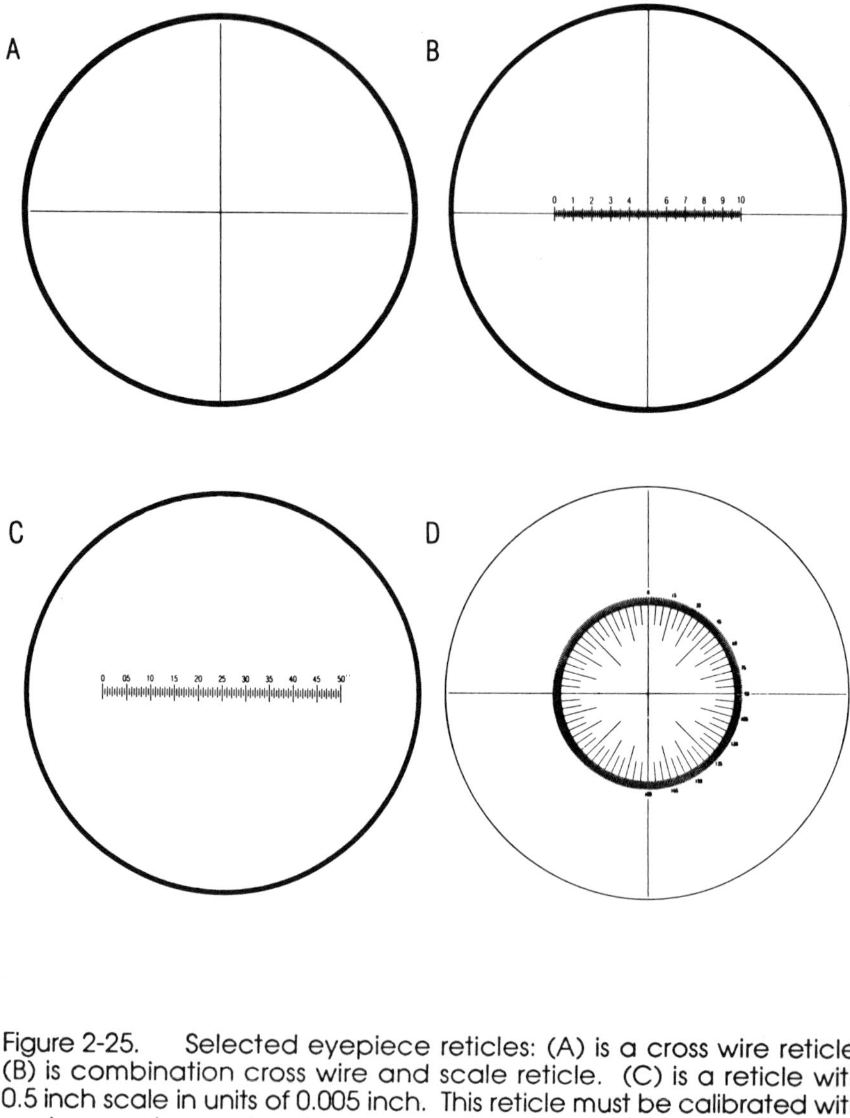

Figure 2-25. Selected eyepiece reticles: (A) is a cross wire reticle. (B) is combination cross wire and scale reticle. (C) is a reticle with 0.5 inch scale in units of 0.005 inch. This reticle must be calibrated with a stage micrometer for a specific objective lens and zoom setting. (D) is a protractor reticle that may be read to 1 degree. Additional reticles are shown in Figures 5-2 and 5-5. Courtesy of Cambridge Instruments Inc., Optical Systems Division.

5. The Widefield Eyepiece.

The viewing angle of a standard modern eyepiece is about 30°; with smaller angles the microscopist has the impression of looking down a tunnel toward a opening. Most of the major manufacturers now produce a series of wide field eyepieces for their instruments which may have viewing angles as large as 60°; however, to accommodate the larger intermediate image, the diameter of the eyepiece is larger than the the standard eyepiece. Figure 2-26 illustrates the increased intermediate image diameter of the wide field eyepiece.

IV. PROCEDURES FOR BRIGHTFIELD TRANSMITTED LIGHT MICROSCOPY

This section will cover the various procedures for setting up a brightfield transmitted light microscope from the occasional selection of an optimum location for the instrument to the many details of daily operation.

A. Instrumentation Location

The room requirements for proper use of the microscope should be carefully evaluated both for the proper performance of the instrument and for operator comfort. The latter may seem to be a luxury, however, fatigue of eye and body can be minimized through effective design of the room and the furniture. The ultimate results of such considerations are both qualitative and quantitative improvements in the microscopical output. These factors are discussed in Chapter 10, Sections III and IV.

B. Alignment of an External Illuminator

An external illuminator is required for older or less expensive microscopes that have no built-in source of illumination. These external illuminators are available in a wide range of complexities. The simplest type has only a lamp housing with a ground glass diffuser and an incandescent lamp that operates directly from the power line. Alignment of this illuminator consists of placing it as close as possible to the sub-stage mirror and orienting it so that the mirror can accept the full beam of illumination. If the illuminator is small enough, it is possible to remove the substage mirror and insert the illuminator in its place with the beam aimed upward and centered on the condenser.

The ideal external illuminator should have, as a minimum, a focusable condenser lens and a fully adjustable iris diaphragm such as shown in Figure 2-7. With this type of illuminator it is possible to obtain either Koehler or Nelson illumination. Since the adjustment for these types of illumination are rather critical it is important to

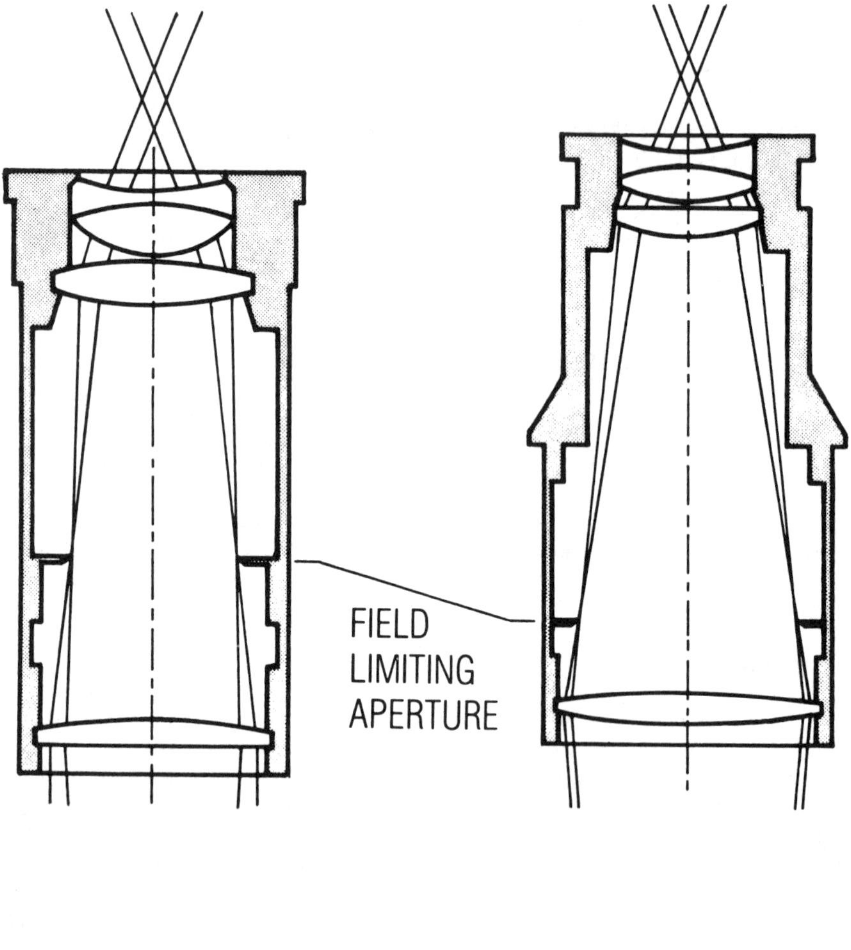

Figure 2-26. A comparison of standard (left) and wide field (right) eyepieces. Note the wider field limiting aperture in the wide field eyepiece. Redrawn courtesy of Wild Leitz USA, Inc.

maintain both the microscope and the the illuminator in a fixed relationship after they have been properly positioned; this can be achieved by clamping or otherwise attaching them to the table.

The first concern in setting up the adjustable illuminator is to make sure that the light exiting the illuminator is exactly normal to the condenser; this is equivalent to saying that the lamp is properly positioned the the illuminator. This is necessary when setting up the condenser for the first time and is especially important after a lamp has been changed. The alignment of the lamp in the illuminator is performed in the following manner:

ILLUMINATOR LAMP ALIGNMENT

1. Adjust the condenser housing to be exactly parallel to the table top.
2. Determine the distance, D, from the center of the condenser lens (or iris diaphragm) to the table top.
3. Focus the light from the illuminator on a sheet of paper that is normal to and about 50 cm in front of the illuminator.
4. Adjust the position of the lamp in the illuminator housing until the spot of light is a distance, D, above the table and is in line vertically with the axis of the illuminator.

Once the illuminator lamp is correctly aligned, the illuminator can then be aligned to the microscope; this is done as follows:

EXTERNAL ILLUMINATOR ALIGNMENT

1. Aim the illuminator at the microscope mirror. The planar side of the mirror should be used for this purpose.
2. Adjust the mirror until the illumination is aimed at the microscope condenser system and an even illumination can be seen through the eyepiece (use the procedure described in C below).
3. Focus the illuminator condenser until the lamp filament is sharply imaged on the aperture diaphragm or the rotating aperture disk of the microscope. This procedure is best accomplished by looking directly at the aperture diaphragm while focusing the illuminator condenser.
4. At this point the external illuminator is properly aligned for Koehler illumination, however it should be emphasized that the illuminator and the microscope should be clamped or otherwise affixed to the table to prevent any relative movement between them.

C. Set-up of a Brightfield Transmitted Light Microscope

Proper set-up of the microscope is necessary to give the best possible image of the sample and to minimize eye fatigue. The truly professional microscopist will take the little extra time to insure that the instrument is optimized before the sample is actually studied.

Koehler illumination provides a very even illumination of the sample for both visual examination and photomicrography, therefore the procedure in this section will emphasize only this type of illumination.

It is worthwhile to consider the concept of conjugate foci to better understand Koehler illumination. Figure 2-27 is a diagram of the brightfield microscope including the illumination and the observer's eye. The source of illumination, in this case the lamp filament (A) is shown as a coil. This filament is focused at various points in the microscope including the aperture diaphragm (B), the rear focal plane of the objective lens (C), and the iris of the observer's eye (D), the latter is coincident with the Ramsden disk above the eyepiece. This lettered set of locations are said to be conjugate foci of the source of illumination. In contrast, the field diaphragm (1), the sample (2), the field limiting aperture of the eyepiece (3) and the retina of the eye (4) constitute another set of conjugate foci. This numbered set constitutes the conjugate foci of the sample.

For those who desire to use "critical" or Nelson illumination, refer to Chapter 5, Section V C. It should be emphasized that with this type of illumination the image of the lamp filament is superimposed on the image of the sample.

This section assumes that the instrument is fully functional. A new instrument should be prepared for use by removing all packing material and shipping hardware. The accessories should also be affixed as shown in the assembly manual. The procedure described here will serve for an instrument with or without a built-in source of illumination. The latter microscope will require an illuminator as described above and when the field diaphragm is discussed that information applies to the illuminator diaphragm. The set-up is performed as given below:

BRIGHTFIELD TRANSMITTED LIGHT SET-UP PROCEDURE

1. If a reticle eyepiece is to be used it first must be focused. Remove the eyepiece from the tube, fully extend the focusing adjustment, gaze through the eyepiece at a broad light source, and focus the eye-piece until the image of the reticle is sharp. If the position of sharp focus is passed, repeat the above procedure. Reinsert the eyepiece.

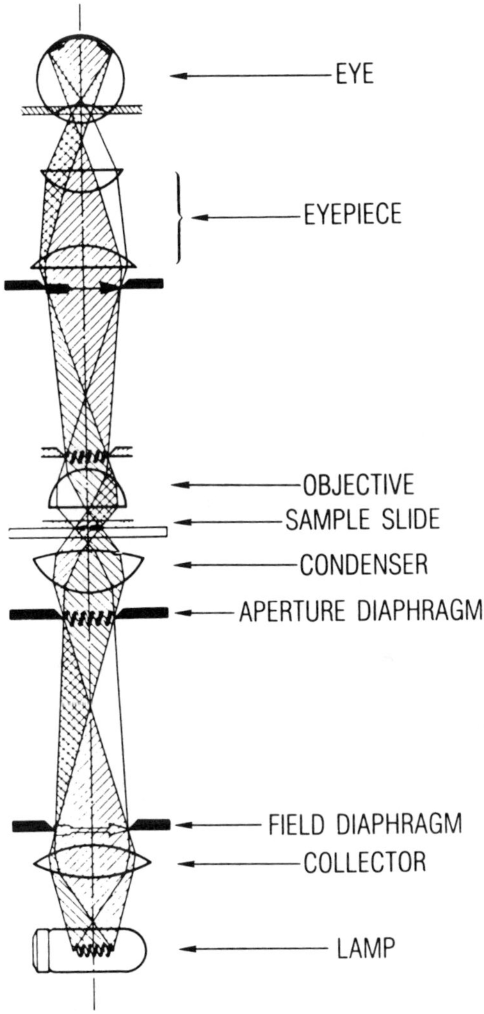

Figure 2-27. A schematic of a transmitted light microscope showing the location of those points in the optical path which are conjugate with the lamp filament (shown as a filament coil) and the sample (shown as an arrow). Thus, the lamp filament is in focus at the aperture diaphragm, the rear aperture of the objective lens and the lens of the eye (the latter is not shown due to the small size of the image). The sample is in focus at the field limiting aperture of the eyepiece and at the retina of the eye. The field diaphragm is focused at the sample plane, hence an outlined arrow is shown at that position. Courtesy of Carl Zeiss, Inc., New York.

2. Turn on the illuminator to a medium brightness, place a sample on the stage and using a low power (10X) objective lens, focus the image of the sample using the coarse focus knob.
   NOTE: The proper procedure for coarse focusing an objective lens, especially the higher magnification lenses, is to move the objective lens close to the sample with the coarse focus knob while viewing the progress from one side, as shown in Figure 2-28. The sample is then viewed through the eyepiece and focused with the fine focus knob. In this way, the motion of the objective lens near the sample is small and generally away from the sample. Thus, there is little chance of damage to either the sample or the objective lens.
3. If a binocular or trinocular body is used it should be adjusted at this point. Adjust the interpupillary spacing so that both eyes can view the image simultaneously. (It is desirable to memorize this spacing number since it is the same for any microscope that you use). If the microscope has threaded focusing collars on both eyepiece tubes, these should both be set at the same numerical value as the interpupillary spacing.
4. Continue the set-up of the binocular or trinocular body by performing the interpupillary acuity adjustment. Refocus the microscope critically using the fine focus knob while observing only with the eye over the eyepiece in the non-adjustable tube (if the microscope has threaded focusing collars on both eyepiece tubes, observe with the right eye only). Then focus for the other eye using only the threaded focusing collar beneath that eyepiece.
5. Close the field diaphragm until its image can be seen in the field of view. Focus the condenser system until that image of the field diaphragm is sharply defined. When the focus is correct there should be no color fringes at the edge of the diaphragm.
6. Center the image of the field diaphragm in the field of view, using the centering adjustment for the condenser.
7. Open the field diaphragm until its image is just outside the field of view. The field diaphragm is now correctly positioned for the present objective lens.
8. The aperture diaphragm adjustment is made by removing an eyepiece and viewing down the tube of the microscope. With the aperture diaphragm wide open, a bright disk of light may be seen, surrounded by the fixed diaphragm of the objective lens rear focal plane. When the aperture diaphragm is closed down, it reduces the diameter of the bright disk. The correct

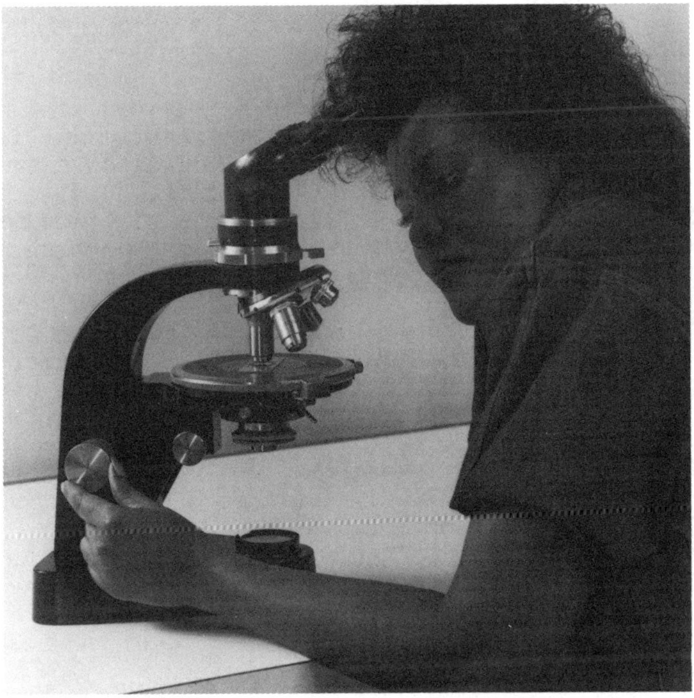

Figure 2-28. The proper procedure to avoid breakage of a sample slide and possible damage to a non-spring-loaded objective lens during focusing is to observe the process from the side. Move the objective lens toward the sample with the coarse focus knob until the objective lens almost touches the sample, then continue focusing with the fine focusing knob while observing through the eyepieces.

setting of the aperture diaphragm for optimum balance of contrast and resolution is achieved when the bright disk of light is 80% to 90% of the full disk diameter, see Figure 2-29. After this adjustment is made and the eyepiece replaced, the sample is ready for study.

9. Critically focus the microscope using the fine focus. If the illumination is incorrect, change the voltage to the lamp or use another neutral density filter in the illuminator filter slot.

10. Unless you are the sole user of the subject microscope, perform steps 1 to 4 each time the instrument is used.

11. Repeat steps 5 through 9 each time the objective lens is changed unless as noted in 12 to 14 below.

12. If a low power (<10X) objective is to be used, it may be necessary to remove or swing-out the final condenser lens in step 5 so that the entire field of view is illuminated. (If the final condenser lens cannot be removed, the entire condenser system should be removed and the sample illuminated with the concave side of the mirror or with a convergent beam of light directly from the built-in illuminator. Steps 6 to 8 are not performed in this case.

13. If an oil immersion objective lens and an oil immersion, final condenser lens is to be used, steps 4a, 4b, 4c, 4d (below) must be inserted between steps 4 and 5.

14. If an oil immersion objective lens and a dry final condenser lens is to be used, steps 4b, 4c, 4d (below) must be inserted between steps 4 and 5. In step 8, the aperture diaphragm should be full open for best resolution.
NOTE: in this case an N.A. of 1.0 is the maximum that may be achieved.

15. When the microscopical observations are completed, remove the sample slide, reduce the illuminator intensity to its lowest level, if possible, and turn off the illuminator. Remove all immersion oil from the objective lens and condenser lens, if used (see Chapter 10, Section VI). Never leave a high power lens in position. Clean the stage to remove any residue of mounting media or immersion oil. Cover microscope to preclude dust.

4a. Move the condenser assembly away from the sample slide, remove the dry final condenser lens and replace it with the oil immersion final condenser lens. For the upright stand, carefully place a drop of bubble-free immersion oil on the final condenser lens then move the condenser assembly toward the sample slide until

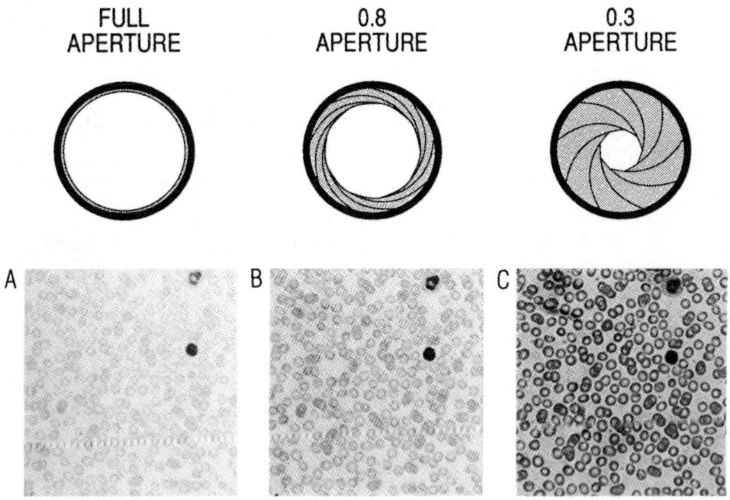

Figure 2-29. The effect of aperture diaphragm setting on the sample image. (A) shows the aperture diaphragm wide open. The accompanying image has maximum resolution but less than optimum contrast. (B) shows the aperture diaphragm at about 80% open. At this setting the resolution in the respective image is only slightly less than in (A) but the contrast is markedly improved. (C) shows the aperture diaphragm closed to about 40%. In the accompanying image the resolution is poorer, contrast higher but artifacts, referred to as diffraction rings, degrade the image .

the oil drop contacts the sample slide. For the inverted stand, place the oil drop on the sample slide.

4b. Move the objective lens away from the sample and replace it with the oil immersion objective lens. For the upright stand, place a drop of bubble-free immersion oil on the cover glass of the sample and carefully move the objective lens toward the sample until oil contact is made (see note in step 2). For the inverted stand, place the oil drop carefully on the front element of the objective lens.

4c. Use the fine focus to focus the sample, then remove an eyepiece and verify the absence of bubbles by viewing down the tube of the microscope. A bright disk will be seen. Any bubbles in the oil film(s) will be imaged as rings of random size and location in the bright disk. If bubbles are present, the oil film(s) must be cleaned off (see Chapter 10, Section VI) and 4a (if used), 4b and 4c repeated.

4d. If no bubbles are present, replace the eyepiece and continue with step 5.

D. Adjustment to Minimize Spherical Aberration

Spherical aberration is that objective lens defect which produces the most pronounced degradation of image. This aberration can be introduced at the higher magnifications by incorrect thickness of the cover glass. The obvious way to minimize this effect is to use the correct cover slip thickness; this can be verified by using a micrometer.

If a sample has a permanently mounted cover slip which is of the incorrect thickness or if the microscopist desires the ultimate of correction, the simplest approach is to adjust the mechanical tube length of the microscope; this is generally possible for the more advanced monocular instruments. The binocular head instruments seldom have a mechanical tube adjustment and this is further complicated by the interpupillary adjustment which also varies the tube length in most binocular systems. Fortunately, in some of the modern instruments, compensation systems are included in the binocular head to correct the tube length for changes introduced by adjustments of the interpupillary spacing and acuity. The adjustment procedure is given below.

The second approach to reducing spherical aberration is to use an objective lens with a correction collar. These objectives are more expensive than the standard ones and are available only at the higher magnifications where correction is most important. The cover glass thickness is measured prior to covering the sample. The collar is then rotated to position a fiducial mark at the value of the cover glass thickness as marked on the objective lens body.

However, the correction collar objective lens can also be used to reduce spherical aberration even if the cover glass thickness is not known; this procedure is given below.

REDUCTION OF SPHERICAL ABERRATION DUE TO COVER GLASS

1. Critically focus the sample using the set up procedure in Section IV C above.
2. Set the adjustment (either correction collar or tube length adjustment) to one extreme of its range and critically focus on a fine structure in the center of the field of view. Note the image quality.
3. Move the adjustment gradually back and forth between the two extremes, focusing continuously on the same fine structure and noting the image quality.
4. Use a gradually decreasing range until the position of best image quality is found. This position will yield an image with the best resolution and image distinction with minimum glare.

E. Problem Correction for Transmitted Light Microscopy

After the above steps have been followed carefully or perhaps during the set up procedure a problem may be noted. Without considerable experience it is often difficult to locate the source of the problem and correct it so that the sample image may be studied. Regardless of the instrument under consideration, Table 2-8 will provide a suitable check list of typical problems, their causes, and solutions. It must be emphasized, however, that it is impossible to include all the problems peculiar to every specific instrument.

TABLE 2-8. TRANSMITTED LIGHT MICROSCOPY PROBLEMS AND THEIR CORRECTION

PROBLEM	CAUSE	CORRECTION
ELECTRICAL SYSTEM:		
No light, pilot light or meter function.	Power not available.* Power supply not plugged in.*	Check power breakers. Plug in power supply.
No light. Pilot light and meter function.	Lamp burned out.* Bad socket contacts.* Defective lamp cord.*	Replace lamp (Cha 10) Repair socket. Repair/replace cord.
Unstable, flickering light.	Lamp defective.* Bad socket contacts.* Defective lamp cord.	Replace lamp. Repair socket. Repair or replace.
Reduced illumination.	Interior of lamp has black evaporated tungston coating.*	Replace lamp and use lower lamp voltage.
	Lamp surface has burned in finger prints.*	Replace with clean lamp.
	Filter in light path.*	Remove filter.
Reduced lamp life.	Selector switch set incorrectly.*	Set selector switch to correct line voltage.
	Incorrect lamp.*	Use proper lamp.
	Transformer setting too high.*	Do not operate lamps at full rated voltage.
OPTICAL SYSTEM:		
Lamp on but no illumination of sample.	A slider for filter or for high/low mangification in mid position.*	Locate offending item and move to correct position.
	Slider for oblique illumination in mid position.*	Move out of light path.
	Mount of swing-in final condenser in light path.	Move fully to correct position.
Sample illuminated but no image.	Revolving nosepiece not at a detent position.*	Rotate nosepiece to desired objective lens.
	Photomicrography slider in camera position.*	Move slider to viewing position.
Insufficient brightness.	Lamp setting too low.*	Increase lamp voltage.
	Neutral density filter in place.*	Remove filter from path.

TABLE 2-8. TRANSMITTED LIGHT MICROSCOPY PROBLEMS AND THEIR CORRECTION (Continued)

PROBLEM	CAUSE	CORRECTION
Field of view cut off or illuminated irregularly.	The high/low magnification slider in mid position.*	Move fully to correct position.
	Auxiliary condenser lens not correctly attached.	Attach correctly.
	Sliding filter partiallty in light path.*	Move fully to correct position.
	Revolving nosepiece not at detent stop.*	Rotate nosepiece slightly until it seats in detent stop.
	Nosepiece or mount for objective lens not fully attached.*	Fully attach nosepiece or objective lens mount,
	Slide for oblique illuminator pulled out partially.*	Move slider out of light path.
	Mount of swing-in final condenser partially in the light path.	Move fully to correct position.
	Illuminator mirror misaligned.	Align external mirror, call repairman for internal mirror.
	Damaged filter.*	Replace filter.
	Eye problem.*	Consult an opthalmologist.
Image of field diaphragm cannot be focused sharply.	Frosted filter in light path between field diaphragm and condenser lens.	Remove frosted filter from light path.
	Glass slide too thick.	Remount sample on thinner glass slide.
Image of field diaphragm blue on one side and orange on the other.	Condenser system misaligned.	Repeat setup procedure and make sure that the condenser is properly centered.
	Swing-in final condenser not fully in place.	Make sure that the swing-in condenser is in the proper position.
Image of field diaphragm blurred or streaked.	Oil film on final condenser lens.	Clean condenser lens per Chapter 10.
	Oil film on front surface of objective lens.*	Clean objective lens per Chapter 10.

TABLE 2-8. TRANSMITTED LIGHT MICROSCOPY PROBLEMS AND THEIR CORRECTION (Continued)

PROBLEM	CAUSE	CORRECTION
Dust or dirt particles visible in field of view.	Dust on light exit in base of stand (for self contained illuminators).*	Clean glass plate per Chapter 10.
	Dust on final condenser lens.	Clean final condenser lens per Chapter 10.
	Dust at level of field diaphragm.*	Clean final condenser lens per Chapter 10. Clean any dirty glass elements near the field diaphragm per Chapter 10, if accessible, otherwise call for service.
	Dust on sample cover glass.	Use clean cover glass.
	Dust on reticle or field lens of eyepiece.*	Dust particles rotate when eyepiece is rotated, clean per Chapter 10.
Excessive image contrast.	Aperture diaphragm closed down too much.*	Aperture diaphragm should be set per Section C 8 above.
	Condenser lens cap or swing-in final condenser not in place.	Attach lens cap or swing-in the final condenser.
Image not sharp, insufficient contrast or poor image detail.	Dirt on condenser lens.	Clean condenser lens per Chapter 10.
	Mounting medium smeared on cover glass.	Use fresh cover glass.
	Aperture diaphragm wide open.*	Aperture diaphragm should be set per Section C 8 above.
	Dirt or old oil film on objective front lens.*	Clean objective lens per Chapter 10.
	Correction collar on objective lens not correctly set.	Set correction collar (if used) per Section D above.
	Incorrect immersion oil used.*	Remove incorrect oil, clean objective lens sample, then reoil with correct oil per Section C above.

TABLE 2-8. TRANSMITTED LIGHT MICROSCOPY PROBLEMS AND THEIR CORRECTION (Continued)

PROBLEM	CAUSE	CORRECTION
Image not sharp, insufficient contrast or poor image detail. (Continued)	Bubbles in immersion oil.*	Remove oil and reoil per Section C above.
	Layer of old oil on oil immersion objective lens, causing schlieren.*	Remove oil old oil per Chapter 10 before adding new oil per Section C above.
	Oil immersion objective lens used without immersion oil.*	Use immersion oil per Section C above.
	Cover glass has wrong thickness.	Replace cover glass with one of correct thickness.
	Sample slide is inverted on stage.	Place sample on stage with cover glass facing objective lens.
	Nosepiece not properly attached.*	Loosen nosepiece and reattach correctly.
	Revolving nosepiece not at detent stop.*	Rotate nosepiece slightly until it seats in detent stop.
	Dust on binocular head prism or windows.*	Clean per Chapter 10.
	Dust, cosmetics, or tears on eye lens of eyepiece.*	Clean the eyepiece per Chapter 10.
	Defective objective lens.*	Test per Chapter 10.
Sample cannot be focused at 40X or greater.	Sample slide is inverted on stage.	Place sample on stage with cover glass facing objective lens.
	Fingerprints, dust, mounting medium, or immersion oil on cover slip.	Replace soilerdcover slip.
Field of view partially out of focus or goes out focus eccentrically.	Nosepiece not correctly attached.*	Loosen nosepiece and reattach correctly.
	Revolving nosepiece not at detent stop.*	Rotate nosepiece slightly until it seats in detent stop.
	Sample not flat on stage.*	Make sample slide level with stage.

TABLE 2-8. TRANSMITTED LIGHT MICROSCOPY PROBLEMS AND THEIR CORRECTION (Continued)

PROBLEM	CAUSE	CORRECTION
Field of view partially out of focus or goes out focus eccentricaly. (Continued)	Slider for oblique illumination partially in light path.*	Move slider completely out of light path.
	Auxiliary condenser lens not correctly mounted.	Correctly position the auxiliary condenser lens.
Various objective lenses are not parfocal.	Affected objective lenses have become partially loosened.*	Tighten the affected objective lenses into the nosepiece.
	Mechanical tube length incorrect.*	Adjust tube length (if possible) per Section D above.
Reticle scale reads backward.	Reticle installed upside down.*	Remove reticle from eyepiece and reinstall correctly.
Large differences in focus between binocular eyepieces.	Eyepieces are not a matched pair.*	Use a matched pair of eyepieces.
	One eyepiece iloosely assembled.*	Carefully tighten threaded parts of eyepiece.
	One eyepiece not correctly assembled.*	Refer to Chapter 10 on repair.
Colorless floating spots in bright image.	"Mouches Volantaes" within eye.*	A normal phenomena with most observers.
STAND:		
Focusing Problems.	Coarse focus does not move easily.*	Refer to professional service personnel.
	Stage slowly drops and sample goes out of focus.*	Refer to professional service personnel.
	Coarse focus cannot be raised far enough to focus sample.*	Stage must be raised by repositioning on the coarse focus assembly.
	Sample slide hits the objective lens when attempting to focus.	Slide upside down.
	Fine focus doens't operate smoothly.*	Refer to professional service personnel.

TABLE 2-8. TRANSMITTED LIGHT MICROSCOPY PROBLEMS AND THEIR CORRECTION (Continued)

PROBLEM	CAUSE	CORRECTION
Focusing problems. (Continued)	Fine focus will not turn in one direction.*	DO NOT FORCE! Move sample away from objective lens with coarse focus knob, rotate fine focus to center of its range, return sample to approximate focus with coarse focus knob, then critically refocus with fine focus knob.
Stage problems.	Image goes out of focus when stage is touched.*	Tighten the stage attachment to its coarse focus assembly.
	Sample stops midway when mechanical stage is translated.*	Adjust sample position on the mechanical stage.

*Entries which pertain to both transmitted and incident light microscopes.

REFERENCES

Abbe, E., Archiv. F. Mikroskopisch Anat., 9, 413, (1873).

Carter, W.H. and Dougal, A.A., The IEEE Southwestern Record, pp18, Dallas, TX April 20-22, (1966).

Hartley, W. G., "Hartley's Microscopy," p53, Senecio Publishing Co. Ltd. (1979).

Needham, G.H., "The Practical Use Of The Microscope", p37, Charles C. Thomas, Publisher, Springfield, IL, (1977).

Osterberg, H., J. Opt. Soc. Amer., 40, 295, (1950).

Royal Microscopical Society Standards, Journ. Roy. Micr. Soc., 56, 377 (1936).

Van Duijn, Jr., C., Microscope, 11, 196, (1957).

3

Brightfield Incident Light Microscopy

I. INTRODUCTION

This chapter discusses the brightfield, incident light microscope, also referred to as an epi or a vertical illumination microscope. These instruments differ from the brightfield transmitted light microscope of Chapter 2 only in the position of the illumination system relative to the sample. In fact, the incident light system may be considered to be arranged like a transmitted light system but with the illumination system rotated 180° about the sample as shown in Figure 3-1. In the latter case the condenser lens becomes coincident with, or in actual practice, is the objective lens. All that is required to make the system workable is the addition of a partially reflecting mirror to move the illumination system out of the path for the image system.

Since there is a considerable similarity between the transmitted light and the incident light instruments, this chapter relies heavily on Chapter 2 and stresses only those areas of difference.

II. THE STAND

A. The Upright Stand.

The Upright stand is not the most common stand for the incident light microscope but is often used when a single instrument must serve for both transmitted and incident illumination investigations. For this reason, the upright stand is often used when a polished thin sections are to be simultaneously examined in both transmitted and incident illumination. A typical upright, incident light instrument is shown in Figure 3-2. One inconvenient feature of the

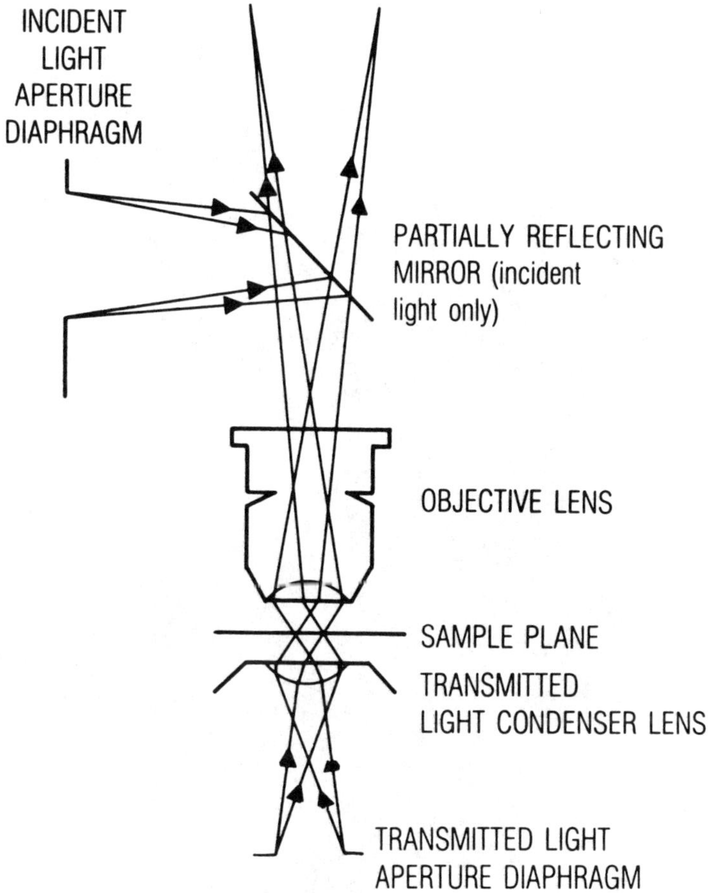

Figure 3-1. A comparison of the illumination paths for brightfield, transmitted light and brightfield, incident light microscopes. In this figure the transmitted light enters from below and the incident light enters from the left, above the objective lens.

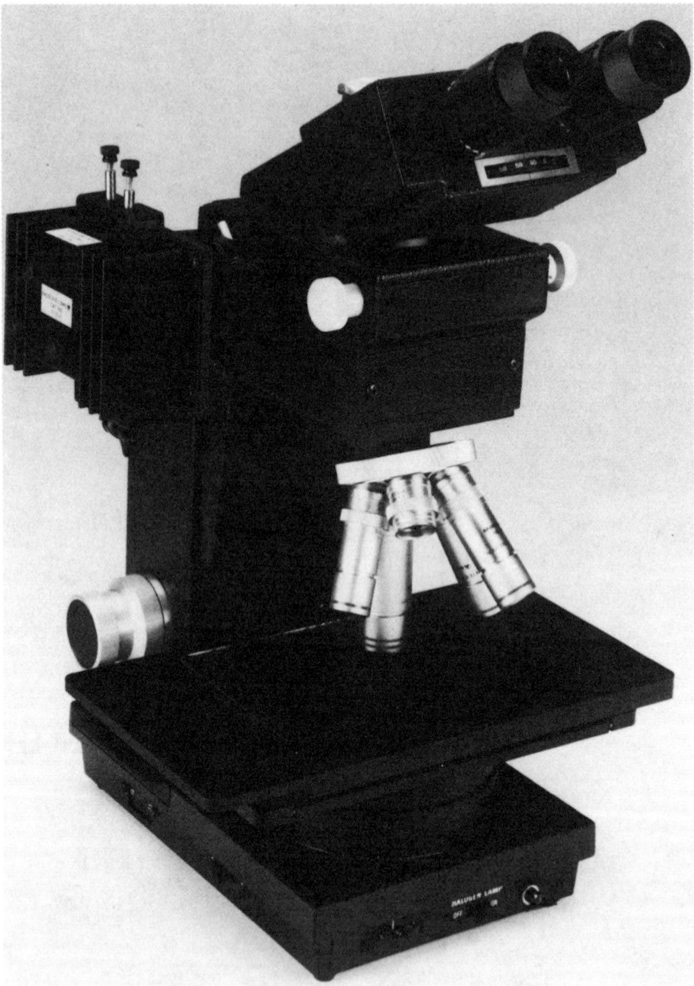

Figure 3-2. An upright, incident light microscope. Courtesy of Cambridge Instruments Inc., Optical Systems Division.

upright, incident light instruments results from the location of the normal incident illuminator, mounted at right angles to the tube of the microscope just above the objective lens. In this location, it often reduces the free working space in the stage area. Also the illuminator adds to the weight of the microscope tube and if the coarse focus assembly is too loose, the tube may slowly move down under the force of gravity, thereby requiring continual refocusing. Methods to remedy this latter problem are given in Chapter 2, Section IV E.

B. The Inverted Stand

The inverted stand for incident light microscopy is usually referred to as a metallograph, see Figure 3-3. This stand is the choice of the metallurgist because of the configuration of the metallurgical sample mount, as described above. In the case of the metallograph, the polished and etched surface of the sample mount rests on the stage; thus it is by definition in focus across its entire surface and therefore it requires little refocusing during an examination.

Leitz has developed an unusual version of the inverted stand which mimics the form of the upright stand, as shown in Figure 3-4. This stand offers the freedom from refocusing of the inverted stand and the comfort of the upright stand's low stage.

C. Upright Vs. Inverted Stand

The basis for selecting one of the two types of stands for an incident light application can be illustrated by some examples. Let us consider first the standard metallurgical mounts. The metallurgical mount as initially prepared consists of a right circular cylinder of plastic with the sample surface of interest exposed on one end of the cylinder. The sample surface is then ground on successively finer grit, abrasive surfaces, polished to a damage free mirror surface, and finally etched to reveal the sample structure. It should be noted that when this process is performed manually, the sample surface and the opposite surface seldom remain parallel.

When a metallurgical mount is examined with an upright stand instrument, the sample rests on the opposite surface with the result that the sample surface is not parallel to the stage. As the sample is examined in various areas, the focus must continually be adjusted to accommodate for the thickness of the mount. This inconvenience may be overcome by placing the opposite surface of the mount on a small quantity of modeling clay or mounting wax which is placed in turn on a glass microscope slide. The sample surface is then made parallel with the bottom surface of the glass slide using a sample leveling press, shown in Figure 3-5.

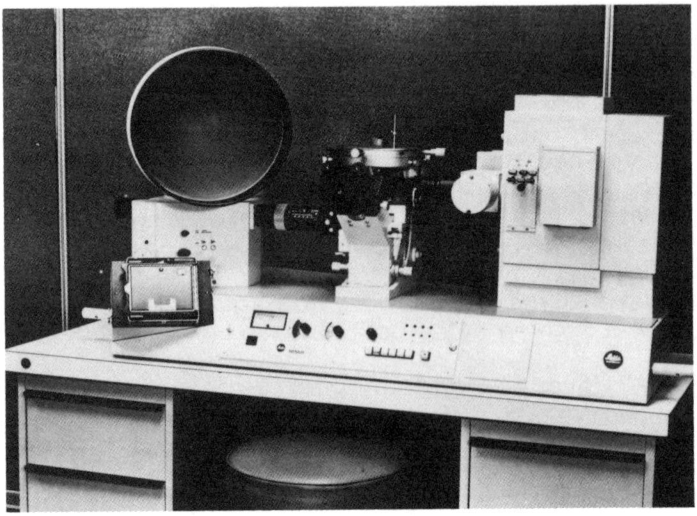

Figure 3-3. An inverted, incident light microscope or metallograph with viewing screen and full photographic capability. Courtesy of Wild Leitz USA, Inc.

A

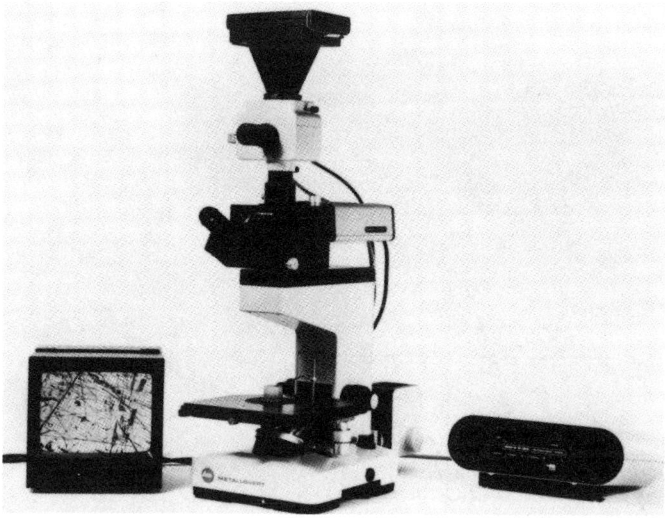

B

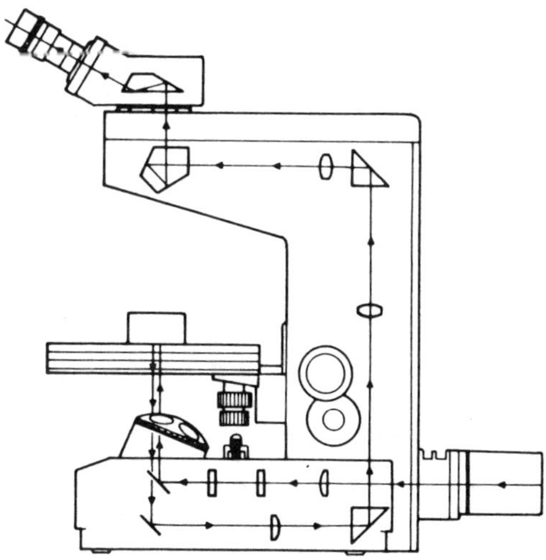

Figure 3-4. A new design of inverted, incident light microscope which retains the convenient and ergonomic, upright stand configuration. A. shows the instrument configured with fixed bellows camera and CCTV camera/monitor. B. shows the unique optical path of this instrument. Courtesy of Wild Leitz USA, Inc.

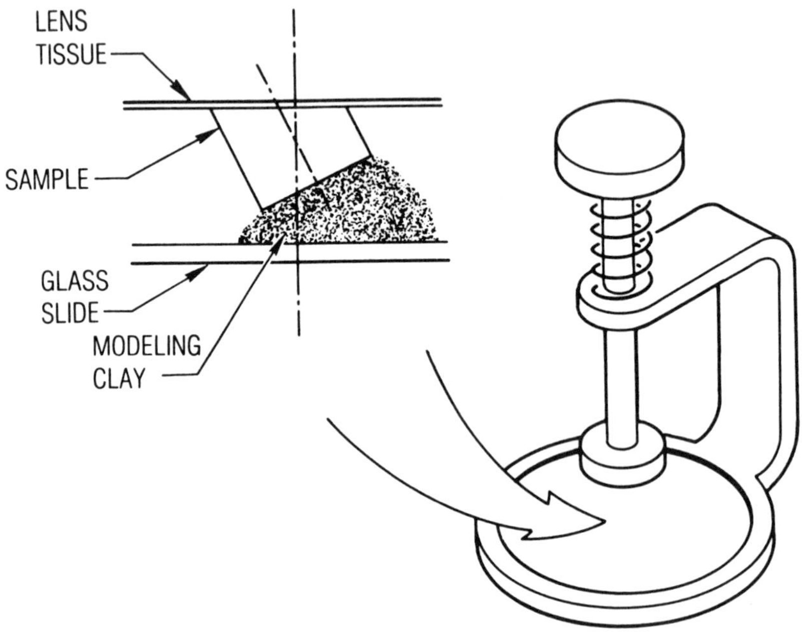

Figure 3-5. The sample leveling press provides a means of making a metallurgical sample surface parallel to the surface of an upright microscope stage. Place a small quantity of modeling clay on a glass slide followed by the sample and a sheet of lens tissue to protect the sample surface. When this entire assemblage is pressed, the sample surface is made parallel to the glass slide. If the press is equipped withan adjustable stop on the shaft, all samples may be pressed to a standard thickness by adjusting the amount of modeling clay used. In this way a measure of parfocality is provided for multiple samples.

After pressing, the lens tissue is removed and the remaining assemblage placed on the microscope stage for easy examination of the sample.

If, however, the metallurgical mount is examined using the inverted microscope, the sample surface rests on the stage and by definition, the sample will remain in focus as the various areas are examined.

A second type of sample might contain contaminant particles in the case of an electronic component such as a transistor or integrated circuit. If such a sample is examined with an upright stand microscope, the particles could be expected to remain properly in the case, barring excessive air movement in the vicinity of the sample. However, if one attempted to use the inverted instrument for this purpose, the particles would be lost, likely coming to rest on the objective lens.

In summary, the type of stand should be selected based on the types of samples to be examined. If both transmitted light and incident light samples are to be examined on the same microscope, the upright is probably the more convenient stand.

D. The Culpeper Stand

In the early eighteenth century England, Culpeper manufactured microscopes that had a tripod stand rather than the more traditional pillar type stand. The modern day version of the Culpeper stand, the Zeiss Axiomat, when configured as an incident light microscope, see Figure 3-6, is a very compact, inverted metallograph. Also, in this configuration the Axiomat instrument provides a large unobstructed stage area. This feature will permit very large samples to be examined or will allow the use of large stage accessories such as described in Chapter 6 Section II.

III. THE OPTICAL SYSTEM

A. Illumination

The preferred source of illumination for the older metallographic instruments was the intense carbon arc source. This source has been largely supplanted by the more reliable and stable halogen and xenon lamps. These are described in Chapter 2, Section III A. Note particularly the cautions mentioned there regarding the use of the xenon lamp.

The illuminator housing is often attached directly to the condenser system which is in turn attached to the side of the body tube for the smaller upright stands. For the larger upright stands and the inverted stands the illumination housing is usually built into the stand. These housings are usually equipped with slots or sliders for the insertion of filters into the light path.

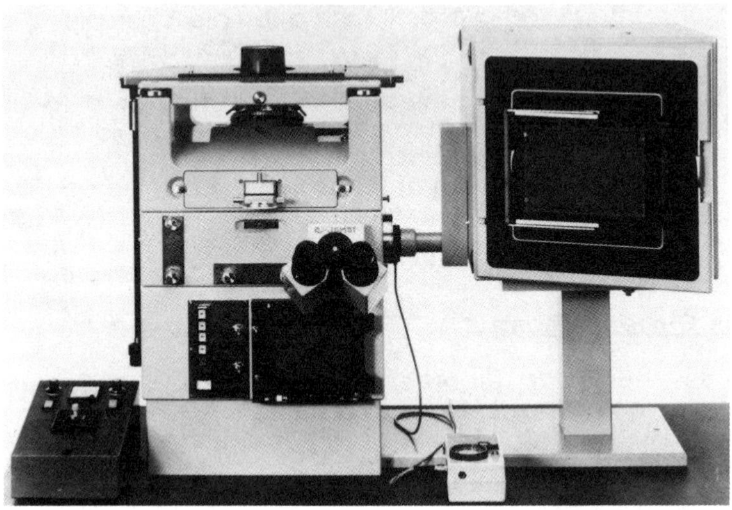

Figure 3-6. The Axiomat microscope, as mentioned in Figure 2-5 is a modern day version of the Culpeper microscope. This picture illustrates this very versatile instrument configured as an inverted, incident light microscope or metallograph. Courtesy of Carl Zeiss, Inc., New York.

B. The Condenser System

The incident light illuminator differs from the transmitted light version in only two respects; (1) a beam diverting prism or mirror is necessary to introduce the illuminating beam into the image path and (2) the objective lens is the final condenser lens in this condenser system.

The procedure for introducing the illuminating beam into the image path has received much attention and there are several methods as shown in Figure 3-7. Each method has specific advantages and disadvantages; these will be discussed in turn.

The plane glass reflector, sometimes called a glass slip illuminator or pellicle, is a thin, flat glass plate positioned at a 45° angle to the axis of the body tube and the condenser assembly. The side of the glass plate adjacent to the condenser assembly is lightly coated with a vacuum evaporated metal coating to produce a surface that is partially reflecting and partially transmitting. The opposite side of the plate is treated with a vacuum deposited anti-reflection coating to minimize double images and glare. It will be recognized from Figure 3-7a that while the desired portion of light from the condenser is reflected toward the objective lens, an appreciable fraction of the light continues undeviated toward the opposite wall of the body tube. Any reflected light from that wall is reflected off of the back side of the plane glass reflector toward the eyepiece as glare, therefore it is imperative that the body tube wall be thoroughly covered with a flat black surface to minimize any reflections.

The advantages of the plane glass reflector are that it is inexpensive and more importantly that it provides full aperture illumination for the sample. It also can be used at any magnification. The most important disadvantage is low image brilliance; this amounts to about 10-20% of that for the prism reflector. Glare can be a problem with poorly reflecting samples. Another disadvantage is some astigmatism or other distortions may be present in the image if the two surfaces of the glass reflector are not flat and parallel, however this is less serious if infinity corrected objective lenses are used. One further problem with this reflector is that it will produce some elliptical polarization if the instrument is also to be used for polarized light investigations, see Chapter 4 Section VI.

The plane glass reflector may require adjustment to provide even illumination of the sample. If this adjustment can be made externally with a knob, it can be accomplished by the microscopist as described below in IV B, otherwise the adjustment should be made by a qualified repairman.

The Smith diverting system, Figure 3-7b, is a modification and improvement of the plane glass reflector. In this illuminator the incoming illumination beam is first totally reflected from a first surface mirror and then reflected from the plane glass reflector toward the

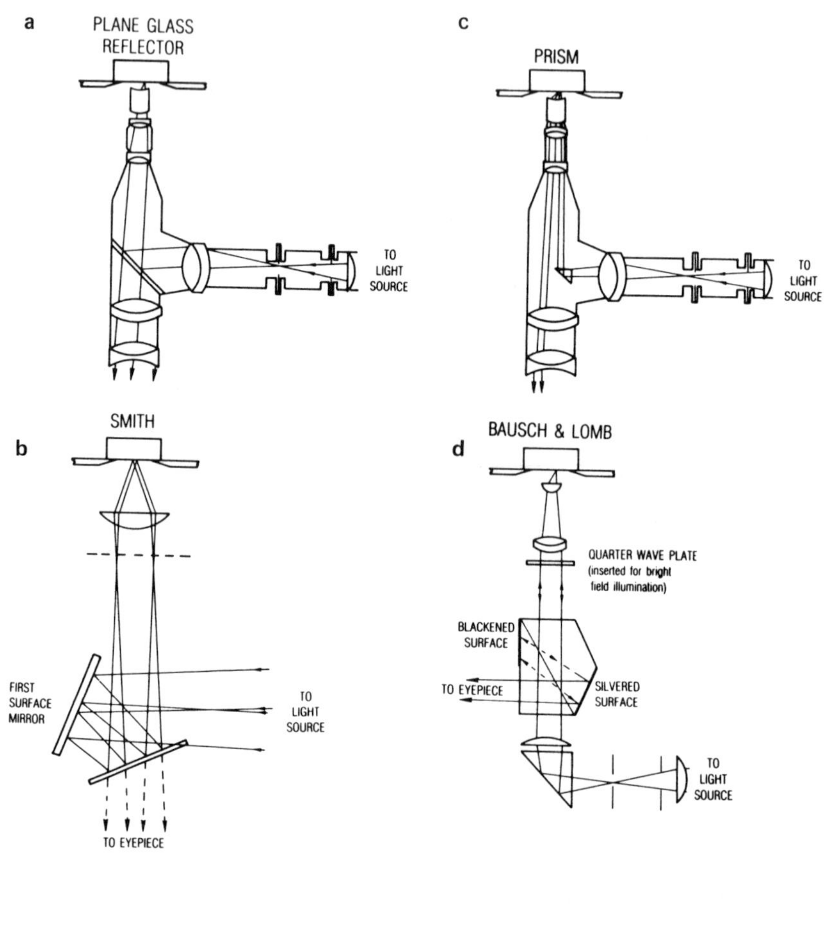

Figure 3-7. Four brightfield, incident light illumination systems.

sample. The image beam passes back through the plane glass reflector and on to the eye. Since the angle of incidence of the returning beam is only 22.5° the amount of elliptical polarization is significantly reduced for polarized light studies.

Any adjustment of the Smith system to correct an uneven illumination condition should be referred to a qualified repairman.

The glass prism diverting system, Figure 3-7c uses a small, glass right prism which obstructs somewhat less than half of the rear focal plane of the objective lens. This system is generally used at total magnifications of 100X or less since the prism physically restricts the aperture of the objective in a unsymmetrical manner and noticeably reduces the resolution of the system. The major advantages of the prism diverting system when compared with the plane glass diverting system and then, only at lower magnifications, are greater light intensity, better contrast and no deleterious effects on polarized illumination.

The glass prism diverting system is almost always equipped with an external adjustment knob. The adjustment of this system is covered in Section IV B below.

The unique prism deviating system shown in Figure 3-7d makes use of the optical properties of the calcite crystal and was developed by Bausch and Lomb for use in their research metallographs. The system consists of a triangular prism and a four-sided prism, both made of calcite, cemented together. Each prism is cut so that the hexagonal C axis of the calcite is both parallel to the upper and lower surfaces of the finished prism and to the cement plane. The illumination entering the prism, in this case, from below is resolved into two plane polarized beams vibrating at right angles to each other. The angle of the cemented surface is such that one beam (the ordinary ray) is totally reflected at the cement interface and is absorbed by the blackened surface, a, while the other beam (the extraordinary ray) passes freely through the interface to the objective lens and the sample. If the light reflected back as an image from the sample has not been rotated or otherwise depolarized, it will pass back through the prism in the extraordinary ray direction. The prism would then direct this light back to the illuminator and no light would be seen in the eyepieces. This situation is analogous to the examination of an isotropic material by polarized light illumination as described in Chapter 4, Section VI. If depolarization takes place, then some light passes through the prism assembly in the ordinary ray direction and can be seen in the eyepieces. This depolarization can occur if the sample is not isotropic or if a quarter wave plate is introduced between the objective lens and the prism assembly. In the latter case the microscope provides the equivalent of a brightfield image.

The advantage of the calcite crystal illuminator is that the light losses are about 4 times less than with the plane glass reflector and that no external alignment is required. The main disadvantage is the high cost of the calcite element.

C. The Objective Lens

Since the incident light illuminator is placed above the sample, the objective lens, in addition to its regular function, serves as the final condenser lens. This is reminiscent of the early transmitted light microscopists who, when they desired the highest quality image, would actually substitute an objective lens for the condenser in their transmitted light microscope.

The details regarding objective lenses for incident light, brightfield microscopy are quite similar to those for the transmitted light instrument, see Chapter 2, Section III C, except for the following factors.

1. Lack of Cover Glass Correction

The most important feature of the incident light objective lens is that it is corrected for use without a cover glass. It is true that most objective lenses of 20X or less generally will produce images of adequate quality with or without cover glasses, however the optimum incident light image is to be obtained with the objective lenses that are corrected for use without cover glasses.

2. Working Distance

Working distance is usually not of much importance to the major user of the incident light microscope, the metallographer, since the samples that they study are prepared with a minimum of surface relief. If however the incident light microscope is to be used with unmounted samples such as electronic components, a large working distance is desired. Since working distance decreases with increasing magnification for a typical objective lens series, it follows that the lowest magnification objective lens that will permit the resolution of the desired structure should be used if working distance is a consideration. It also follows that for a given magnification of objective lens (and a given lens diameter) the resolution decreases as the working distance increases.

Appendix II gives the optical parameters for a number of long working distance (LWD) incident light objective lenses. These, however, cannot be thoughtlessly substituted for the lenses on a given microscope without paying a penalty in image quality. Every attempt should be made to accommodate such factors as tube length and eyepiece corrections. It should be emphasized that LWD objective lenses often are not strain free, therefore, they cannot be used with differential interference contrast or polarized light techniques, See Chapter 4, Sections V and VI, respectively.

3. Plano Corrected Objective Lenses

Perhaps the most desirable feature for an incident light objective lens is plano correction since the sample surface of a typical mount, when properly prepared, is a planar surface. If large areas of a sample are to be scanned using the mechanical stage, the plano objective lens will undoubtedly reduce eye fatigue.

4. Radiation Resistant Objective Lenses

Radiation resistance is a design consideration which affects only a small portion of the scientific community, i.e., nuclear science and engineering, however to them it is an exceedingly important property. When ordinary glasses are subjected to intense ionizing radiation and high energy particles they tend to darken or "brown". Prime subjects for this browning are the optical glasses in the objective lenses used in nuclear studies. These objective lenses are a crucial part of remote microscopes used in glove boxes or hot cells, as described in Chapter 6, Section IV, because they are continually in close proximity to highly radioactive materials. To minimize this browning special radiation resistant glasses are used.

D. Viewing Screen

A feature of the incident light microscope, particularly of the metallograph is the viewing screen, see Figure 3-3. This screen ranges in size from 100 X 125 mm (4 X 5 in.) to about 300 mm (12 in.) in diameter. The screen may be in a fixed position that is convenient for the operator or may be on a swivel pedestal so that several may view the image. To make the image as bright as possible a shallow shroud often surrounds the screen.

Optically the screen is a ground glass onto which the image is projected from an eyepiece. A Fresnel lens may also be affixed to the ground glass to produce a brighter image. Since the screen generally suffers from low light intensities is is necessary to use the brightest illumination possible, such as a xenon lamp. A word of caution is necessary at this point; if either the sample or the mounting medium will soften or distort as a result of the intense xenon illumination, a heat filter, see Chapter 7, Section VI C, should be inserted in one of the filter slots of the condenser system during extended sessions with the viewing screen.

IV. PROCEDURES FOR BRIGHTFIELD INCIDENT LIGHT MICROSCOPY

A. Instrument Location

The special considerations for a proper microscopy laboratory are discussed thoroughly in Chapter 10, Section IV. Since an incident light microscope, particularly the metallograph, makes use of high intensity, xenon arc illumination, the caution about ventilation of the lamp housing to preclude the buildup of toxic or irritating ozone is quite important.

B. Alignment of the Incident Light Condenser System

Correct alignment of the condenser system is desirable in visual examinations and crucial in photomicrography. The right angle relationship between the condenser axis and the image path in the incident light microscope is established by the manufacturer, however the position of the aperture diaphragm and/or the angle that the beam diverting system makes with the image beam may be adjustable. The following procedure is necessary if either or both of the above are adjustable to assure that the condenser system is properly aligned, i.e., the illuminating beam and the image beam are coaxial at the sample.

PROCEDURE FOR ALIGNMENT OF CONDENSER SYSTEM

1. Turn on the illuminator to medium brightness and position a low power objective for viewing.
2. Focus the microscope carefully on the metallic surface of a small first surface mirror. If the microscope is an upright instrument, the front and back surfaces of the mirror must be parallel.
3. Remove an eyepiece and view down the tube of the microscope. Open the aperture diaphragm fully, note the diameter of the fixed aperture at the rear focal plane of the objective lens and then close down the aperture diaphragm until its diameter is slightly smaller than the aperture of the rear focal plane of the objective lens.
4. Center the image of the aperture diaphragm in the aperture at the rear focal plane of the objective by moving the centering adjustment for the aperture diaphragm and/or rotating the beam diverting system.
5. Return the eyepiece to the tube of the microscope and store the first surface mirror for future use.

C. Setup of a Brightfield Incident Light Microscope

Proper setup of the microscope is necessary to give the best possible image of the sample and to minimize eye fatigue. The truly professional microscopist will take the little extra time to insure that the instrument is optimized before the sample is actually studied.

Since Koehler illumination provides a very even illumination of the sample for both visual examination and photomicrography, the procedure in this section will provide Koehler illumination.

This section assumes that the instrument has been made ready for use by removing all packing material and shipping hardware, and by attaching the appropriate accessories. The setup is performed as given below.

BRIGHTFIELD INCIDENT LIGHT SETUP PROCEDURE

1. If a reticle eyepiece is to be used, it first must be focused. Remove the eyepiece from the the tube, fully extend the focusing adjustment, gaze through the eyepiece at a broad light source, and focus the eyepiece until the image of the reticle is sharp. If the position of sharp focus is passed, repeat the above process. Once focus is achieved, reinsert the eyepiece.
2. Turn on the illuminator to a medium brightness. If a xenon lamp is used, insert a medium neutral density filter into the illuminator filter slot.
3. Place a sample on the stage and using a low power objective lens, focus the image of the sample using first the coarse and then (if necessary) the fine focus knob. NOTE: The proper procedure for focusing an objective lens, especially the higher magnification lenses, is to move the objective lens close to the sample while view viewing the progress from one side, as shown in Figure 2-28. This procedure assures that the sample and objective lens will not be damaged by their inadvertent contact.
4. If a binocular or trinocular body is used it should be adjusted at this point. Adjust the interpupillary spacing so that both eyes can view the image simultaneously. (It is desirable to memorize this spacing number, since it is the same for any microscope that you use.) If the microscope has threaded focusing collars on both eyepieces tubes, these should both be set at the same numerical value as the interpupillary spacing.

5. Continue the setup of the binocular or trinocular body by performing the interpupillary acuity adjustment. Refocus the microscope critically using the fine focus while observing only with the eye over the eyepiece in the nonadjustable tube (if the microscope has threaded focusing collars on both eyepiece tubes, observe with the right eye only.) Then focus for the other eye using only the threaded focusing collar beneath that eyepiece.
6. If the beam diverting system has an external adjust ment, its proper setting is shown in Figure 3-8 for both the plane glass reflector and the prism diverting systems. Remove an eyepiece and makes the adjustment by viewing down the tube of the microscope. After the adjustment, replace the eyepiece.
7. Close the field diaphragm until its image can be seen in the field of view. It should be centered as shown in Figure 3-8 using the plane glass reflector or prism diverting system. Then open the field diaphragm up until its image is just outside the field of view. The field diaphragm is now correctly positioned for the present objective lens.
8. The aperture diaphragm adjustment is made by removing an eyepiece and viewing down the tube of the microscope. With the aperture diaphragm wide open, a bright disk of light may be seen, surrounded by the fixed aperture of the objective lens rear focal plane. When the aperture diaphragm is closed down, it reduces the diameter of the bright disk. The correct setting of the aperture diaphragm for optimum balance of contrast and resolution is achieved when the bright disk of light is 80 to 90% of the full disk diameter. After this adjustment is made and the eyepiece replaced, the sample is ready for study.
9. If the illumination is incorrect, change the voltage to the lamp or use another neutral density filter in the illuminator filter slot.
10. Unless you are the sole user of the subject microscope, perform steps 1 to 6 each time the instrument is used.
11. Repeat steps 7 through 9 each time the objective lens is changed unless as noted in 12 below.
12. If an oil immersion lens is to be used, steps 5a, 5b, 5c (below), and steps 6 through 8 must be performed.

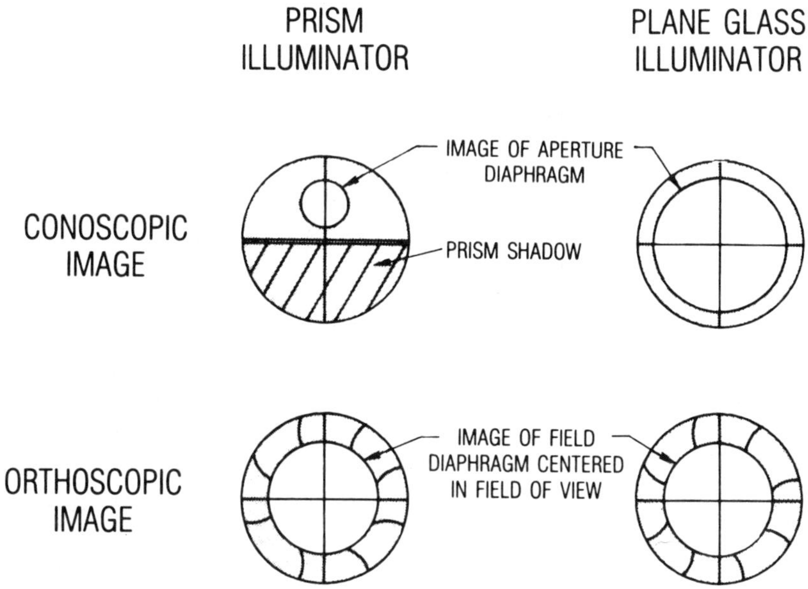

Figure 3-8. The prism and the plane glass illuminators shown in Figure 3-7 are sometimes mounted on an adjustable pivot. The proper adjustment of the pivot knob is achieved by focusing on a clean first surface mirror and observing the image as shown in this figure. The conoscopic image is achieved by removing the eyepiece and viewing down the tube of the microscope or by replacing the eyepiece with an eyepiece telescope or by inserting the Bertrand Lens into the light path, if present on the instrument. The orthoscopic image is the normal image.

13. When the microscopical observations are completed, remove the sample, reduce the illuminator intensity to its lowest level, if possible, and then turn off the illuminator. Remove the immersion oil from the objective lens, if used (see Chapter 10, Section VI). Clean the stage if it has any residue of immersion oil. Cover the instrument to preclude dust.

5a. If an oil immersion objective lens is to be used, move the objective lens away from the sample and replace it with the oil immersion objective lens. For the upright stand, place a drop of bubble-free immersion oil on the sample in the area immediately below the lens and carefully move the objective lens toward the sample until oil contact is made (see note in step 2). For the inverted microscope apply the drop of oil on the front lens element; avoid touching the lens element with the oil applicator.

5b. Focus on the sample, remove the eyepiece and verify the absence of bubbles in the oil by viewing down the tube of the microscope. A bright disk will be seen. Any bubbles present will be imaged as ring(s) of various size and location in the bright disk. If bubbles are present, carefully clean the oil off the sample and the objective lens (see Chapter 10, Section VI) and repeat steps 5a and 5b.

5c. If no bubbles are present replace eyepiece.

D. Problem Correction for Incident Light Microscopy

After the above steps have been followed carefully or perhaps during the set up procedure a problem may be noted. Without considerable experience it is often difficult to locate the source of the problem and correct it so that the sample image may be studied. Even though there are a large number of different instruments to consider, Table 3-1 in combination with Table 2-8 will provide general check list for typical problems, their causes and solutions. Table 3-1 is arranged to include problems unique to the incident light microscope and Table 2-8 lists with an asterisk all those problems which are common to both transmitted and incident light microscopy. It must be emphasized again that it is impossible to include all of the instrument specific problems.

TABLE 3-1. INCIDENT LIGHT MICROSCOPY - PROBLEMS AND THEIR CORRECTION[1]

PROBLEM	CAUSE	CORRECTION
OPTICAL SYSTEM		
Field of view cut off or illuminated irregularly.	Beam diverting system not aligned.	Align per Section C 6.
Dust or dirt particles visible in field of view.	Dust on sample surface.	Clean sample.
	Dust on beam diverting system.	Refer cleaning to qualified repairman.

[1]Also see Table 2-8, Chapter 2 for additional information on problems associated with incident light microscopy. Items listed with an asterisk (*) in the CAUSE column also apply to incident light microscopes.

SELECTED READING

Vander Voort, G. F. "Metallographic Principles and Practice," McGraw Hill (1984).

Vander Voort, G. F. "Applied Metallography," Van Nostrand (1986).

4

Enhancement of Contrast

I. INTRODUCTION

The concept of contrast as it pertains to a microscopical image was discussed in Chapter 1; furthermore, the microscopy of samples with brightness or color contrast was discussed in Chapters 2 and 3. It must be emphasized, however, that light is changed in other ways as it interacts with the sample; these do not produce much, if any, contrast and are not normally visible in the brightfield microscope. Table 4-1 lists most of the optical properties of a sample and their effects on the illuminating light. One or more of these properties must be present in every sample, otherwise no structural information about the sample can be obtained using the optical microscope.

This chapter discusses the various modifications to the basic brightfield microscope so that these normally non-visible effects of the sample also can be observed as brightness and color contrast images. Numerous modifications have been developed as the result of continuous research for a number of years. Figure 4-1 illustrates these various modifications for several types of samples. The techniques that produced these images are documented on the following pages in more or less chronological order.

Perhaps the first indication that the sample influenced the illuminating light in a way that did not produce brightness or color contrast was noted very early in the history of microscopy. It was observed during focusing of very low contrast samples that very slight changes from optimum focus improved the contrast of the image. This effect was more recently explained as a slight difference in path length for the background and image rays

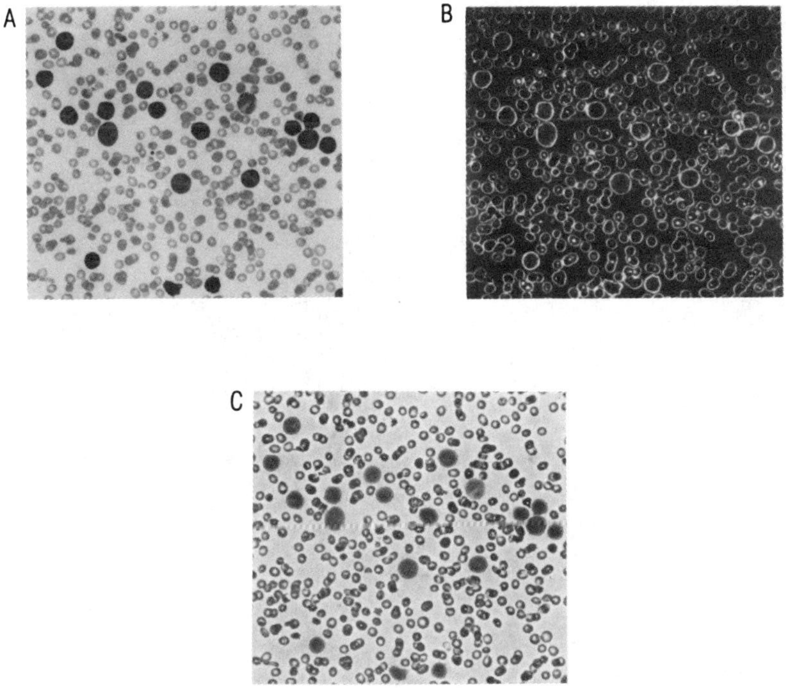

Figure 4-1. Photomicrographs of three samples, representative of biomedical, electronic manufacturing and materials analyses using the various contrast enhancing described in this chapter.

(A) Blood sample - brightfield illumination
(B) Blood sample - darkfield illumination
(C) Blood sample - phase contrast illumination

(continued)

Figure 4-1: (continued)

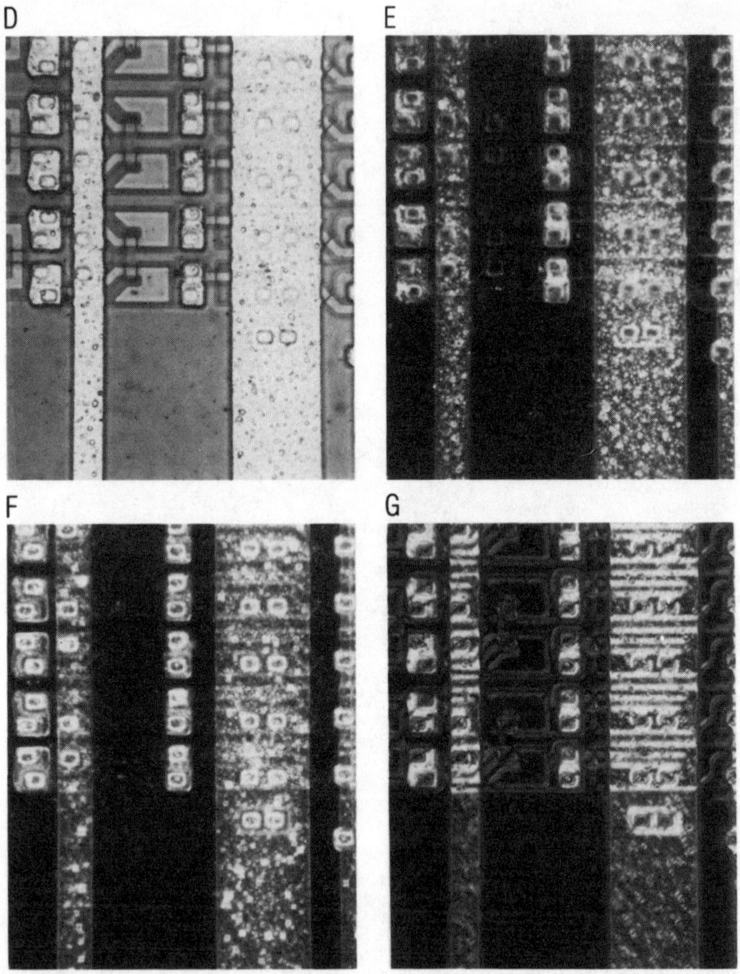

(D) Integrated electronic circuit - brightfield illumination
(E) Integrated electronic circuit - darkfield illumination
(F) Integrated electronic circuit - crossed polarized light illumination
(G) Integrated electronic circuit - differential interference contrast illumination

(continued)

Figure 4-1: (continued)

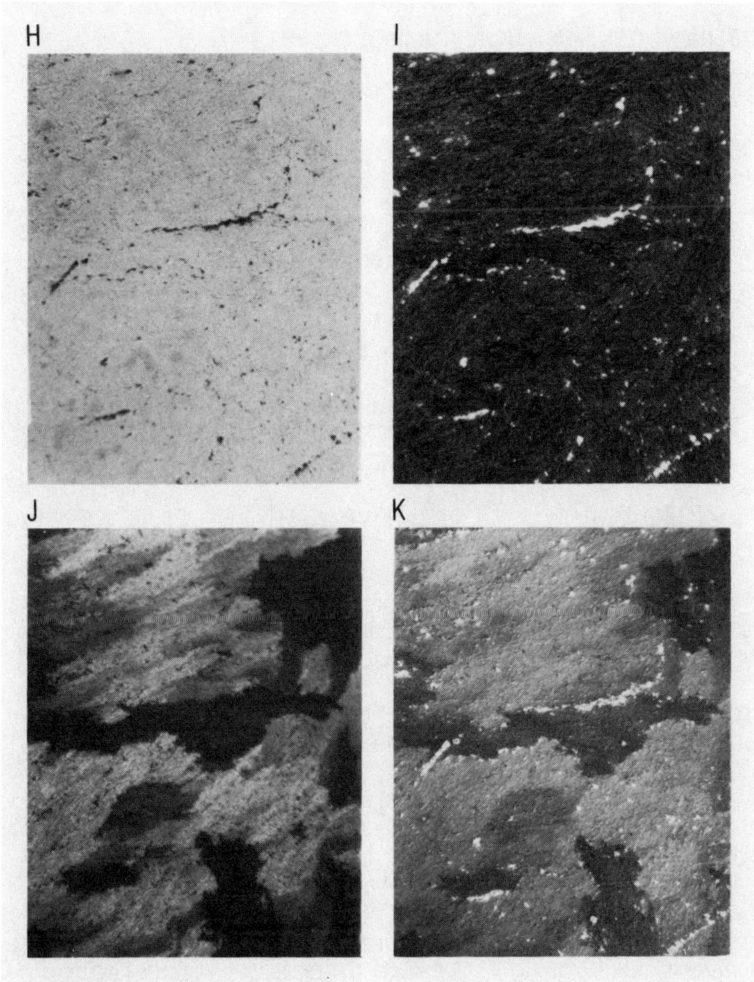

(H) Titanium diboride intermetallic compound (unetched) brightfield illumination
(I) Titanium diboride intermetallic compound (unetched) darkfield illumination
(J) Titanium diboride intermetallic compound (unetched) crossed polarized light illumination
(K) Titanium diboride intermetallic compound (unetched) differential interference contrast illumination

thereby giving a rudimentary phase contrast image. This technique, however gives contrast at the expense of image sharpness.

II. OBLIQUE ILLUMINATION MICROSCOPY

A. General

Oblique illumination may be considered to be any illumination which is not coaxial with the image beam. The oblique illumination may be inclined at any or all angles to the image beam axis ranging from almost parallel, to grazing incidence with the stage. The direction of the illumination may be from any or all azimuths around the sample. The most common oblique illumination is however, from one angle of inclination and one azimuth.

TABLE 4-1
SAMPLE PROPERTIES AND THEIR EFFECTS ON THE ILLUMINATING LIGHT

TRANSMITTED LIGHT	INCIDENT LIGHT	EFFECT ON IMAGE LIGHT
Differing opacity, large refractive index differences	Variable reflecttivity and etched grain boundaries	Brightness contrast
Interference or absorption colors	Reflection or interference colors	Color Contrast
Changes in sample thickness or small refractive index changes	Differences in sample to objective lens distance	Phase changes (independent of the sample orientation)
Differences in refractive index with orientation on the stage	Differences in sample reflection with orientation on the stage	Phase changes (orientation dependent)
Fine, dispersed second phase or fractures or cleavages	Fine, dispersed second phase or pits or surface particulates	Diffraction effects

If the angle of inclination for the oblique illumination with respect to the image beam is less than α, the half angle of the most oblique rays entering the objective lens, the resulting image will be essentially a brightfield image. This illumination, which we shall refer to as low angle, oblique illumination can be achieved by altering the brightfield illumination as described below. The image is essentially a modified brightfield image, i.e., the sample is observed against a bright background. If the angle of oblique illumination with respect to the image beam is greater than α, the illumination is usually achieved with an external illuminator; this results in what we shall refer to as high angle, oblique illumination. The image in this case is darkfield, i.e., the subject is bright against a dark background.

Generally, oblique illumination has been superseded by more advanced illumination techniques which provide more uniform, axial illumination, but for the microscopist equipped with only a brightfield instrument this is the simplest option for increasing contrast.

B. Procedure For Transmitted, Oblique Illumination

Oblique illumination is useful for transmitted light samples when it is desirable to accentuate periodic structures or linear structures such as fibers, fractures or cleavages. In these cases the azimuth of the illumination should be at right angles to the periodic or linear structure. This method of illumination is also the basis for one method of refractive index determination as described in Chapter 5, Section V C.

SET-UP FOR A TRANSMITTED, LIGHT OBLIQUE ILLUMINATION MICROSCOPE

1. Adjust the microscope for correct brightfield illumination per Chapter 2, Section IV C, except that the aperture and field diaphragms should be wide open. Remember to readjust these diaphragms to their original positions if it is necessary to return to brightfield microscopy.
2. Obtain oblique illumination using one of the following methods. These are arranged in order of increasing complexity and may not be suitable for every instrument. Methods a through c generally give low angle oblique illumination
   a. For simple instruments with external illuminator, swing the mirror to one side while examining the image. If the image intensity becomes too low move the mirror back until suitable intensity is restored. This technique provides variable azimuth and angle of inclination.
   b. Insert a cardboard sector stop partially across the illumination path at the aperture diaphragm. See Figure

4-2 for details of the sector stop. The angle of inclination is fixed by the position of the opening cut in the sector stop. Some variation of azimuth is possible by rotating the sector stop but excessive adjustment should be avoided to prevent possible damage to the aperture diaphragm.

c. Move the rotating sector stop into the illumination path. This device is an integral part of some condenser systems. The azimuth of the illumination can be changed by rotating the sector in its housing, however the angle of inclination is fixed.

d. Decenter the condenser system, either with the centering adjustments or in the case of certain Abbe condensers by translating the aperture diaphragm. The azimuth of the illumination can be changed by rotation of the condenser in the translated condition. The angle of inclination is determined by the amount of decentering.

e. Turn off the brightfield illumination and illuminate the sample from the side with a separate, focused micrscope lamp to give high angle oblique illumination. The lamp is focused on the sample from below the stage. The azimuth and the angle of inclination is determined by the placement of the illuminator.

C. Procedure For Incident, Oblique Illumination

Oblique illumination in the case of the incident light sample is also suitable for the examination of periodic or linear structures. In these cases, the azimuth of the illumination should also be at right angles to the periodic or linear structures. Perhaps the most important use for incident light microscopy is the production of relief or three-dimensionality especially at lower magnifications. In this way it can be demonstrated whether a structure on the surface of a sample is a cavity or a protrusion by the position of shadows relative to the illumination azimuth.

To maximize the effect of the oblique illumination the microscopist should explore the complete range of inclination angles and azimuth direction; the latter may be varied by changing the position of the illumination or by rotating the stage, if that option is available.

Finally a word of caution is in order because oblique illumination can produce artifacts such as color fringes and directional contrast, thus it is always wise to compare the resulting image with that of the sample in brightfield illumination before attempting an interpretation based on the oblique illumination case.

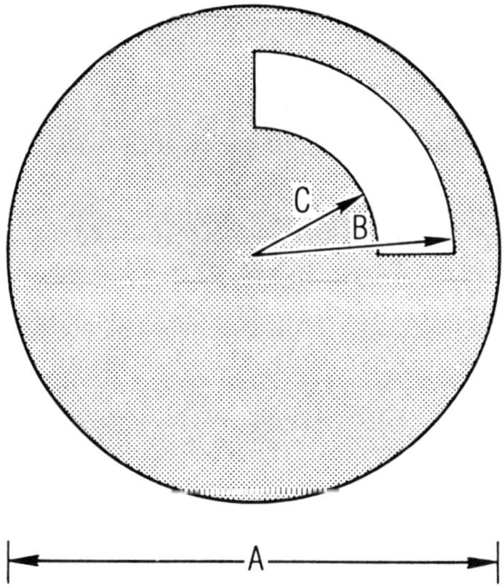

Figure 4-2. Pattern for a temporary cardboard sector stop. A is slightly larger than the total diameter of the completely open aperture diaphragm. The value of C determines the obliquity of the illumination and the difference B-C determines the amount of illumination. NOTE: do not attempt to close the aperture diaphragm with this stop in place as the diaphragm could be damaged.

SET-UP FOR AN INCIDENT LIGHT, OBLIQUE ILLUMINATION MICROSCOPE

1. Adjust the microscope for correct brightfield illumination per Chapter 3, Section IV C, except that the aperture and field diaphragms should be wide open. Remember to readjust these diaphragms to their original positions if it is necessary to return to brightfield microscopy.
2. Obtain oblique illumination using one of the following methods. These are arranged in order of increasing complexity and may not suitable for every instrument. Methods a through d generally give low angle oblique illumination.
   a. Insert a cardboard sector stop partially across the illumination path at the aperture diaphragm (be careful not to damage the aperture diaphragm). See Figure 4-2 for details of the sector stop.
   b. Move an existing, rotating sector stop into the illumination path. This device is an integral part of some condenser systems. The azimuth of the illumination can be changed by rotating the sector in its housing.
   c. Decenter the condenser system by translating the aperture diaphragm out of the illuminating beam (the azimuth of the illumination is established by the direction that the aperture diaphragm is translated). In some instruments the position of the correctly centered aperture diaphragm is located by a detent.
   d. Turn off the brightfield illumination and illuminate the sample from the side with a separate, focused microscope lamp to give high angle oblique illumination.

D. Problem Correction For Oblique Illumination

The microscope should be set up initially for brightfield microscopy, Chapter 2, Section IV C. If any problems occur during that procedure consult the associated problem correction section in Chapters 2 and 3, and eliminate that problem before proceeding to oblique illumination. If a problem is noted during or after one of the above oblique illumination procedures, Table 4-2 will provide a general check list for typical problems, their causes and solutions. Due to the wide variety of instruments and techniques it is impossible to include all of the instrument specific problems.

TABLE 4-2
OBLIQUE ILLUMINATION MICROSCOPY PROBLEMS AND THEIR CORRECTION

PROBLEM	CAUSE	CORRECTION
No illumination or uneven illumination of sample.	Mirror moved too far from image beam axis.	Correctly position mirror.
	Slider for rotating sector not fully in position.	Move slider into correct position.
	Condenser system moved too far from image beam axis.	Move condenser back toward image beam axis.
	External illuminator not focused on sample.	Correctly position external illuminator.
Insufficient oblique illumination effect.	Mirror or condenser not moved far enough from image beam axis	Move mirror or condenser farther from image beam axis.
	Angle between image beam axis and external illuminator too small.	Make angle between external illumination and stage larger.
Desired structure on sample not emphasized.	Azimuth of illumination not correct.	Rotate sample or stage.
		Rotate sector in its slider.
		Rotate cardboard sector at aperture diaphragm. Use caution to avoid damage to aperture diaphragm.
		Move external illuminator to different place around stage.

III. DARKFIELD MICROSCOPY

A. General

In brightfield illumination a majority of the light incident on the sample enters the objective lens and contributes to forming an image. The image may be characterized as dark or colored in an otherwise bright field of view. Darkfield illumination, however utlizes a condenser system in which the sample is uniformly illuminated with light that has a numerical aperture greater than that of the objective lens, i.e., with a high numerical aperture cone of illumination. In this latter case the image is characterized as bright or colored in a field which is otherwise dark. Darkfield illumination may also be considered to be a special case of oblique illumination in which the sample is illuminated from all azimuths with high angle illumination.

Darkfield microscopy can be accomplished easily at low magnifications, i.e., with objective lenses of low N.A. by using a high N.A. condenser with a central stop which excludes the direct illumination to the objective lens; such a system is shown in Figure 4-3a. The construction of such a stop is given below.

For higher N.A. objective lenses, a number of cleverly designed condenser systems have been devised; most of which employ reflecting rather than refracting elements. One of these condensers is shown in cross section in Figure 4-3b. Since the critical angle for glass to air is only about 41° it is absolutely imperative to oil these higher N.A. condenser lenses to the sample slide. The objective lenses used with these condensers are usually equipped with an adjustable aperture stop or with removable fixed stops, see Figure 4-4. These stops reduce the N.A. of the objective to less than that of the hollow cone of light from the condenser, otherwise any direct illumination entering the objective would destroy the darkfield effect. It should be emphasized that once a darkfield condenser system is set up for an objective lens of a given N.A., any objective lens with a lower N.A. may also be used if properly centered to the condenser system.

Darkfield illumination is created in a completely analogous manner in the incident light case. However, the special condenser system is built around the objective lens using either reflective or refractive elements, as shown in Figure 4-5. When the darkfield stop is inserted in the field of view a darkfield image exists, when it is out of the light path an ordinary brightfield image is obtained. The higher N.A. objective lenses require either oil or water immersion and are so marked.

B. Procedure For Transmitted, Darkfield Illumination

If a perfectly clear, clean sample is observed in transmitted darkfield illumination or a clean specular surface is studied with

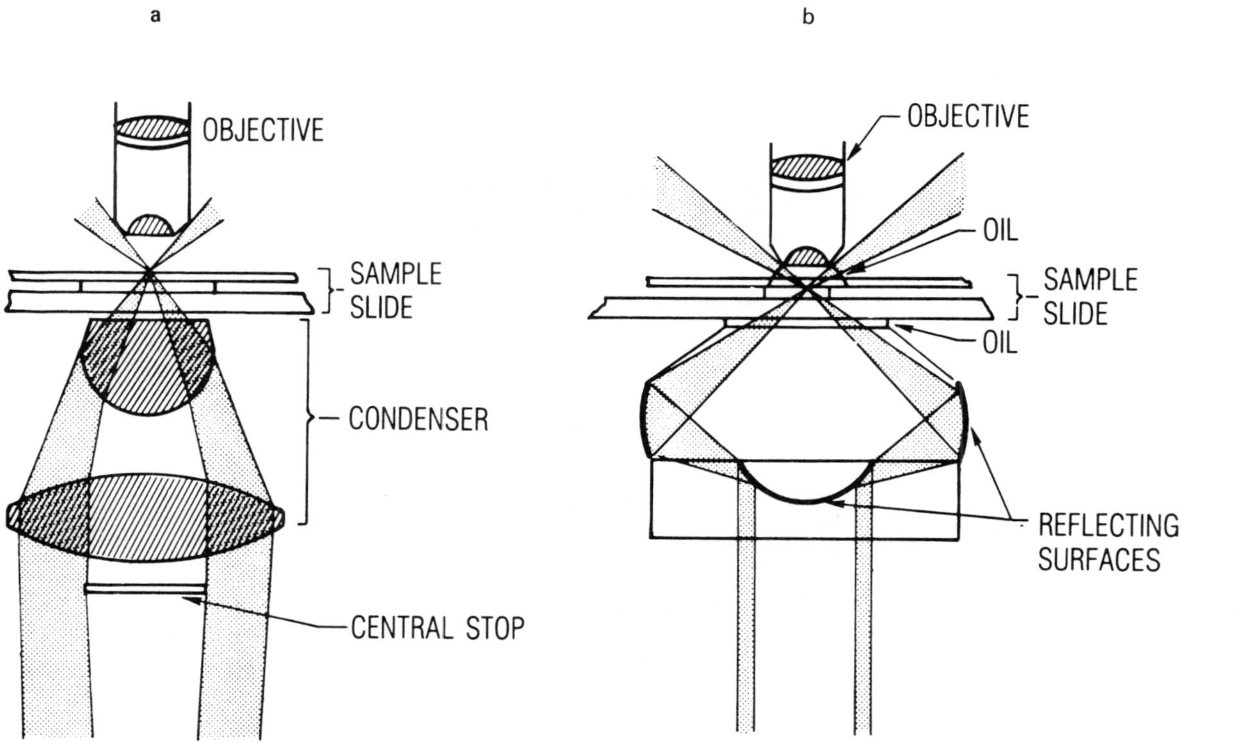

Figure 4-3. Two types of transmitted light darkfield condensers. Redrawn courtesy of Wild Leitz USA, Inc.

Figure 4-4. Removable funnel stops are available for reducing the numerical aperture of certain medium and high power objective lens for use with darkfield condenser systems. Courtesy of Alan DeHaas.

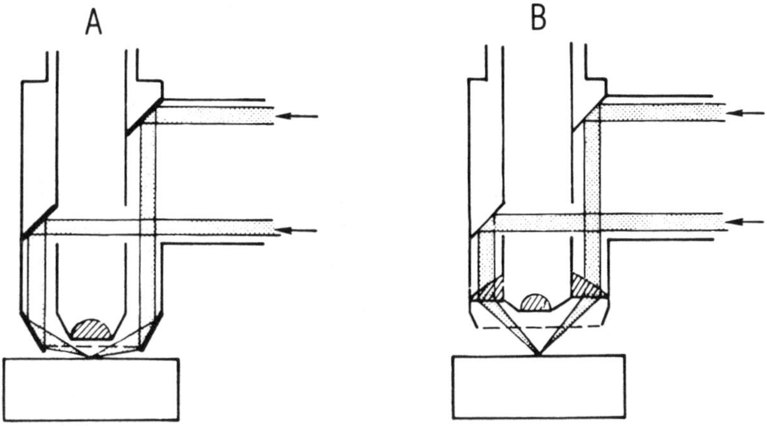

Figure 4-5. Two forms of incident light darkfield objective lenses.

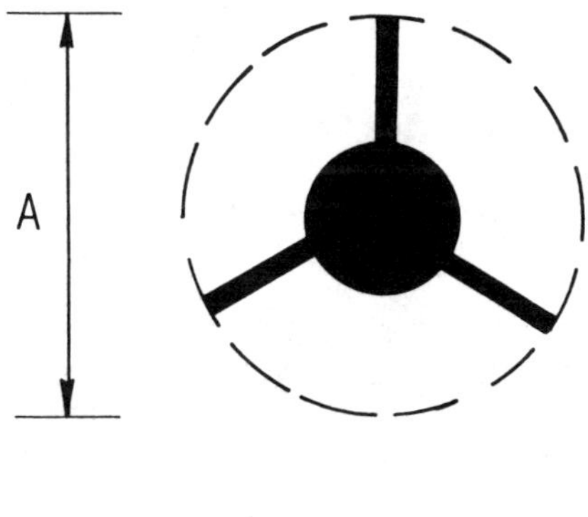

Figure 4-6. Pattern for a central stop to provide simple darkfield illumination. Redrawn from Proceedings of the Royal Microscopical Society.

incident darkfield illumination the image would be completely dark. When discontinuities, or second phases are present light is diffracted, refracted or reflected such that a small portion is accepted by the objective and produces the bright image of the object. Thus, the darkfield technique is a powerful technique for looking for traces of second phases such as films, contamination, or fine structure.

In transmitted darkfield illumination many unstained objects may be seen in striking detail, for example colloidal particles, bacteria, protozoa and yeasts are vividly displayed against the dark background. In the non biological areas, this illumination is desirable for the examination of water or permanent mounts of small particles or thin sections of ceramics or plastics containing small inclusions, porosity, or areas of differing refractive index.

A disadvantage of the darkfield method is that it is not too effective with large amounts of second phase or high contrast structures; these can be seen as well with brightfield microscopy. Perhaps a more serious disadvantage of this technique when the slower photographic emulsions are used is that very long exposures are required to record the structures. This disadvantage is of little consequence if high speed films are used.

Bradbury (1980) has described how a simple central stop can be fabricated in the laboratory that is suitable for transmitted light investigations using objective lenses of 10X or less.

CONSTRUCTION OF A SIMPLE CENTRAL STOP ACCORDING TO S. BRADBURY

1. Set up microscope for proper brightfield illumination as described in Chapter, Section IV C using the desired objective lens ($\leq$ 10X).
2. Remove the eyepiece and adjust the aperture diaphragm until its image just enters the rear focal plane. Replace the eyepiece.
3. Remove the condenser and measure the diameter of the opening of the aperture diaphragm. Add 15% to the measured value (to allow for possible eccentricity of the filter holder): this is the diameter, D, of the central opaque disk required to produce darkfield illumination with the objective lens selected in 1 above.
4. Figure 4-6 shows the form of the stop which may be fabricated from black cardboard or thin metal. The three arms support and position the central stop in the condenser filter carrier; the measurement, OD, is the diameter of the filter required for that filter carrier.

Transmitted light darkfield microscopy can be accomplished at the lower objective magnifications by inserting a central stop in the brightfield condenser system as is shown in Figure 4-3a. Some student instruments have central stops mounted on a rotating disk that is a part of the condenser system. The stop described in C above, commercially available stops of various diameters, or an adjustable darkfield stop available from either R. & J. Beck or W. Watson and Sons may be inserted in the filter carrier beneath the condenser system.

SET-UP FOR A LOW POWER TRANSMITTED LIGHT DARKFIELD MICROSCOPE

1. Place sample on stage and using an objective lens of 10X or less set up the microscope for brightfield transmitted light as described in Chapter 2, Section IV C.
2. Open both field and aperture diaphragms fully, place a central stop in the filter carrier of the condenser and move the filter carrier into the optical path of the microscope. Otherwise, if available, rotate the aperture disk containing a central stop until it is in the optical path of the microscope.
3. Remove an eyepiece and observe the rear focal plane of the objective lens by viewing down the tube. Unless the condenser is perfectly aligned a bright disk of light will be seen which is partially obscured with a dark disk (the central stop). The central stop should be only slightly larger in diameter than the bright disk of light. The condenser is then centered so that the bright disk of light is completely covered by the dark central stop.
4. Replace the eyepiece and observe the sample in darkfield. If the image is too dark, increase the intensity of the lamp.

SET-UP FOR A HIGHER POWER TRANSMITTED LIGHT DARKFIELD MICROSCOPE

1. Place the sample slide on the stage and using an objective lens of 10X or less set-up the microscope for brightfield transmitted light as described in Chapter 2, Section IV C.
2. Lower the condenser using its focus knob and replace the brightfield condenser with the darkfield condenser. Adjust the condenser so that it just touches the sample slide.
3. Focus carefully on the bright ring of light projected onto the sample by the condenser and center this in the field of view using the centering screws on the condenser.
4. If the condenser is an oil immersion type, lower the condenser, place a drop of bubble-free immersion oil on the top lens of the condenser, then raise the condenser slowly until oil contact is made with the sample slide. If any bubbles should be noted in the oil film, clean the oil from both the condenser and sample slide as described in Chapter 10, Section VI, then repeat step 4 until no bubbles are present.

5. Replace the 10X objective with the desired objective lens. If the objective lens is an oil immersion type, apply a drop of bubble-free immersion oil to the cover glass of the sample, carefully move the objective lens toward the cover glass until oil contact is made. Focus the sample then maximize the intensity of illumination by very carefully adjusting the focus knob of the condenser.
6. If the sample does not appear bright against a black field remove an eyepiece and examine the rear focal plane of the objective lens.
   a. If a large ring of light is seen, the aperture of the objective is too large for the condenser. For those objective lenses equipped with a built-in iris diaphragm, use it to reduce the N.A. until the bright ring is just outside the field of view. For those objective lenses equipped with funnel stops, rotate the objective lens to one side, and depending on its type either remove the objective lens, drop the funnel stop into the back of the objective lens and reinstall or remove the lower half of the objective lens, insert the funnel stop, and reinstall that segment to the remainder of the objective lens. Finally reposition the objective toward the sample (when the sample is correctly focused, no bright ring should be observed in this step). If neither the built-in iris diaphragm or the funnel stops are available for a given objective lens, an objective lens of lower N.A. must be used.
   b. If only a portion of a bright arc can be seen, the condenser must be more carefully centered so that the arc disappears.
7. Return the eyepiece and focus critically. The sample may now be studied in darkfield illumination.
8. If another objective lens is to be used repeat steps 5 through 7.

C. Procedure For Incident, Darkfield Illumination

Darkfield illumination is also impressive for use in incident light microscopy. In this case, the surface irregularities such as grain boundaries, cracks or cleavage traces stand out in bold relief. It is often unnecessary to etch a metallurgical specimen to bring out detail in this way. The color rendition of materials is particularly good with incident darkfield illumination.

Since the objective lens is also the condenser lens of the incident light microscope, any changes to the instrument to permit darkfield illumination must include the objective lens. In fact, the objective lens for darkfield illumination includes a peripheral illuminating system that is physically isolated from the imaging portion of the objective but is an optical extension of the imaging portion. That is to say, the peripheral portion focuses at the same point as the imaging portion. This peripheral illuminating system is not as well corrected optically as the imaging portion of the objective and may contain either refracting or reflecting elements as shown in Figure 4-5.

Although it may not be immediately apparent, these objective lenses can, as well, be used as brightfield objectives. To change from brightfield to darkfield a central stop is inserted in the condenser system which permits illumination only from the peripheral illuminating portion of the objective lens.

SET-UP FOR AN INCIDENT LIGHT, DARKFIELD MICROSCOPE

1. Adjust the microscope for correct normal incident brightfield illumination per Chapter 3, Section IV C, and verify that darkfield objective lenses are being used. The aperture and field diaphragms should be wide open. Remember to return these diaphragms to their original positions if it is necessary to return to brightfield microscopy.
2. Insert the darkfield stop into the light path. The sample can now be observed in darkfield illumination.
3. Objective lenses may be changed without readjustment of the illumination.

D. Problem Correction For Darkfield Illumination

If any problems occur during the brightfield illumination set-up procedure, consult the associated problem correction section in Chapters 2 or 3 and eliminate that problem before proceeding to darkfield illumination. If a problem is noted during or after one of the above darkfield illumination procedures, Table 4-3 will provide a general check list for typical problems, their causes and solutions. Due to the wide variety of instruments and techniques it is impossible to include all of the instrument specific problems.

IV. PHASE CONTRAST MICROSCOPY

A. General

In the years immediately after the second world war, the field of optical microscopy made great advances through the work of Frits Zernike (1935) on phase contrast microscopy. This technique when applied to samples with little or no brightness or color contrast permitted the evaluation of structure by their phase effects alone. The impact of this capability can be appreciated in the biological field alone, when it is realized that live specimens could be examined without applying potentially lethal stains. For his efforts, Zernike was awarded the Nobel prize in physics in 1953.

TABLE 4-3
DARKFIELD MICROSCOPY-PROBLEMS AND THEIR CORRECTION

PROBLEM	CAUSE*	CORRECTION
Insufficient brightness.	lamp voltage too low. 1,2,3.	increase lamp voltage but lower it after the observation.
	Interior of lamp has black evaporated tungsten coating. 1,2,3.	Replace lamp.
	Condenser not focused correctly. 1,2.	Carefully focus condenser on sample.
	Filter in light path. 1,2,3.	Remove filter.
Reduced lamp life.	High lamp voltages required for darkfield studies. 1,2,3.	Reduce lamp voltage after observation.
Image not darkfield	Central stop diameter too small. 1.	Use larger diameter central stop.
	Condenser not centered. 1,2.	Carefully center condenser per above instructions.
	Objective lens N.A. too large for condenser system. 2.	Reduce N.A. with adjustable aperture stop or removable fixed stop. Use lower N.A. objective lens.
	Darkfield stop not fully inserted. 3.	Insert darkfield stop completely.
	Revolving nosepiece not at detent position. 2,3	Rotate nosepiece slightly until it seats in detent stop.
Image not clear.	Sample has too much structure. 2,3.*	Sample not suitable for darkfield study.

*Comments apply to: 1. Low power transmitted light darkfield
2. Higher power transmitted light darkfield.
3. Incident light darkfield.

So that we might appreciate the various components and the complexity of the phase contrast system, it is worthwhile to consider the process by which phase contrast is achieved. Let us, for the sake of a simple discussion, assume that we are going to examine with a microscope, a sample that consists of a single, completely transparent cubic grain with a R.I. of n_s in a mounting medium with a R.I. of n_b where $n_s > n_b$. We will illuminate the sample with essentially coherent, monochromatic illumination of wavelength, λ. This arrangement is shown in Figure 4-7. When the illumination interacts with the sample a portion of it is diffracted. The diffracted light is accepted by the objective lens and is imaged at the primary image plane. The background or undiffracted illumination however, is imaged as a small spot, X, at the rear focal plane of the objective lens. The background light passes through point X and illuminates the primary image plane uniformly, assuming that we have correctly set up Koehler illumination in the condenser system.

Now let us consider the various ray bundles passing through the assemblage made up of the glass slide, mounting medium, sample grain and the cover glass. Since our illumination is essentially coherent, the rays incident on the glass slide are in phase as shown in Figure 4-8a. Likewise, all the rays forming the background illumination are coherent since the thickness of glass slide, mounting medium and cover glass is constant. However, the light rays passing through anisotropic sample grain, R_s, are slightly retarded with respect to the background light rays, R_b due to the grain's higher R.I.; this is shown in Figure 4-8b. The actual retardation, Δ, of R_s is

$$\Delta = (n_s - n_b) = t \qquad (19)$$

where t is the thickness of the grain. The phase shift, ϕ, between R_s and R_b is given by

$$\phi = 2\pi\Delta/\lambda. \qquad (20)$$

The intensities of R_s and R_b are proportional to the square of their amplitudes, A_s and A_b, respectively. Since the sample is transparent, R_s has been retarded but not reduced in amplitude. Thus, $A_s = A_b$ and their corresponding intensities are equal. R_s may be resolved into two rays, one of which has the same phase and amplitude as R_b and a new ray, R_c; this is shown in Figure 4-8c. When d is small, as is the case with sample suitable for phase contrast observation, R_c is shifted by 90° or $\pi/2$ in terms of phase angle. Furthermore, the amplitude, A_c, of R_c for small values of ϕ is proportional to ϕ and is a function of $(n_s - n_b)$. An obvious consequence of the latter is that if the refractive index of the sample is

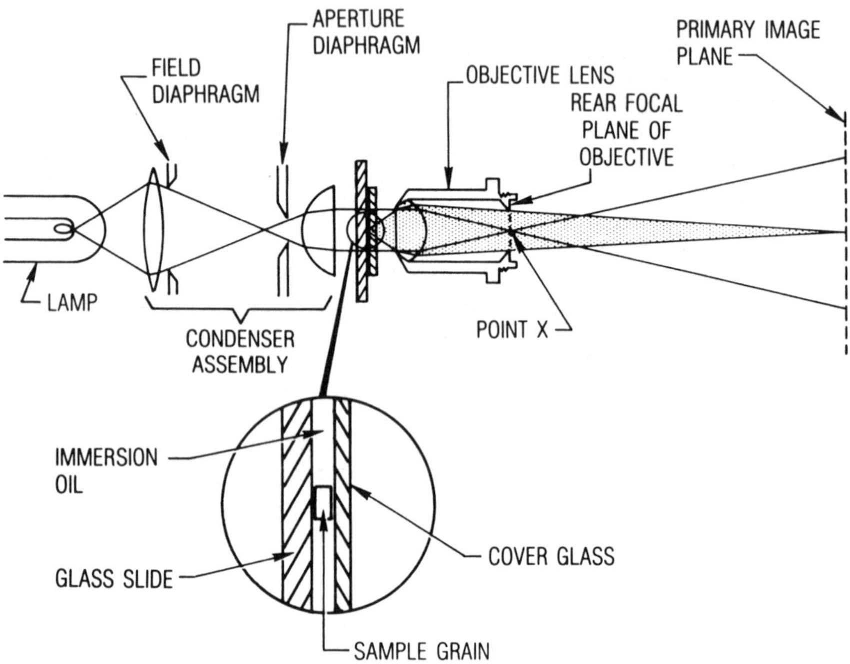

Figure 4-7. Illumination and image paths for a transmitted light microscope. The illumination source is imaged at X in the rear focal plane of the objective lens while the sample grain is imaged at the primary image plane (the sample image rays are shaded).

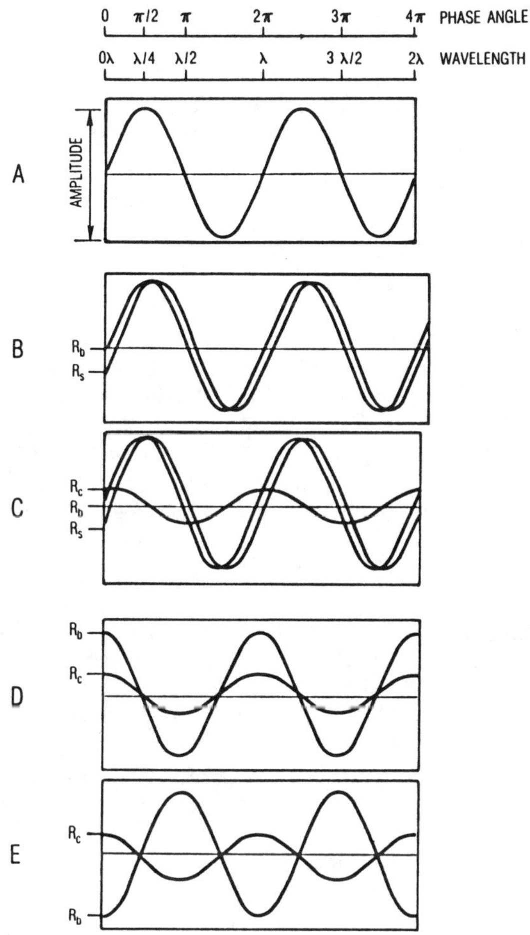

Figure 4-8. Simplified graphical presentation of the ray phase relationships in phase contrast microscopy:

A. Phase of ray incident on sample.
B. Relationship between Rb, the unretarded ray adjacent to the sample and Rs, the ray as retarded by passing through the sample.
C. The sample ray, Rs, may be resolved into two rays, one of which is equivalent in phase to Rb and another, Rc; these two rays have a phase difference of $\pi/2$ or $\lambda/2$.
D. Rb has been selectively retarded so that it is now in phase with Rc (bright contrast).
E. In another case, Rs has been retarded so that Rb is 180° out of phase with Rc (dark contrast).

exactly matched by the refractive index of the immersion medium no phase contrast is possible.

The intensity of the background at the primary image plane is $R_b = A_b^2$; if we set A_b at unity then R_b is also 1. The image of the sample at the primary image plane at this time can be evaluated in the same way. The image of the grain is formed by two rays, one of which is in phase with R_b and another, R_c which differs from R_b in phase angle by 90° or $\pi/2$. The intensity of light in the sample image is $(A_b + A_c)^2$. Since A_c is proportional to m which is small, the intensity of the sample image is ~1. If the intensity of the sample and the background are identical, then the sample grain is not visible as an intensity contrast image.

Consider now the effect of adding a small optical element known as a phase plate at point X in Figure 4-7. The function of the phase plate is to retard the background ray, R_b that is focused on that spot. If the properties of that phase plate are such that R_b is retarded by 90° or $\pi/2$, then R_b and R_c are in phase and the intensity of the image of the sample grain is given by:

$$I = (1 + \phi)^2 = 1 + 2\phi + \phi^2 \sim 1 + 2\phi \qquad (21)$$

It will be recalled that the intensity of the background is nominally 1, therefore the sample grain with a greater intensity of $(1 + 2\phi)$ will have brightness contrast and can be seen. This condition, known as negative phase contrast, is shown in Figure 4-8d.

If the phase plate has sufficient thickness that R_b is retarded by 270° or $3\pi/2$, the rays R_b and R_c are completely out of phase. The intensity of the image of the grain is given by:

$$I = (1 - \phi)^2 = 1 - 2\phi + \phi^2 \sim 1 - 2\phi \qquad (22)$$

In this case the intensity of the grain is less than that of the background and thus, appears darker than the background. This condition, known as positive phase contrast, is illustrated in Figure 4-8e.

In practice, the small circular aperture used in the condenser system to produce the coherency required for the above illustration limits the resolution of the system to an unacceptable value. Furthermore, the small, circular phase plate in the objective also produces an image which lacks uniformity of the phase effect. A number of aperture configurations and the corresponding phase plate shapes have been explored. For example, a slit has been used in the condenser with a narrow rectangular phase plate in the objective lens. The English firm, C. Baker, has designed a phase system which used a cross shaped aperture plate that was used with a corresponding cross shaped phase plate. However, theory suggests that the optimum configuration for these elements is an annulus in the condenser and an annular phase ring in the rear focal plane of the objective lens, as shown in Figure 4-9. Since the phase ring is the most inaccessible part of the system it's dimensions are fixed and the annulus is adjusted to match the phase ring. Typical dimensions of the phase ring are as follows; the ring is located about one half of the distance between the center and the edge of the rear focal plane aperture of the objective lens and has a thickness of about one-fifteenth the diameter of that aperture.

Some phase rings retard the background rays while others retard the image rays. A further improvement in contrast may be realized by the addition of a vacuum deposited, absorbing layer either to the ring or to the background ray area. Depending on the specific arrangement of the retarding and absorbing layers the phase contrast objective lens is referred to as a positive (dark contrast) objective lens or a negative (bright contrast) objective lens. Various phase ring arrangements are shown in Figure 4-10.

Wild has developed a "Varicolor" variable phase contrast system which makes use of special phase rings in combination with an interference filter in the illumination path. For certain conditions of sample preparation either positive or negative phase contrast can be obtained through varying the color of the illumination; this is accomplished by changing the tilt of the interference filter.

With an ideal phase ring, a path difference of about 1 in the sample has been observed using an absorption of 96% in the phase ring. To achieve such sensitivity a superior instrument, properly adjusted must be used.

Since the phase ring blocks a portion of the central aperture of the objective lens it is reasonable to suspect that it may reduce the resolution of the system. This has been thoroughly investigated and the result is that the observer resolution is somewhat better than anticipated for high contrast samples and somewhat poorer than expected for low contrast samples. Therefore, the criteria for lateral resolution given in Chapter 2 may be used for most phase contrast applications.

B. Transmitted Light Applications

In its simplest form the phase contrast microscope consists of a brightfield microscope which has an annulus in the condenser system and a phase ring in the objective lens. Both Wild and Unitron have student versions of the phase microscope in which the annulus is pressed into or screwed onto the bottom of the brightfield condenser. All that is needed to complete these instruments is the installation of objective lenses with phase rings.

In more advanced instruments the condenser system includes either a turret arrangement with several annuli of various diameters or a single annulus with a pancratic system that varies the effective diameter of the annulus. A typical phase contrast

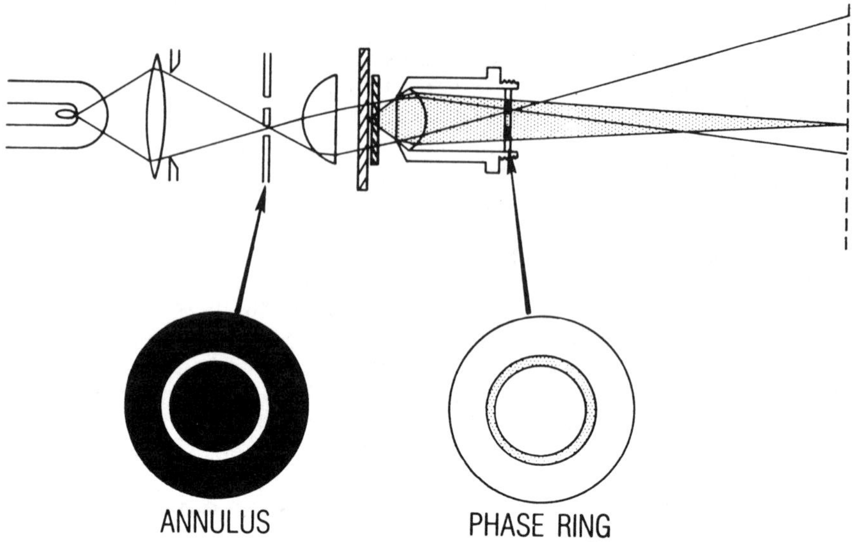

Figure 4-9. Illumination and image paths for a transmitted light, phase contrast microscope. The annulus is imaged on the phase ring at the rear focal plane of the objective lens. The sample is imaged at the primary image plane (the sample image rays are shaded).

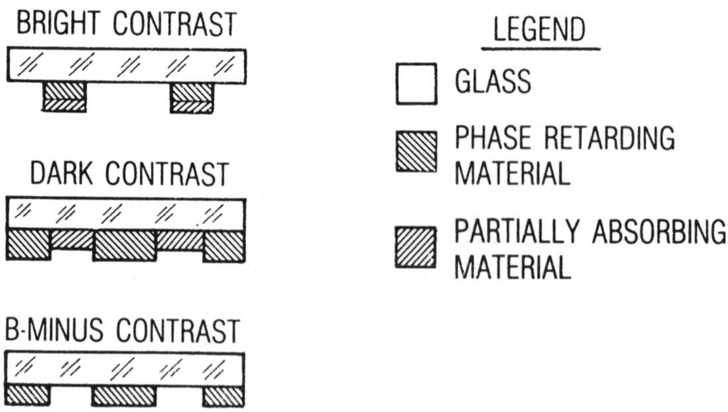

Figure 4-10. Cross sections of three types of phase rings showing the arrangement of the various coatings. Courtesy of Cambridge Instruments Inc., Optical Systems Division.

condenser system with a revolving turret, shown in Figure 4-11, includes an overall centering system, an aperture diaphragm and individual centering for each of the annuli. Some manufacturers design their systems to use only three different diameters of annuli, one for objective lenses of 10 X or less, one for the intermediate magnifications and the final one for the high dry and oil immersion objective lenses. Other manufacturers use one annulus per objective lens. One turret position is usually without an annulus and serves for standard brightfield observations and for initial alignment purposes. Often an additional turret position is equipped with a central stop for low power darkfield illumination.

The pancratic phase contrast condenser has been developed by several manufacturers including, Leitz, Unitron, Carl Zeiss and Zeiss Jena. These condensers rely on a pancratic system which varies the N.A. of the condenser over a wide range, thereby projecting the image of a single fixed annulus of correct diameter onto the phase rings of the corresponding set of phase contrast objective lenses. Furthermore, by simply adjusting the pancratic system, as shown in Figure 4-12 , one can also have brightfield or darkfield illumination. Since there is only one annulus in this condenser, it is only equipped with an overall centering adjustment.

The phase ring is the other crucial item required to produce phase contrast images; these take a number of forms as described above. The phase ring is nearly always located in the rear focal plane of the objective lens. Therefore, a special set of objective lenses is required for phase contrast investigations. It should be emphasized that most phase contrast objectives lenses can be used for casual brightfield examinations during a phase contrast study without severe degradation of the image by using an open turret position or adjustment of the pancratic assembly in the condenser.

Reichert, has used a remote phase contrast assembly for a number of years which eliminates the need for special phase contrast objective lenses; this is shown in Figure 4-13. This system which mounts on the side of the instrument forms an image of the rear focal plane of the objective lens within its confines. A turret is present at that point with a series of both positive and negative phase rings as well as an open position for brightfield examinations. Using a built-in pancratic system, the single fixed annulus in the condenser system can be matched with any objective lens magnification. The phase contrast image can be viewed only through an eyepiece on the phase assembly. This eyepiece is mounted in one tube of a two position turret assembly; the other turret position is for a focusing telescope as described below.

Correct functioning of the phase contrast microscope requires careful superposition of the annulus on the phase ring images. While one can accomplish this procedure by removing an eyepiece and viewing down the open tube of the microscope, critical adjustment requires some magnification.

Figure 4-11. Typical phase condenser and objective lenses. The revolving disk in the condenser contains a phase annulus to match the phase ring in each objective lens. Courtesy of Nikon, Inc.

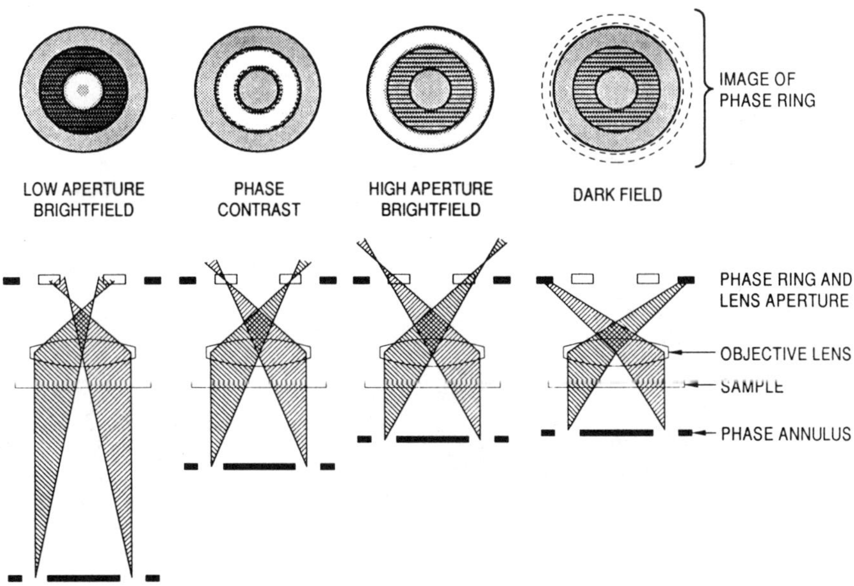

Figure 4-12. The pancratic phase contrast condenser in which the annulus diameter is adjustable to match the phase rings of a complete set of phase contrast objective lenses. This condenser also permits a complete range of viewing conditions for a single phase contrast objective lens. Adjusted to position I the condenser yields a low aperture brightfield image. When the condenser is at position II the sample is seen in phase contrast. At position III the condenser yields a high aperture brightfield image and at position IV the sample is seen in darkfield illumination. Redrawn courtesy of Unitron Instruments, Inc.

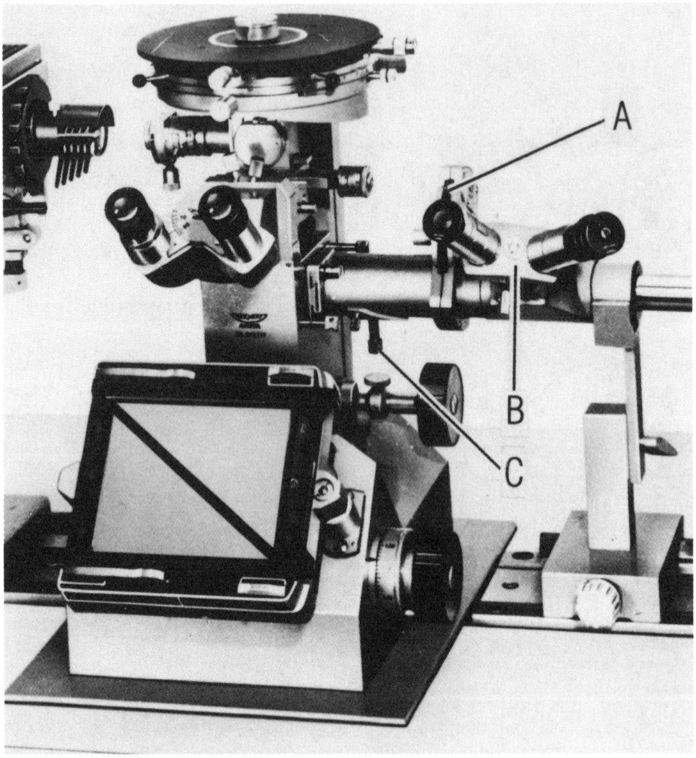

Figure 4-13. Reichert has used a pancratic phase ring arrangement for some time; this system is shown as the horizontal attachment to the microscope body. In this system the phase annulus is focused on the phase ring which is separate from the objective lens and at a point between the objective lens and the eyepiece. The pancratic system is adjusted with lever C so that the annulus image is matched to a phase ring in turret A for any given objective lens. The eyepiece through which the phase contrast image is viewed and the eyepiece telescope are mounted on the turret B. The phase effect may be photographed using a camera mounted on the tube to the right of the pancratic system.

Some research grade instruments have an insertable lens, known as a Bertrand lens, see Chapter 2, Section III D, located between the objective lens and the eyepiece. The Bertrand lens gives sufficient magnification of the phase ring that centering adjustment is easily accomplished. For the instruments not equipped with a Bertrand lens, an accessory known as a eyepiece telescope or centering telescope, see Chapter 7, Section VII, can be substituted for an eyepiece to provide a magnified view of the phase ring alignment process.

The ideal sample for transmitted light, phase contrast microscopy is one which is thin, has low contrast and some difference in refractive index from the mounting medium. The effective retardation of the sample compared with the mounting medium is determined by the difference in refractive indices between sample and mounting medium and the sample thickness. For example, a sample 1 micrometer thick, can be contrasted and seen if the difference in refractive indices between it and the mounting medium is only 0.0002. The appearance of the sample in phase contrast depends on the refractive index of the sample relative to its mounting medium and on the type of phase contrast objective lens used. This is illustrated in Table 4-4.

TABLE 4-4
CONTRAST BETWEEN SAMPLE AND BACKGROUND IN TRANSMITTED LIGHT PHASE CONTRAST

REFRACTIVE INDEX OF THE SAMPLE	POSITIVE OR DARK PHASE OBJECTIVE LENS	NEGATIVE OR BRIGHT PHASE OBJECTIVE LENS
Greater than that of the mounting medium	Sample is darker than the background	Sample is brighter than the background
Less than that of the mounting medium.	Sample is brighter than the background.	Sample is darker than the background.

The effect of increased sample thickness is to cause contrast reversal. The nominal contrast type of a given objective lens applies for samples that have a retardation less than $\pi/2$. As the thickness increases the contrast will suffer a contrast reversal at a retardation of $3\pi/2$ and then back to the original contrast at $5\pi/2$ and so on.

Contrast reversal can occur in yet another way. The nominal contrast effect of a given objective lens is based on the refractive index of the sample being greater than that of the mounting medium. If the situation is reversed, the contrast is also reversed.

An effect that is observed, particularly in nominal positive contrast, i.e., when the sample is darker than the background, is that the sample is surrounded with a bright halo. This halo effect makes size measurements very difficult since the actual edges are not easy to discern.

In summary, the phase contrast technique is excellent for observing very subtle phase structures in samples that do not have brightness contrast, color contrast or discernible effects in polarized light. In general, the thinner the sample, the better the effect. The technique produces images that have much greater brightness than darkfield therefore, very satisfactory photomicrographs can be produced.

On the negative side, the phase contrast technique requires special objectives that must be installed before phase contrast observations. Furthermore, the halo effect limits the quantitative use of the phase contrast technique. As a result the technique is gradually giving way to the interference contrast technique. Finally, as with any phase sensitive technique, one must always remember that the contrast observed is in the image and not in the sample.

Phase contrast microscopy is widely used in the biological field to this day. Transmitted light, phase contrast microscopy has also found limited application in the examination of rocks and minerals in thin section (Smithson,1948).

SET-UP OF A TRANSMITTED LIGHT, PHASE CONTRAST MICROSCOPE

1. Adjust the microscope for correct brightfield illumination per Chapter 2, Section IV C.
2. Carefully remove the brightfield condenser system and replace it with the phase contrast condenser.
3. Place a suitable brightfield sample, i.e., one that has observable brightness contrast, on the stage and focus it carefully with the 10X phase contrast objective lens in place.
4. Close field diaphragm to about 1/2 the field and focus its image in the the plane of the sample image using the condenser focusing knob. If the condenser has an aperture diaphragm, closing it down will improve the sharpness of the field diaphragm image.
5. If the field diaphragm is a part of the illuminator, center the field diaphragm in the field of view with the illuminator assembly. If the field diaphragm is a part of the condenser, center the image of the field diaphragm in the field of view using the condenser centering adjustments. Then open the field diaphragm until its image is just outside the field of view. If the aperture diaphragm was adjusted in Step 4, open that diaphragm fully.
6. Replace the brightfield sample with a thin unstained sample. Refocus the sample critically.

7. Observe the phase annulus and phase ring in one of three ways:
    a. Remove an eyepiece an look down the tube of the microscope. This is not too effective because of the small image. A pinhole cap, Chapter 7, Section VI maybe used to improve the contrast of this image.
    b. Replace an eyepiece with a eyepiece telescope. Focus the telescope to produce a sharp image of the phase ring. The Reichert remote system has a rotating assembly which holds both the eyepiece and the centering telescope so that either one can be rotated into viewing position as desired, see Figure 4-13.
    c. Move the Bertrand lens lever or slider to place the lens in the image path. If the Bertrand lens is focusable, sharply focus the image of the phase ring. Some advanced instruments also provide centering adjustment for the Bertrand lens, if this is available, center the image of the phase ring in the field of view.
8. Focus the image of the annulus using the condenser focus knob.
9. Adjust the phase annulus so that its image falls exactly on the phase ring by proper adjustment of the phase condenser centering.
    a. If a turret condenser is supplied, rotate the turret to the annulus for the 10X objective then center the annulus using the condenser centering adjustments so that the annulus image falls exactly on the phase ring. If the annuli are not so identified, try each annulus in succession until the one is found that matches the diameter of the phase ring.
    b. If a pancratic condenser is supplied, adjust the condenser centering until the annulus is coaxial with the phase ring, then adjust the pancratic assembly knob until the diameter of the phase annulus image exactly matches that of the phase ring. Repeat Step 8 if the latter adjustment affects the focus of the annulus. Correctly and incorrectly adjusted annuli are shown in Figure 4-14.
10. Reinstall the eyepiece or remove the Bertrand lens from the light path and observe the sample in phase contrast.
11. If the objective lens is changed repeat Steps 7 to 10.

C. Incident Light Applications

Incident light phase contrast microscopy is accomplished in the same manner as with transmitted light, using phase objective lenses and matching phase annuli in the condenser system, as shown in Figure 4-15. The isolated phase assembly of Reichert, discussed in B. above, not only serves well for the incident light

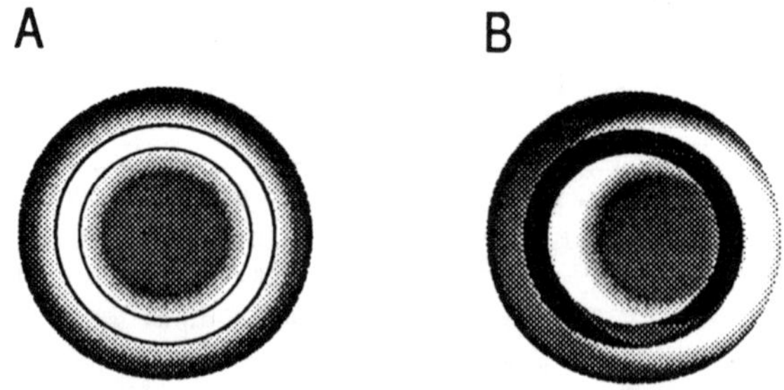

Figure 4-14. Diagrams showing the correct (A) and incorrect (B) positioning of the phase annulus image on the phase ring in the objective lens. This can be seen by removing the eyepiece, using an eyepiece telescope or using a Bertrand Lens, if present on the instrument.

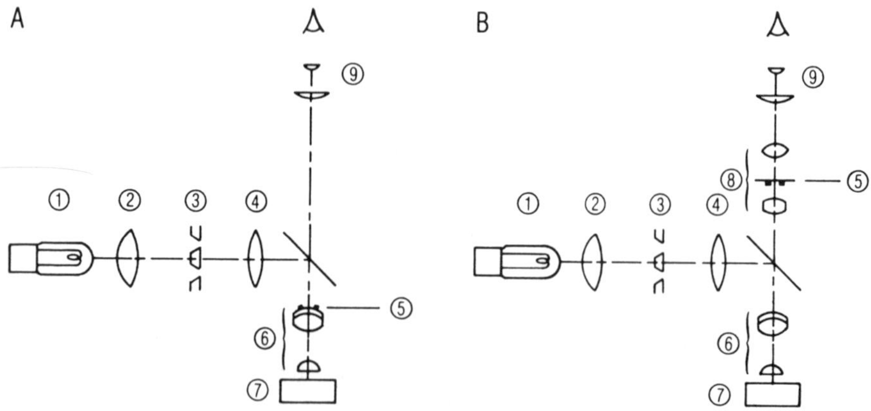

Figure 4-15. Cross sectional view of an incident light phase contrast systems; (A) is a standard system with phase rings in each objective lens and (B) is the Reichert system as shown in Figure 4-13.

1. Lamp
2. Lamp collector lens
3. Phase annulus
4. Condenser lens
5. Phase ring
6. Objective lens
7. Sample
8. Pancratic lens arrangement
9. Eyepiece

microscope but also minimizes any disturbing reflections from the phase rings in the objective lenses.

It will be recalled that the phase effects in transmitted light result from slight differences in the refractive index of the sample with respect to the mounting medium. In incident light illumination, the phase effect results from differences in the phase of the light reflected from the sample due to local variations in the flatness of the sample, this is illustrated in Figure 4-16. For example, the aim of metallographic preparation is to produce a smooth flat sample that is subsequently etched to accentuate the grain boundaries. However even after the finest polishing a small "relief", or deviation from flatness, at the edges of grains on the order of 10 to 100Å occurs due to differences in hardness of the individual grains. This small relief can be clearly observed in phase contrast if it exceeds 50Å. The effect is often improved however, with a light etch.

The contrast of a given grain depends on whether a grain is elevated or depressed relative to its neighbors and on the contrast type of the phase rings; this is illustrated in Table 4-5. As the relief between grains increases the retardation increases and the resulting contrast undergoes reversal as described in B above.

TABLE 4-5
CONTRAST BETWEEN SAMPLE AND BACKGROUND IN INCIDENT LIGHT PHASE CONTRAST

RELATIVE HEIGHT OF SAMPLE GRAIN	POSITIVE OR DARK PHASE OBJECTIVE LENS	NEGATIVE OR BRIGHT PHASE OBJECTIVE LENS
Greater than the surounding grains.	Sample is darker than the background.	Sample is brighter than the background.
Less than the surrounding grains.	Sample is brighter than the background.	Sample is darker than the background.

Incident light, phase contrast microscopy has served as a powerful tool in the investigation of structures that would not etch well. It clearly delineates such structures as grain boundaries, precipitates, twinning, slip bands and other deformation structures.

The major problem with incident light phase contrast, is the halo effect, mentioned in B above. Since quantitative measurements are so important in metallurgical studies, the halo is a severe drawback and the recent trend is to use interference contrast techniques for this purpose. To the author's knowledge no incident light phase contrast accessories are available on new instruments. However, due to the number of existing instruments, set-up procedures is given below.

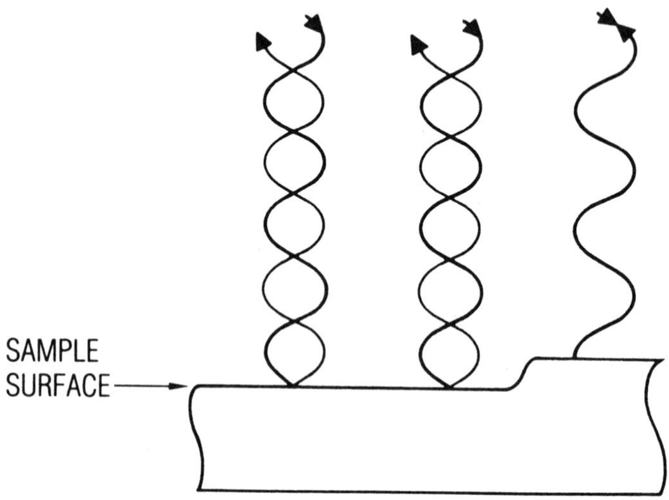

Figure 4-16. Cross section of sample surface through a step. The effect of the step on the phase of the light rays reflected is shown schematically. All of the rays incident on the sample are shown to be in phase. The two reflected rays at the left are in phase with each other because their path length is the same. However, the ray reflected from the step in the surface at the right is out of phase with the other reflected rays because its path length is shorter.

SET-UP OF AN INCIDENT LIGHT, PHASE CONTRAST MICROSCOPE

1. Adjust the microscope for correct incident light, brightfield illumination per Chapter 3, Section IV C.2. Place a lightly etched or unetched sample on the stage and focus it carefully. The sample must be leveled as described in Chapter 2, Section II A, if an upright stand microscope is used.
2. Adjust the field diaphragm until its image is just outside the field of view. Open the aperture diaphragm fully.
3. Insert the correct phase annulus for the desired objective lens into the condenser system by rotating the turret containing the annuli. The single phase annulus for the Reichert system is mounted onto the end of the condenser system; its one annulus is also rotated into position.
4. Observe the phase annulus and phase ring in one of three ways:
   a. Remove an eyepiece an look down the tube of the microscope. This is not too effective because of the small image. A pinhole cap, Chapter 7, Section VI may be used to improve the contrast of this image.
   b. Replace an eyepiece with an eyepiece telescope. Focus the telescope to produce a sharp image of the phase ring. The Reichert remote system has a rotating assembly which holds both the eyepiece and the eyepiece telescope so that either one can be rotated into viewing position as desired, see Figure 4-13.
   c. If the instrument has a Bertrand lens move its lever or slider to place the lens in the image path. If the Bertrand lens is focusable, sharply focus the image of the phase ring. Some advanced instruments also provide centering adjustment for the Bertrand lens, if this is available, also center the image of the phase ring in the field of view.
5. Adjust the phase annulus so that it falls exactly on the phase ring by proper adjustment of the annulus centering. If the Reichert remote phase ring system is used, rotate the phase ring turret until the diameter of the phase ring matches that of the annulus. This system has adjustments for annulus diameter, focusing, and centering as shown in Figure 4-13. Correctly and incorrectly adjusted annuli are shown in Figure 4-14.
6. Reinstall the eyepiece or remove the Bertrand lens from the light path and observe the sample in phase contrast.
7. If the objective lens is changed repeat Steps 4 to 6.

D. Problem Correction For Phase Contrast Illumination

If any problems occur during the brightfield illumination set-up procedure, consult the associated problem correction section and eliminate that problem before proceeding to phase contrast ilumination. If a problem is noted during or after one of the above phase contrast illumination procedures, Table 4-6 will provide a general check list for typical problems, their causes and solutions. Due to the wide variety of instruments and techniques it is impossible to include all of the instrument specific problems.

V. DIFFERENTIAL INTERFERENCE CONTRAST MICROSCOPY

A. General

Interference microscopy is one of the newer techniques for evaluation of samples. The interference microscope simply consists of a brightfield microscope to which an interferometer is attached. Interference microscopes may be divided into two basic types, the two beam instruments and the multiple beam instruments. Either of these instruments may be used to obtain quantitative information about the sample but under certain conditions they produce contrasting of the sample structure in the same manner as darkfield and phase contrast instruments. The contrasting capability, specifically of the two beam instrument is the topic of this section; this technique is referred to as differential interference contrast microscopy or simply, DIC. Interference microscopy as a quantitative technique is discussed in Chapter 5, Section 6.

The operation of the two beam interference microscope is illustrated in Figure 4-17 beam splitter system (1) between the illumination and the sample divides the illuminating rays into two coherent beams, one of which, the sample beam, interacts with the sample and the other, the reference beam interacts with the reference; the two beams are then recombined at a second beam splitter system (2). Since the beams are coherent they interfere with each other and produce a pattern of Newtonian color fringes in the image of the sample that are dependent on the structure of the sample. In the DIC microscope the fringe system is expanded either by manual adjustment or by instrument design to fill the field of view with any single fringe. The actual location of this interferometer within the framework of the microscope varies from instrument to instrument.

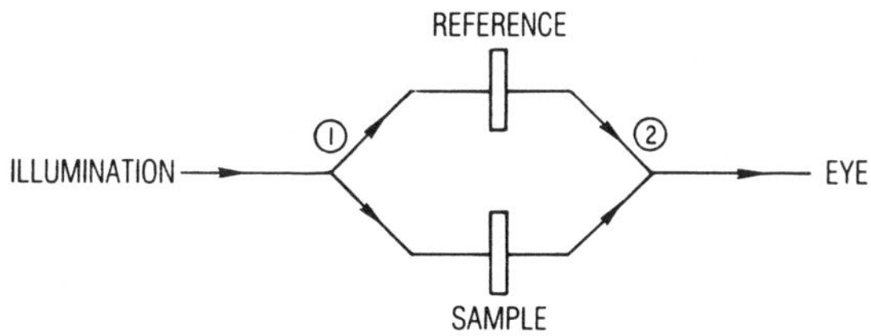

Figure 4-17. Schematic drawing of a two beam interferometer. (1) is a beam splitter which divides the illuminating beam so that both reference and sample are illuminated. (2) is a second beam splitter in reverse that recombines the beams after undergoing some retardation in both reference and sample. The combined beam gives rise to a useful interference image of the sample if the reference has uniform retardation throughout the field of view.

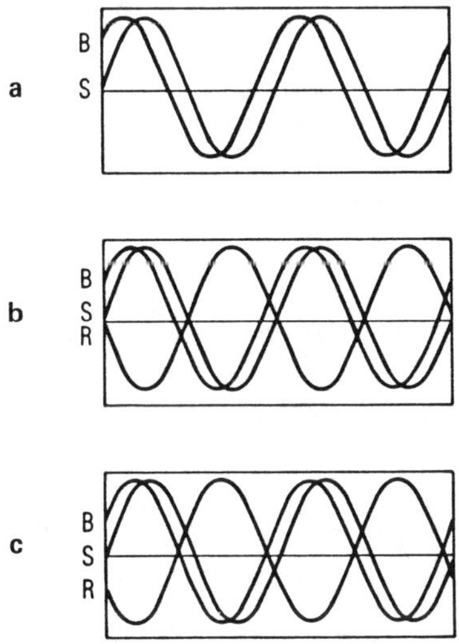

Figure 4-18. Simplified graphical presentation of the relationship of the sample and background rays to the reference rays in interference contrast microscopy:

a. The sample ray, S, is slightly retarded with respect to the background ray, B.
b. The reference ray is made exactly out of phase with the sample ray; this makes that sample area dark with respect to the background.
c. The reference ray is made exactly out of phase with the background ray: in this case that sample area is bright with respect to the background.

TABLE 4-6
PHASE CONTRAST MICROSCOPY-PROBLEMS AND THEIR CORRECTION

PROBLEM	CAUSE	CORRECTION
Phase ring not visible when viewed without eyepiece or with Bertrand lens or with eyepiece telescope.	Brightfield objective lens in place.	Replace the bright-field lens with the correct phase contrast objective.
	B-minus phase objective lens in place.	Phase ring cannot be seen for a B-minus phase contrast object. Set-up first with a standard phase contrast objective lens then change to the B-minus objective lens and carefully adjust annulus centering to give optimum contrast.
Annulus cannot be centered on the phase ring.	Annulus (or phase ring) turret not at detent position.	Rotate turret slightly in either direction until detent seats.
	Revolving nose-piece not at detent position.	Rotate nosepiece slightly in either direction until detent seats.
	Centering adjustment inoperative.	Refer to professional service personnel.
	Sample not level (only for upright stand incident light instrument.	Level sample per Chapter 3, Section IV C 3.
Poor phase contrast image.	Wrong size of annulus or annulus incorrectly centered.	Readjust instrument per instructions in Sections D or E above.
	Inappropriate phase contrast sample.	For transmitted light use thinner sample and minimize R.I. differences between sample and mounting medium. For incident light use flat, highly polished samples with little or no etching.

Figure 4-18 illustrates graphically the relationships between the sample and the reference beams and how they interact. It will be recalled that any variations in the sample produce phase changes in the sample image relative to its background. In Figure 4-18a, a sample beam, (S) and a reference beam, (B) are shown to be out of phase with respect to each other. If the reference beam is adjusted so that it is exactly out of phase with the background, as shown in Figure 4-18b, the background will be dark and the sample will be bright or colored. However, if the reference beam is made exactly out of phase with a portion of the sample, it will be dark and the remainder of the sample and the background will be bright or colored, this is shown in Figure 4-18c.

The resulting interferometric image is rich in contrast and apparent relief and may be in color. It is this latter feature that makes the technique so popular in a time when color photomicrography is so easy to accomplish.

As mentioned above, the actual physical arrangement of the various elements of an interference microscope is quite diverse and a number of variations for both transmitted and incident light will be discussed. Even the reference material can vary significantly; in the self reference instrument, the reference material is some portion of the sample slide or mount itself; whereas, in the case of the independent reference instrument the reference material may be of an entirely different composition that is separated from the sample by a large distance on the microscopic scale.

B. Transmitted Light Applications

The DIC technique provides excellent imaging of subtle differences in refractive index in a sample with about the same sensitivity as the phase contrast technique. Brightness or color contrast does not seriously interfere with the technique, however the DIC effect is most dramatic in essentially transparent, colorless samples. DIC microscopy is not used quite as much for transmitted light applications as phase contrast microscopy because of the somewhat higher cost of the DIC accessories. Perhaps, more importantly, the images of the phase contrast microscope are well understood and are familiar. It should be emphasized, however that for particle or cell size measurements, interference contrast microscopy is to be preferred over phase contrast microscopy because the halo effect that tends to confuse the boundaries of structures in the phase contrast image are not present the DIC image.

The most popular configuration for transmitted light DIC is the self referencing system, of which the Nomarski system is an excellent example. The Nomarski system consists of a basic polarizing microscope, see Section VI, to which special objectives and condenser system is added. The operation of the Nomarski DIC is illustrated by Figure 4-19a. The light from the lamp is linearly polarized by the polarizer (see Chapter 6, Section IV). The linearly polarized light passes through the condenser and then through a Wollaston prism. The Wollaston prism consists of two wedges of calcite so cut and arranged that their vibration directions are diagonal to the plane of the light from the polarizer. The Wollaston prism effectively divides that light into two linearly polarized beams

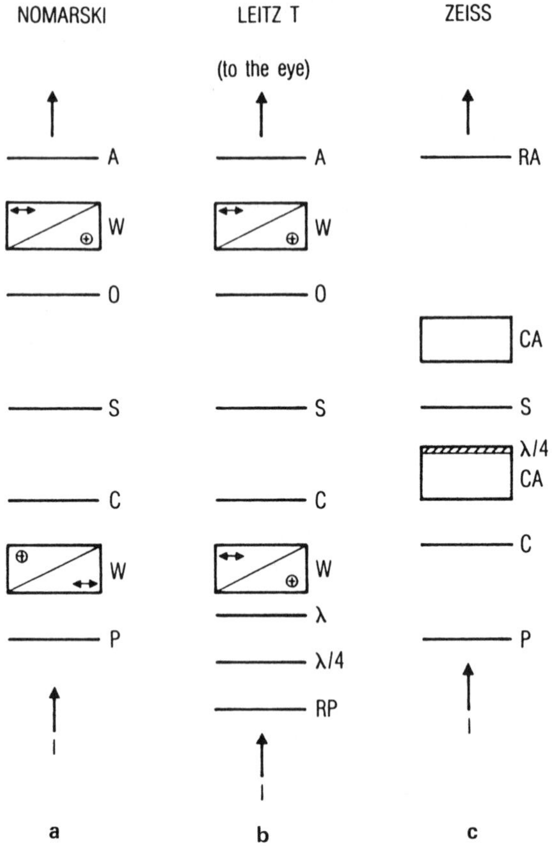

Figure 4-19. Schematic arrangement of the components for the various transmitted light DIC microscopes.

A - Analyzer		RA - Rotary analyzer	
C - Condenser		RP - Rotary polarizer	
CA - Calcite plate		S - Sample	
I - Illumination		W - Wollaston prism	
O - Objective Lens		λ - 1st order plate	
P - Polarizer		λ/4 - 1/4 wave plate	

of about equal intensity and vibrating perpendicularly to each other; these are referred to as the ordinary and extraordinary rays. The ordinary ray passes through the Wollaston prism without deflection, while the extraordinary beam is laterally displaced. In operation the sample slide is positioned so that one of the rays pass through the sample and the other through a clear area of the slide.

The vibration direction of both beams is rotated 90° upon passing through the half-wave plate. This procedure equalizes the optical path length of the two beams as they pass through the entire interferometer.

After passing through the sample slide the two separate beams are recombined in the second Wollaston prism as axially parallel beams. The combined beam exits the interferometer system through the analyzer, a second polarizing filter, that is positioned so that its transmission direction is oriented 90° to that of the polarizer. In the analyzer the components of the combined beams that are vibrating perpendicular to the transmission direction of the polarizer are absorbed, whereas the components vibrating parallel to the polarizer transmission direction pass through but are displaced by one half-wave to each other. Since the two beams are now vibrating in a single plane they are now able to interfere with each other.

If no structure is present in the sample slide the path length of both beams is equal. In this case, the beams that recombine at the analyzer are exactly out of phase, i.e., there is one half-wave difference between them; thus no light exits the interferometer. If, however, a sample slide is placed so that the sample beam passes through the area of interest and the reference beam passes through an empty area of the slide, the two beams upon recombination at the analyzer will not, in general, be exactly out of phase in all areas. Thus, the image will have varying degrees of brightness and color depending on the difference in retardation between the sample and reference beam at any specific point.

One advantage of DIC microscopy given by Allen, et al.(1969) is that it can be used effectively at large condenser apertures; this gives a shallow depth of field. The shallow depth of field combined with the mode of producing the interference images has yet another unique advantage. Objects that are only somewhat above or below the plane of focus do not produce significant contrast, i.e., they are not seen. Therefore, each plane of focus is essentially a crisp optical section with little or no interference by structure above or below that plane. They point out that this feature is one of the most important features of the technique for biological and medical microscopy.

The Jamin/Lebedeff, the Smith, and the Leitz Interference Contrast Device T are similar to the Nomarski system. The optical arrangement for several of these systems is shown in Figure 4-19.

The actual physical location of the interferometer parts vary significantly for the various DIC types and for the various manufacturers. The condenser system houses the polarizer and the first beam splitter. Since each objective lens requires a specific matched beam splitter in the condenser, a separate condenser may be provided for each objective lens or a single condenser assembly will have a revolving turret equipped with an individual beam splitter for each objective lens. This turret often has an open position for true brightfield observations and a position with an annulus for phase contrast comparisons.

The recombining beam splitter may be an integral part of each objective lens, or it may be part of an insert that fits between each objective and the nosepiece or it may be mounted in the body tube.

The analyzer is located in the body tube. It is always arranged so that it can be inserted in the beam path at will. It is necessary that the analyzer be rotatable to optimize the interference effect. Also a brightfield-like image may be easily obtained by rotating the analyzer to a ±45° position from the fully crossed position.

Some caution must be given with regard to suitable samples for examination with self reference DIC. The distance between the sample and reference beams is relatively small even for low power objective lenses and becomes progressively smaller with increasing magnification. It will be recalled that the reference beam must be in a clear area of the sample slide, i.e., a structureless area. Therefore the ideal samples of this instrument is one containing isolated cells, grains or particles separated by areas of clear mounting medium. The actual minimum distance between the two beams is, thus, an inverse function of the objective lens magnification. If a large continuous structure is to be examined, it can be accomplished near the margins of the mount where clear areas are available for passage of the reference beam.

For critical examinations it must also be noted that a pale, secondary image occurs next to the primary image. It results from the inability of the half-wave plate to rotate the two beams exactly 90° for any color except the one color for which it was computed. All other colors are elliptically polarized. The additional component produced by these elliptically polarized rays produce this secondary image. This effect can be eliminated by using monochromatic light that matches the the design wavelength of the half-wave plate, but this negates the impressive color contrast of the method. Since the brightness or color contrast of the DIC image does not relate to actual absorption effects in the sample it is often desirable to make comparisons with the sample in brightfield illumination. It is not necessary with the self reference systems to change to another instrument or to remove the DIC accessories, one must only rotate the analyzer plus or minus 45°

from its normal position, i.e., 90° to the polarization plane of the polarizer. This procedure actually allows the viewing of either the sample beam or the reference beam individually; each of these beams by itself yields a brightfield image of the sample in plane polarized illumination.

The advantages of the self reference, transmitted light DIC instruments are:

1. Excellent contrast exists even at high condenser and objective lens apertures.*
2. High N.A. ensures high lateral resolution and extremely shallow depth of field (the latter is useful for optical sectioning without disturbing halo or scattered light from sections above or below area of interest).
3. While phase contrast is optimum only for samples with small retardations, DIC is suitable for samples with small to large retardations.
4. Abrupt changes in retardation causes halos in phase contrast, DIC images have no halos.
5. DIC renders phase differences in either brightness or color contrast.
6. The reference and sample beams have nearly identical paths, therefore vibration does not produce "flutter" of the interference effects (see discussion of the Mach-Zehnder DIC below)
7. It is easy to compare the DIC and brightfield images of the sample by merely rotating the analyzer ±45°.

* 1-5 from Lang (1970).

The disadvantages of the self reference, transmitted light DIC instruments are:

1. Elements of the sample, grains, cells etc., must have a certain minimum distance from each other to provide sufficient free area for the reference beam .
2. In tissue sections or other extensive sections only the marginal areas can be unambiguously examined.
3. Critical examination of the image will reveal a pale, secondary image that occurs next to the primary image.
4. Cost of a DIC instrument is somewhat higher than for a phase contrast instrument.

The Mach-Zehnder instrument, as developed by Leitz, was the only commercially produced example of the independent reference interference microscope for transmitted light. This very complex microscope was primarily designed for quantitative measurement of phase differences in biological research. The determination of these phase differences permits the measurement of object thickness as well as the refractive indices and dry weight mass of biological structures. With careful adjustment the Mach-Zehnder instrument could be used to provide DIC images, as well. The chief advantage of these measurements is that they can be carried out without damage to living specimens.

C. Incident Light Applications

Incident light DIC has become very popular as a contrasting techniques in recent years and has, for all practical purposes, superseded the incident light phase contrast instrument. DIC is very useful in metallurgical examinations but undoubtedly the most important application is the examination of semiconductor wafers in various stages of their fabrication.

The incident light DIC microscope is, as with other methods of contrasting, the direct incident light analog of the transmitted light instrument. Generally speaking, the incident light instrument is simpler in both construction and operation. Examples of both the self reference and the independent reference instruments are available and they each serve a useful function.

The Nomarski system is the most popular of the self reference DIC devices for incident light microscopy, as well. Due to its simplicity of operation, it is probably the most popular of all DIC systems for this purpose. A typical incident light Nomarski system is shown schematically in Figure 4-20a. The illumination is plane polarized by the polarizer, P, and is reflected from the plane glass reflector to the modified Wollaston prism where it is divided into two rays of linearly polarized light. The principal direction of the prism is oriented at 45° to the vibration direction of the polarized light to equalize the intensities of the two rays; and thereby, maximize the intensity of the resulting image. The axis of the Wollaston prism is also slightly inclined to the optical axis of the microscope to avoid glare reflected from the plane surface of the prism.

The two rays intersect at the rear focal plane of the objective lens; this is the plane of fringe localization. The rays pass through the objective lens. These rays which are then parallel and slightly displaced to each other, impinge on the sample.

When the two rays are reflected back by the sample they are focused on the plane of localization by the objective lens and then recombined by the Wollaston prism. These recombined rays do not produce an interference image until they pass through the analyzer, A. As with the transmitted case, an image similar in appearance to a brightfield image is formed if the analyzer is rotated ±45° from its normal position, i.e., 90° to the polarization plane of the light from the polarizer.

The actual placement of the Wollaston prism varies according to the manufacturer. Reichert's system uses a Wollaston prism that is mounted either in the body tube or in an accessory that fits between the body tube and the objective lens. This

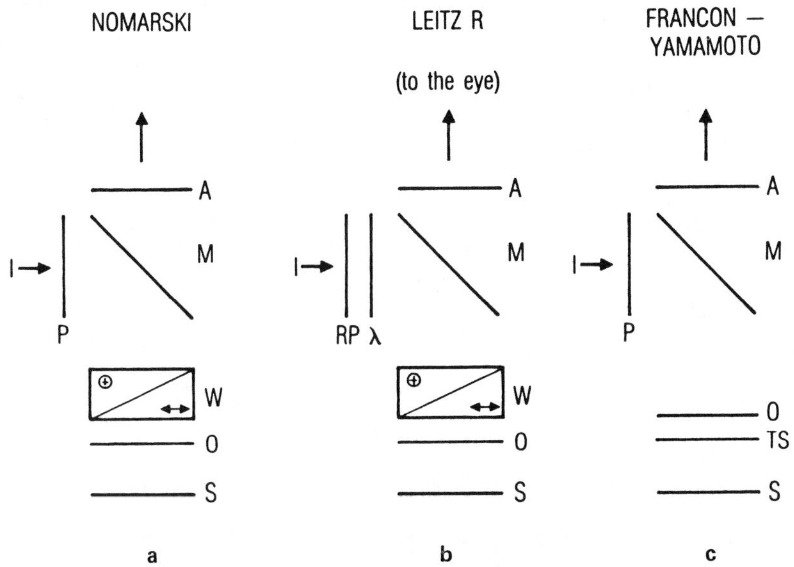

Figure 4-20. Schematic arrangement of the components for various Incident light DIC microscopes. In these drawing the illumination enters from the left, is directed to the sample by the partially reflecting mirror; the returned illumination passes upward through partially reflecting mirror to the eye.

A - Analyzer
I - Illumination
O - Objective lens
λ - 1st order plate
M - Partially reflecting mirror
P - Polarizer
RP - Rotary polarizer
S - Sample
TS - Tilting Savart plate
W - Wollaston prism

arrangement permits DIC studies using any of their strain-free objective lenses. By adjustment of the Wollaston prism the relative retardation of the system can be adjusted so that a give step on the sample can be imaged in a variety of contrasts ranging from a darkfield effect to a wide range of colors.

The Francon-Yamamoto system of Nikon, and the Leitz Interference Contrast R system are variants of the polarizing type systems. The arrangement of their optical elements is shown in Figure 4-20b and c. The placement of the Nikon DIC elements is similar to that of the Reichert instrument so that these instruments too, only require standard, strain-free objective lenses. Other manufacturers generally use a Wollaston prism that is an integral part of each DIC objective lens.

The advantages of the self reference, incident light DIC instruments are:

1. Excellent contrast is obtained even at high condenser and objective lens apertures.*
2. High N.A. ensures high lateral resolution and extremely shallow depth of field.
3. While phase contrast is optimum only for samples with small retardations, DIC is suitable for samples with small to large retardations.
4. Abrupt changes in retardation causes halos in phase contrast, DIC images have no halos.
5. DIC renders phase differences in either brightness or color contrast.
6. Through comparison with scanning electron microscopy, Bluhm and Haller (1976) found Nomarski DIC measurements to be valid, extremely precise ($\pm$ 0.2 - 0.3 mm), rapid, and highly accurate.
7. The reference and sample beams have nearly identical paths, therefore vibration does not produce "flutter" of the interference effects (see discussion of the independent reference, incident light DIC systems below).
8. It is easy to compare the DIC and brightfield images of the sample by merely rotating the analyzer $\pm 45^\circ$.

* 1-5 from Lang (1970).

The disadvantages of the self reference, incident light instruments are:

1. Elements of the sample, e.g., grain boundaries, second phases, particulates etc., must have a certain minimum distance from each other to provide sufficient free area for the reference beam.
2. Critical examination of the image will reveal a pale, secondary image that occurs next to the primary image.

3. Cost of a DIC instrument is somewhat higher than for a phase contrast instrument.

Watson supplies two independent reference, interference objective lenses for incident light observations. These lenses, a 10X and a 20X, attach directly to almost any microscope. Other independent reference DIC devices that are no longer produced are the Wild interference attachments for their M12 and M20 microscopes and the Leitz Mirau interference attachment for the 10X, 20X and the 32X objectives of their Metallux instrument.

TABLE 4-7
TYPES OF SELF REFERENCE DIC MICROSCOPES

MANUFACTURER	MODEL*	ILLUMINATION	TYPE
Americal Optical	21 DIC	Trans	Nomarski
	10 DICV	Incident	Nomarski
Bausch & Lomb	Research II Met	Incident	
Leitz	Various	Trans	DeviceT (Smith)
	Orthoplan & -Pol	Trans	Nomarski
	Various	Incident	Device R (Smith)
Nikon	Apophot	Trans	Francon - Yamamoto
	Apophot	Incident	Francon - Yamamoto
	Microphot	Trans	Nomarski
	Microphot	Incident	Nomarski
Olympus	Vanox	Trans	Nomarski
	Vanox	Incident	Nomarski
Reichert	Various	Trans	Nomarski
	Various	Incident	Nomarski
Vlickers	M17	Trans	Smith or Nomarski
	M17	Incident	Smith or Nomarski
Zeiss	Various	TRANS	Jamin - Lebedeff
	Various	Trans	Nomarksi
	Various	Incident	Nomarski

*It should be emphasized that some of these instruments are no longer manufactured but are included because many are currently in use.

Although these independent reference systems were designed for quantitative, fringe measurements, careful adjustment can provide an acceptable DIC image. The main advantage of these instruments is that image quality is completely independent of grain size and distribution of the structure in the sample because the sample and reference beams are independent.

Flutter of the interference effect is a most disconcerting problem with the independent reference instruments, however the use of a microscope with a sturdy stand and a table equipped with vibration isolation will do much to minimize this problem.

D. General Comments On The Set-up Of A DIC Microscope

The DIC microscope, as mentioned above, is the combination of an interferometer and a microscope. There are a large number of interferometer types and their placement in the optical train of the microscope can vary to some extent, thus it is not possible to discuss a generic instrument and describe its set-up as with the previous instruments. The best and most explicit information about the set-up of a specific instrument is to be found in its instruction manual, however these typically have a short lifetime. The several instruments that are representative of the major classes of DIC microscopes, are shown in Table 4-7. No attempt will be made to provide initial assembly and installation information for the DIC elements. If the original instruction manual is not available for this purpose, these procedures should be performed by an authorized service organization for the specific microscope.

E. Procedure For Transmitted, DIC Illumination

The most common forms of the transmitted light DIC microscopes are the Nomarski and the Smith versions, therefore the following procedure embodies the essentials of both systems. These systems have a special condenser that includes a revolving turret which has an individual prism for each DIC objective lens. The turret may also include an open position and phase ring positions for brightfield and phase contrast comparisons, respectively.

The sample for DIC examination should be $\leq$10 μm in thickness and with or without color or brightness contrast. The DIC image gives a three-dimensional character to the sample image, but it must be remembered that this effect is due to either refractive index or thickness differences in the sample.

SETUP OF A TRANSMITTED LIGHT DIC MICROSCOPE

1. Insert polarizer, if not already in the optical path.
2. Place a suitable sample on the stage, turn on the illuminator and focus the sample carefully.
3. Select the desired objective lens and rotate the condenser turret to the corresponding prism for that objective lens.
4. Close field diaphragm, if present, to about 1/2 the field and focus its image in the the plane of the sample image using the condenser focusing knob; otherwise go to Step 6.

5. Open field diaphragm to just beyond the field of view.
6. Remove an eyepiece to observe the aperture diaphragm. Close the aperture diaphragm until the opening remaining is 75% of the full aperture setting. Replace the eyepiece.
7. Insert the analyzer into the optical path. If the analyzer has more than one possible orientation, place it in the crossed position.
8. If the objective lens prism is in the body tube of the microscope, insert it into the optical path. Otherwise, the prisms are an integral part of the objective lenses.
9. Change the contrast effect by: (a) Adjusting knob on the body tube prism assembly, or (b) by inserting the accessory lambda plate into the condenser optical path.
10. Observe the sample in DIC illumination.
11. To change magnification, repeat Steps 3 - 9.
12. To compare the DIC image with the brightfield image, rotate the analyzer to 45° position, if possible, otherwise, remove the upper prism and analyzer from image path.

F. Problem Correction For Transmitted, DIC Microscopy

If the DIC image is not satisfactory upon completion of the above procedure, repeat the steps carefully. If a satisfactory image is still not present refer to Table 4-8 for possible problems.

G. Procedure For Incident, DIC Illumination

The procedure for incident light DIC is generally easy, however the wide variety of accessories makes it difficult to give a detailed procedure that will accommodate all instruments.

The sample for DIC examination should be flat and, if a metallurgical sample, only lightly etched. The DIC image gives a three-dimensional character to the sample image; this effect is due to differences in sample height and is greatly exaggerated.

SETUP OF AN INCIDENT LIGHT DIC MICROSCOPE

1. Adjust the microscope for correct incident light, brightfield illumination per Chapter 3, Section IV C.2. Place a lightly etched or unetched sample on the stage and focus it carefully. The sample must be leveled as described in Chapter 2, Section II A, if an upright stand microscope is used.
2. Insert the polarizer and analyzer in crossed positions
3. Insert the DIC prism assembly into the optical path. The location of the prism is described below:
   (a) In body tube: American Optical, B&L Research II, newer Reichert, Unitron, newer Zeiss
   (b) Slider in objective lens: Nikon.
   (c) Fixed in objective lens: Leitz, Olympus.

(d) Accessory between body tube and objective lens: older Reichert, older Zeiss.

4. Observe sample in DIC.
5. Contrast can be changed in the following manner.
   (a) Adjust prism control: American Optical, B&L Research II,Olympus, Unitron, Reichert.
   (b) Insert lambda plate: Leitz, Nikon, Zeiss.
6. When objective lens is changed, the prism must also be changed for: American Optical, B&L Research II, Zeiss.

TABLE 4-8
DIC MICROSCOPY PROBLEMS AND THEIR CORRECTION

PROBLEM	CAUSE	CORRECTION
Field diaphragm image not sharp.	Glass slide too thick or too thin.	Use glass slides 0.9-1.4 mm thick.
	Sample too thick.	Prepare thinner sample.
No interference effect.	Polarizer and analyzer not crossed.*	Cross polarizer and analyzer per Table 4-12.
	Polarizer destroyed by illuminator heat.*	Refer instrument to service organization for repair.
	Sample too thin.	Prepare thicker sample.
	Difference in R.I. between sample and mounting medium too small.*.	Prepare new sample with different mounting medium.
	Sample with too little relief.*	Sample may require light etching.
	Sample with too much relief.*	Sample inappropriate for DIC or needs repolishing.
	DIC prism not perfectly oriented.*	Refer instrument to service organization for adjustment.
Interference effect not uniform.	Polarizer damaged by illuminator heat.*	Refer instrument to service organization for repair.
	Dirty slide, cover glass, objective lens or condenser lens.*	Clean per Chapter 10, Section VI.
	Sample in wrong azimuthal orientation.*	Rotate stage or sample on stage.
	DIC prism not correctly oriented.*	Refer instrument to service organization for adjustment.

* Entries which pertain to incident light DIC

H. Problem Correction For Incident DIC Microscopy

If the DIC image is not satisfactory upon completion of the above procedure, repeat the steps carefully. If a satisfactory image is still not present refer to the CAUSES marked with an asterisk in Table 4-8 for possible problems.

VI. POLARIZED LIGHT CONTRAST MICROSCOPY

A. General

The first recorded application of polarizers to the microscope is attributed to Henry Fox Talbot in 1834. This addition of plane polarizers to the brightfield light microscope has produced a very powerful tool for the microscopist that provides both quantitative information about the sample and a very powerful contrasting technique. The emphasis in this section is on the contrasting of sample structures; the quantitative aspects of polarized light microscopy are covered in Chapter 5, Section V.

Since the unique capabilities of the polarizing microscope rest with the added plane polarizers, it is worthwhile to learn what they are, examine the various types and evaluate their merits. A plane polarizer is a device that allows light vibrating in only one direction to pass through it; this is illustrated in Figure 4-21.

Prism polarizers were one of the earlier types to be used in microscopy. These consist of two or three carefully oriented, cut and polished calcite prisms which are cemented together to form a single prism. A typical prism polarizer is shown in Figure 4-22. These prisms are very effective polarizers with an extinction ratio of $<1X10^{-5}$ and an individual transmission of about 37%. Extinction ratio is the ratio of the light intensity transmitted through two polarizers when exactly crossed, to the light intensity through the same two polarizers in the parallel position. Prism polarizers, however are expensive, and have a comparatively small optical aperture. Since heat can cause decementing, a heat absorbing filter, Chapter 7, Section VI C, should precede these polarizers if a high energy illuminator such as a xenon lamp is used.

Filter polarizers are made of a polymeric material in which the molecules are oriented by stretching and to which strongly dichroic pigment molecules are selectively attached. In the direction of transmission the pigment molecules are relatively transparent, in the direction at right angles to the transmission direction these same molecules are highly absorbing. Thus the light that passes through is strongly polarized. Filter polarizers have a typical individual transmission of about 30% and an extinction ratio of about $1X10^{-4}$. Filter polarizers are widely used on modern polarizing microscopes because they are relatively inexpensive and can have large apertures. The main disadvantages to filter

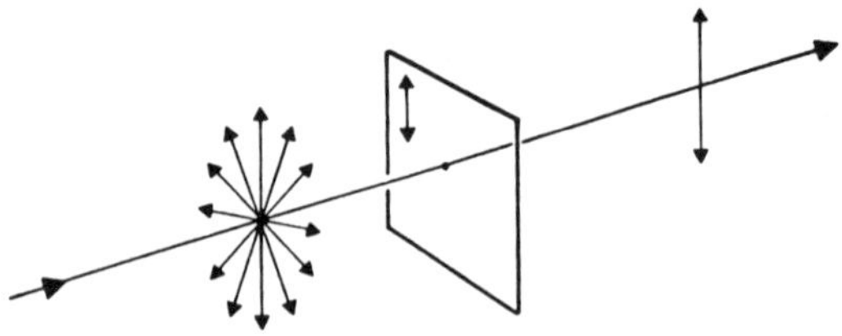

Figure 4-21. Schematic drawing showing the production of polarized light as a beam of randomly polarized light passes through a polarizer. The exiting beam is polarized in one direction.

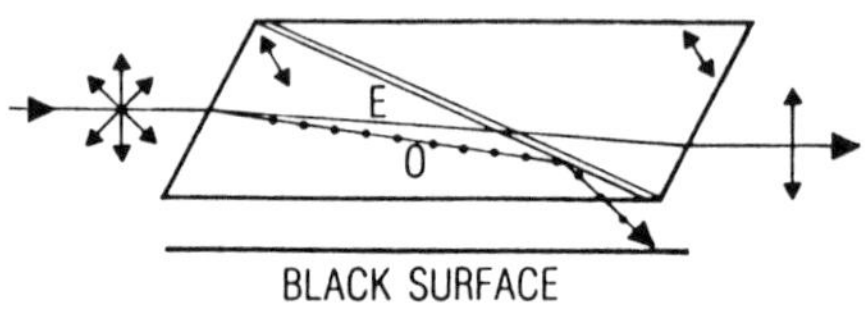

Figure 4-22. Cross section of a Nicol polarizing prism (Halle form) showing the polarization of a beam of light. The prism consists of two pieces of clear calcite cemented together with Canada balsam (η = 1.52). The light upon entering the calcite is resolved into two beams, polarized at right angles to each other. The refractive index of the calcite for the E beam is about 1.49, whereas, the refractive index for the O beam is 1.66. The E beam passes through the entire prism essentially unaffected. However, the angle of the cemented surface is such that it exceeds the critical angle for the O beam, thus the O beam is reflected at the cement surface and subsequently absorbed by the black interior surface of the prism cell.

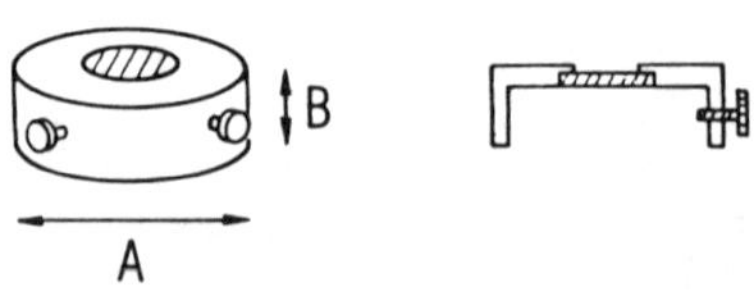

Figure 4-23. Analyzer cap which may be placed over the eyepiece of a brightfield microscope. Used in conjunction with a sheet polarizer in the condenser system, this analyzer cap will provide some polarizing contrast capability. The polarizer and analyzer are crossed by rotating the analyzer cap to a minimum transmission position with a blank sample in place. Dimensions A and B are not critical but must be sufficient for the cap to fit over the eyepiece. Redrawn by permission from the Proceedings of the RMS.

polarizers are a somewhat less than optimum extinction ratio and heat sensitivity. Since even a brief heating above 100°C will destroy the filter, a heat absorbing filter, Chapter 7, Section VI C, should be used even with 50 watt lamps. With even more intense sources a second heat filter or a prepolarizer should be used.

Reflection polarizers make use of a stack of obliquely positioned glass plates or an inclined, vapor-deposited metal film. These reflection devices are not efficient polarizers (their extinction ratio is only about $2X10^{-2}$), therefore they are suitable only as prepolarizers.

Two procedures are used for contrasting with plane polarized light. If the sample is illuminated with a plane polarizer in the condenser system, this is referred to as a plane polarized examination. If, in addition a second polarizer, oriented at 90° to the previous polarizer is introduced at a point beyond the sample, this is referred to as cross polarized illumination. This second polarizer is known as the analyzer.

B. Construction Of Simple Polarizing Accessories

Bradbury (id.) has described the construction of a simple polarizer and analyzer for use with a standard brightfield instrument. The polarizing elements are cut from sheet of filter polarizing material (available from: Edmund Scientific, 101 E. Gloucester Pike Barrington, N.J. 08007). The polarizer should be of a size and shape to accept the full aperture of the light source and should be affixed firmly to the microscope at any point between the illuminator and the sample; although the closer the filter is to the sample the less likely that extraneous polarization effects from the optics will be observed. The analyzer is in the form of a cap that fits over the eyepiece containing a smaller piece of filter polarizer. The cap may be machined from brass, as shown in Figure 4-23. Alternately, the analyzer can be made from a small cardboard pillbox with a central hole formed by a sharp cork borer. The polarizing filter is carefully glued in position over the central hole.

In use, the polarizer is mounted in the illumination path and the cap placed over the eyepiece. The transmitted light instrument is focused on a blank slide while an incident light instrument is focused on a first surface mirror. The analyzer is then rotated until the vibration directions of the two polarizing elements are crossed and the field of view is dark. If the brass analyzer is used it may be locked in the crossed position with the set screw provided. Since brightfield optics are not selected for minimum strain it is possible that the field will not be completely dark, however this should not seriously affect most observations.

C. Transmitted, Polarized Illumination

The transmitted light, polarizing microscope is useful for the examination of transparent materials, i.e., those that transmit visible light. Even some very dark materials fall in this category since small grains or thin slices do transmit some light. The refractive index is an important and characteristic optical property of transparent solids. Liquids also have a refractive index and they have an important role in polarized light microscopy. For the purpose of contrasting, solids may be grouped in either of two categories; isotropic materials having only one refractive index, and birefringent materials having 2 or 3 refractive indices. Table 4-9 lists the two classes of materials and examples of the materials that belong in them.

TABLE 4-9
CLASSES OF TRANSPARENT MATERIALS

CLASS OF MATERIAL	EXAMPLES
ISOTROPIC	CUBIC CRYSTALLINE SOLIDS. GLASSES. MOST LIQUIDS.
BIREFRINGENT	MECHANICALLY STRAINED ISOTROPIC SOLID TETRAGONAL CRYSTALLINE SOLIDS. HEXAGONAL CRYSTALLINE SOLIDS. ORTHORHOMBIC CRYSTALLINE SOLIDS. MONOCLINIC CRYSTALLINE SOLIDS. TRICLINIC CRYSTALLINE SOLIDS. OPTICALLY ACTIVE LIQUIDS AND SOLIDS. LIQUID CRYSTALS. FERROMAGNETIC SOLIDS. SOME LIVING CELLS OR ORGANISMS.

Cross polarized illumination provides contrasting only for birefringent materials. Thus, those samples which are made up of some fraction of birefringent materials can be profitably studied using cross polarized illumination. Another property that can produce contrasting for some birefringent materials is their selective absorption of certain colors of light in certain specific directions as seen with a polarizing microscope. This effect is known as dichroism if the material has two refractive indices or pleochroism if the material has three refractive indices. This effect may be observed in plane polarized illumination.

A third way in which contrast may be observed in cross polarized, transmitted light is by rotation of the plane of the polarized light by the sample. This rotation occurs in optically active liquids and solids, see Richardson (1978); liquid crystals, see Nicastro (1982); and ferroelectric materials, see Richardson, (id.). The latter materials exhibit in a magnetic field what is known as the Faraday effect. Extinction of the light transmitted by these materials can only be achieved by rotation of the analyzer in one direction or the other from the normal crossed position.

A typical transmitted light, polarizing microscope, usually referred to as a petrographic microscope, is shown in Figure 4-24. The accessories that are necessary to convert the brightfield, transmitted light microscope into a petrographic, microscope are discussed below. For critical polarized light investigations it is necessary to maintain the sample object of interest exactly coincident with the axis of both the illumination and image optics; the various methods for accomplishing this are described in Chapter 10, Section VI.

The first plane polarizer, called the polarizer, is mounted in the condenser system. It is usually rotatable and in the research instruments the rotation is calibrated. The older instruments used a Nicol prism polarizer, while the more recent instruments make use of the filter polarizer.

The condenser system for a polarizing microscope is unique, in that it requires strain free lenses in the optical path after the polarizer. It is equipped with an aperture diaphragm and often a field diaphragm. The final high N.A. lens in the condenser is usually designed as a swing out element to permit rapid changes between high and low N.A. illumination. This latter feature is valuable in quantitative applications.

The stage should be of the revolving type with precision ball bearings and should be calibrated to 1/10 of a degree for the full 360° of rotation. A worthwhile feature of the stage is a system that may be engaged as desired and that will provide a detent every 45° of stage rotation. Some microscopes are equipped with centering adjustments to precisely place the rotation axis of the stage on the optical axis of the instrument.

The objective lenses are unique only in the sense that they must be strain free. Achromatic lenses are generally sufficient for this purpose and they are relatively easy to produce with strain free optics. As indicated in Table 4-9 even isotropic materials such as optical glass exhibit birefringence if strained, therefore objective lenses for polarized light applications are carefully tested and selected to have minimum birefringence. To preserve the strain free character of these objective lenses they must not be dropped or subjected to other sharp blows. An evaluation test for strain free objectives is given in Chapter 10.

The typical, polarizing objective lens is mounted in either an individual, centerable, quick-change clutch as shown in Figure 4-25 or in a revolving nosepiece. At least one manufacturer, Zeiss, provides each polarizing objective lens with two individual centering adjustments, located toward the sample end of the lens.

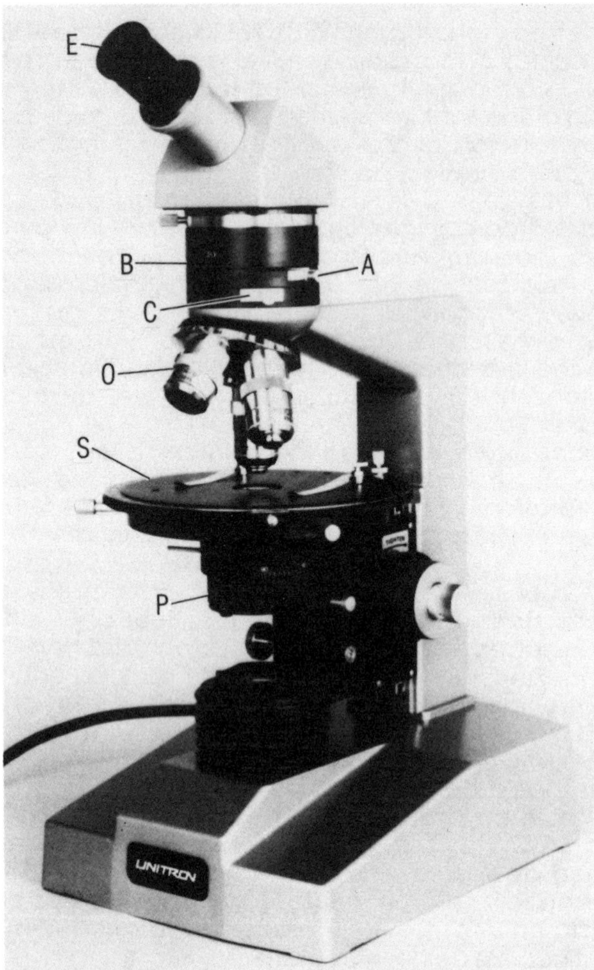

Figure 4-24. Polarizing microscope suitable for optical crystallography and petrography as well as for contrasting. Courtesy of Unitron Instruments Inc.

A Analyzer
B. Bertrand lens
C. Compensator slot
E. Focusable, cross wire eyepiece
P. Graduated, rotatable polarizer
S. Revolving calibrated stage
O. Centerable objectives on revolving nosepiece

Figure 4-25. Quick change clutch with attached objective lens and a separate objective lens. The clamp on the front of the clutch is compressed to remove and replace an objective lens and its mount. Note the centering adjustments on either side of the objective lens mounts.

Leitz supplies a quadruple or a quintuple revolving nosepiece for this purpose with individual centering adjustments for each objective lens. If neither the objective lens nor the revolving nosepiece is equipped to permit centering the objective lenses, the stage must be centerable.

Just above the objective lens is the accessory slot that allows the insertion of various compensator accessories into the optical path. This slot is positioned 45° to the polarizer direction; this is usually the northwest-southeast orientation. These compensators are crucial to the determination of crystal properties as discussed in Chapter 5, Section V, however, they find some use in the contrasting mode for altering or improving the color contrast of the sample. A few older instruments may also have an accessory slot in the condenser. The accessory slot should not be used for ordinary filters since they are not usually selected for low strain birefringence.

A second plane polarizer, called the analyzer is located in the body tube. The analyzer is insertable into the optical path at will and is usually rotatable about the optic axis. The rotation is calibrated in the research grade instruments. The normal orientation of the analyzer is "crossed" or at 90° to the polarizer direction.

To complete the description of the petrographic microscope it is necessary to mention two additional features although they are not generally used in the contrasting mode of operation; these are the Bertrand lens, see Chapter 2, Section III C, and the grain isolating diaphragm.

The grain isolating diaphragm is located between the Bertrand lens and the eyepiece. It is used only in conjunction with the Bertrand lens and there only to restrict observations to a desired single grain or particle. The grain isolating diaphragm serves no function in normal viewing and should remain completely open.

Binocular and trinocular heads used for polarized light microscopy have two unique features. The first of these is a locating pin that maintains alignment of the eyepieces to the polarizer. The second feature is a depolarizer plate that is mounted either in the head or immediately before it in the body tube. The function of the depolarizer is to eliminate image differences in the two eyepieces resulting from interaction of the polarized light with the beam splitter and prisms in the head.

The eyepieces are of the standard types, but at least one is equipped with a crosswire reticle. A locating pin on the side of the eyepiece mates with a slot in the body tube to provide an accurate alignment of the crosswire relative to the polarizer direction.

As a contrasting device, the petrographic microscope is useful to distingush isotropic from anisotropic materials in the same sample and for increasing the contrast of the anisotropic materials. The sample may consist of individual grains or a thin section of a consolidated material in a mounted slide. If the microscope is used without the analyzer in the light path, the observation is said to be in "plane polarized illumination"; this technique is useful for contrasting due to dichroism or pleochroism of the sample.

If the analyzer is inserted into the optical path and rotated until its polarization direction is exactly 90° to the polarizer direction, the observation is said to be in "cross polarized illumination". This technique gives an image somewhat like darkfield illumination with the birefringent materials illuminated, often in bright colors. Rotation of the stage can dramatically affect the contrast in either plane polarized or cross polarized illumination.

The method by which transmitted light, cross polarized microscopy produces contrast can be explained by reference to Figure 4-26. A sample consisting of a transparent, anisotropic crystal in a general orientation resolves the plane polarized beam of light from the condenser into two plane polarized beams which are vibrating at right angles to each other. The refractive indices are different in the two directions, thus one beam will be retarded with respect to the other. The amount of retardation will depend on the composition of the crystal, its orientation to the incident beam of light and its thickness. To simplify the discussion we shall assume the light to be monochromatic. The two beams exiting the crystal are magnified by the objective lens and are resolved into the vibrational direction of the analyzer. The resolved beams will either reinforce, as shown in Figure 4-26a, or cancel, as shown in Figure 4-26b, depending on the phase relationship of the two beams. If the beams reinforce, light will be see from that crystal, however if the beams, cancel the crystal will appear dark. If we now allow white light, the different colors of light will have different retardations, thus there will be cancellation for some colors and reinforcing for other colors; the observed color of the grain will be due to the latter. The maximum intensity of the color will be with the grain's vibration directions midway between the polarizer and analyzer directions.

If the grain is not in a general orientation but rather with its vibration directions parallel to the polarizer and analyzer directions, it can be seen from Figure 4-27, that no light of any color reaches the eye. Therefore, when a given grain is observed as the stage is rotated a full 360°, it will be dark four times, at 0°, 90°, 180°, and 270° with the reinforcement interference colors increasing to a maximum at the 45°, 135°, 225° and 315° positions.

If the analyzer is inserted into the optical path and rotated until its polarization direction is exactly parallel to the polarizer direction, the observation is said to be in parallel polarized illumination. The field in this case is bright because the polarizer and analyzer are parallel. The colors of birefringent materials differ from those in crossed polarized light and are generally weaker. This techniques is not extensively used for contrasting purposes. The normal interference color observed for a given grain will depend on its total retardation, This is show in Table 4-10 for cross polarized light and for parallel polarized light. Dichroism or pleochroism of a

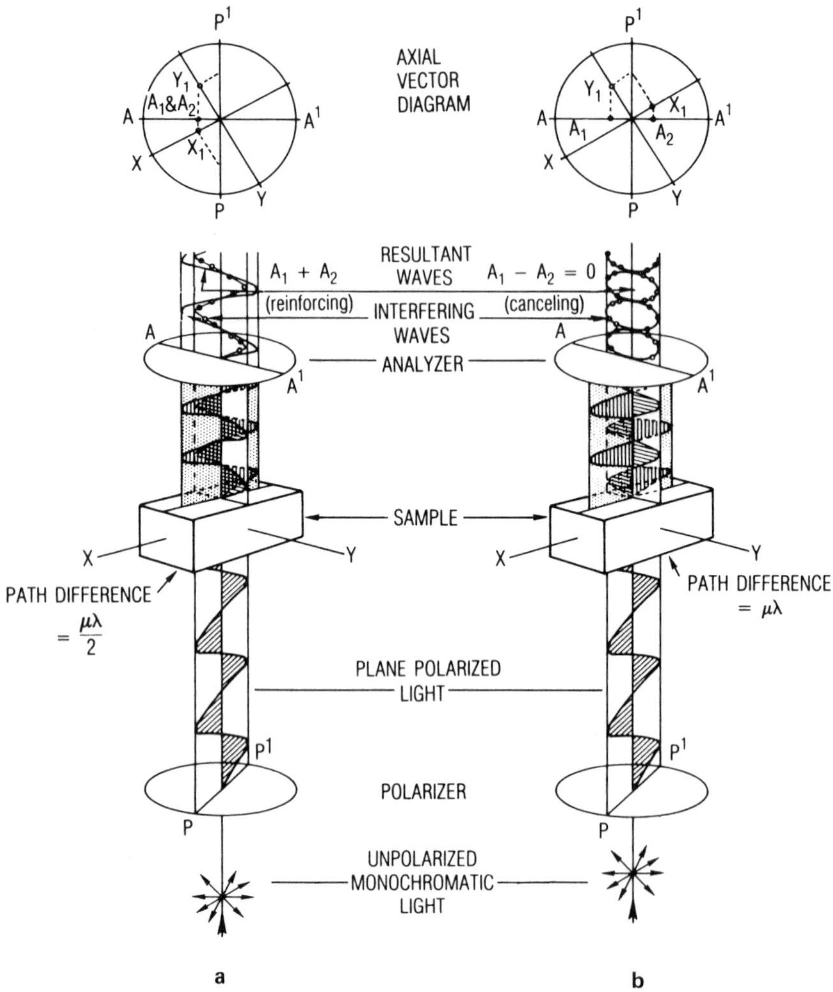

Figure 4-26. Perspective drawings showing the effect of an anisotropic sample in a general orientation, on the crossed polarized, monochromatic illumination of a polarizing microscope. The upper drawings are vector diagrams that summarize the resultant waves in the drawings below them. The left diagram shows that the rays Y_1 and X_2 add vectorially to produce visible brightness, whereas in the right diagram Y_1 and X_1 cancel to produce a dark image.

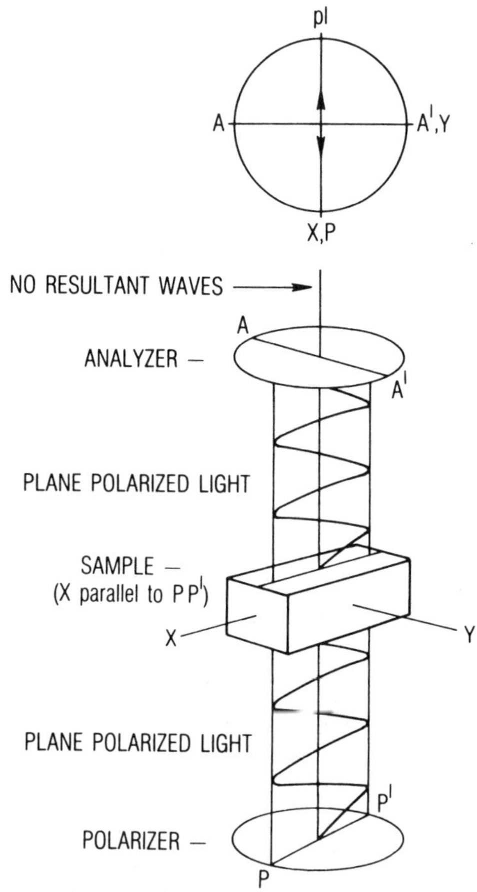

Figure 4-27. Perspective drawing showing the effect of an anisotropic sample in parallel orientation, on the crossed polarized, monochromatic illumination of a polarizing microscope. The upper vector diagrams shows that with no rays resolved onto the Y axis of the sample, no resultant waves are resolved onto the analyzer and hence a dark image is produced.

sample grain will alter these colors producing what is referred to as anomalous interference colors.

TABLE 4-10
TRANSMITTED LIGHT INTERFERENCE COLORS FOR ANISOTROPIC MATERIALS*

ORDER OF COLOR	CROSS POLARIZED LIGHT RETARDATION (in nm)	CROSS POLARIZED LIGHT COLOR	PARALLEL POLARIZED LIGHT RETARDATION (in nm)	PARALLEL POLARIZED LIGHT COLOR
FIRST	0	BLACK	0	BRIGHT WHITE
	40	WHITE		
	120	BLACKISH BLUE	97	YELLOWISH WHITE
	165	MOUSE GRAY	158	BROWNISH WHITE
	180	DARK BLUE	218	BROWNISH YELLOW
			234	BROWN
			259	LIGHT RED
			267	CARMINE RED
			275	DARK REDDISH BROWN
	285	WHITE	281	DEEP VIOLET
			306	INDIGO
	367	YELLOW	332	BLUE
	390	STRAW YELLOW	430	GRAYISH BLUE
	435	BROWNISH YELLOW	505	BLUISH GRAY
	464	ORANGISH YELLOW	536	PALE GREEN
	521	REDDISH PURPLE	551	YELLOWISH GREEN
SECOND				
	570	VIOLET	575	GREENISH YELLOW
			589	GOLDEN YELLOW
	624	BLUE	664	ORANGE
	687	BLUISH GREEN	728	BROWNISH ORANGE
	720	GREEN	746	LIGHT CARMINE RED
	765	WHITE	826	PURPLISH RED
	780	YELLOWISH GREEN	855	VIOLET
	834	YELLOW	910	INDIGO
	894	YELLOWISH ORANGE	948	DARK BLUE
	974	ORANGISH BROWN	998	GREENISH BLUE
	1065	REDDISH VIOLET	1101	GREEN

TABLE 4-10
TRANSMITTED LIGHT INTERFERENCE COLORS FOR ANISOTROPIC MATERIALS* (Continued)

ORDER OF COLOR	CROSS POLARIZED LIGHT RETARDATION (in nm)	COLOR	PARALLEL POLARIZED LIGHT RETARDATION (in nm)	COLOR
THIRD				
	1155	BLUE	1151	DIRTY YELLOW
	1224	BLUISH GREEN	1258	YELLOWISH PINK
	1299	GREEN	1334	BROWNISH RED
	1350	PALE GREEN	1376	VIOLET
	1380	WHITE	1426	GRAYISH BLUE
	1434	PALE YELLOW	1495	SEA GREEN
	1524	YELLOWISH RED	1534	GREEN
	1587	WINE RED	1621	DULL SEA GREEN
	1644	REDDISH BLUE	1652	YELLOWISH GREEN
FOURTH				
	1856	GREEN	1711	YELLOWISH GRAY
	1965	PALE YELLOW	1744	PALE GRAYISH RED
	1994	ORANGE	1811	CARMINE
	2040	ORANGISH RED	1927	GRAYISH RED
	2094	WINE RED	2007	BLUISH GRAY
	2138	PURPLISH RED	2048	GREEN

*General orientation in cross or parallel polarized illumination.

SET-UP OF A TRANSMITTED LIGHT, POLARIZING MICROSCOPE FOR CONTRAST ENHANCEMENT

1. Set the orientation of the polarizer, in the condenser system, to zero. Remove the analyzer in the body tube from the light path. If high illumination levels are used, place a heat absorbing filter in the illumination path prior to the polarizer (see Chapter 7, Section VI C).
2. Adjust the microscope for correct brightfield illumination per Chapter 2, Section IV C. The microscope is now set up for observations in plane polarized light.
3. For examinations in cross polarized illumination, insert the analyzer in the light path. For maximum contrast the angular position of the analyzer should be carefully adjusted so that it is completely crossed with respect to the polarizer, i.e., the field of view should be completely dark with no sample on the stage.

4. Contrast may be altered by rotation of the stage, especially if the sample has preferred orientation. The contrast may also be changed by the insertion of a compensating accessory inserted in the tube slot, see Chapter 7, Section VI F.

D. Incident, Polarized Illumination

The incident light microscope or the metallograph may be outfitted with a polarizer and analyzer in a completely analogous manner to the transmitted light instrument, as described above. The Bertrand lens and the grain isolating diaphragm, however are not usually available, or necessary, on a purely incident light instrument. A typical arrangement of the optics for incident polarized illumination is shown in Figure 4-28.

The incident light polarizing microscope can be used profitably whenever anisotropic phases are present in opaque materials. The samples for these examinations are the typical polished and etched metallurgical mounts, see Mott and Haines (1951). The anisotropy that provides the contrast for opaque samples can be intrinsic to the material or can be an applied or naturally formed coating, the character of which is indicative of the structure of the material on which it rests. Typical areas where the method has proved useful are the examination of metals and their alloys, ore minerals, ceramic materials, coal, and cement clinker.

As with the transmitted light microscope, the incident light instrument may be used to view a sample in plane polarized light, using only the polarizer or in cross polarized light using both the polarizer and analyzer.

If only isotropic grains are present in the sample mount, their appearance in incident, plane polarized light is similar to that in brightfield illumination. However, variations in the brightness of anisotropic grains may be noted in plane polarized light as the stage is rotated; this effect is known as bireflection. Every anisotropic grain in the sample in a general orientation reflects light back to the objective lens, polarized in two mutually perpendicular directions. The light incident on the grain is resolved into the two directions with equal amplitudes. The amplitudes of the light beams reflected back into the objective, however are reduced as a result of their individual reflection coefficients. The reflection characteristics for a given grain in each of these directions may be observed by rotating the stage until they are, in turn, parallel to the polarizer.

The reflection characteristics for the two principal directions may be strongly a function of wavelength. In this case, different colors will be observed in the two unique directions of such a crystal when the unique directions are aligned with the polarizer; this is known as "color".

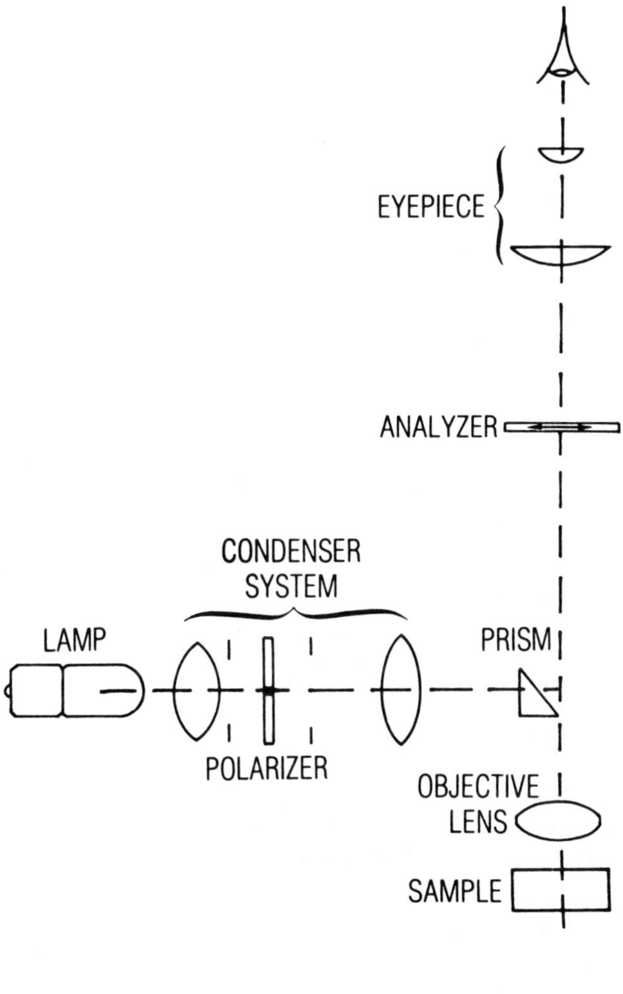

Figure 4-28. Arrangement of an incident light polarizing microscope.

In incident, cross polarized illumination, the empty field is dark. The structure of isotropic materials is also dark, however, the structure of anisotropic samples is shown in brightness or color contrast that varies in intensity as the stage is rotated.

Contrast for anisotropic grains in incident, cross polarized light is produced as shown in Figure 4-29. Every anisotropic grain in the sample that is in a general orientation reflects light back to the objective lens, polarized in two mutually perpendicular directions. It will be recalled from above that these grains are dark when either of the principle directions of the grain, XX' and YY' are parallel to the polarizer direction, PP'. As the grain is rotated away from either of these two unique positions that grain will show increasing brightness or color, with a maximum at the 45° position. In this vector diagram, the analyzer direction is indicated by the line AA'. The light incident on the grain, having an intensity of OP_i, is resolved onto XX'and YY" with equal amplitudes. OX_i and OY_i, the amplitudes of the light beams reflected back into the objective, however are reduced as a result of their individual reflection coefficients. The amplitude of the XX' beam is $OX_r = OX_i \cdot R$, where R is the reflection coefficient. The amplitude of the YY' beam is similarly, OY_r. These two reflected components recombine if there has been no retardation of either beam to yield a plane polarized beam with the direction and amplitude OR. OR is at an angle of ε from the polarizer direction. When OR is resolved onto the analyzer a plane polarized beam of intensity OA_r results, i.e., light may be seen coming from that grain.

If an opaque, anisotropic grain has an appreciable absorption coefficient, the light is elliptically polarized. The light from these grains can be completely extinguished by rotating the analyzer by the angle ε while using a variable, elliptical compensator (see Chapter 7, Section VI G) in the accessory slot.

Contrast can sometimes be observed in incident, cross polarized illumination as a result of multiple reflections from scratches or small wedge shaped etch pits in the surface of the sample. This is shown in Figure 4-30. Since etch pits are generally made up of simple, low index crystal faces they are an indication of a given grain's orientation, and thus they can provide a convenient means for studying preferred orientation in that sample. By rotating the stage so that the etch pits for a given grain are in the 45° position, the effect is maximized for that grain.

The incident light, polarized light microscope may also be used to contrast magnetic domains. The plane of polarized light is rotated upon reflection from the electrolytically polished surface of a ferromagnetic material that is magnetized in the direction parallel to the optical axis of the microscope; this is known as the Kerr effect. The direction of rotation depends upon the polarity of the surface viewed. The Kerr effect is generally smaller than the Faraday effect mentioned for transparent materials above.

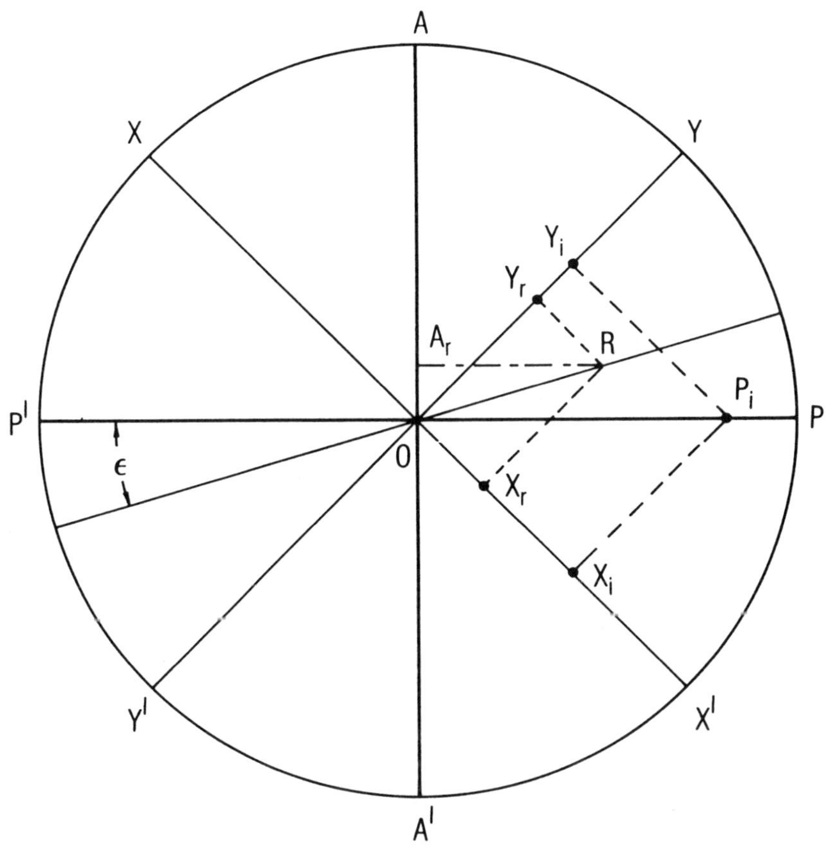

Figure 4-29. Vector diagram illustrating the production of visible light by a polished sample of an opaque, anisotropic material in crossed polarized light. The sample is oriented such that the grain under observations has crystal directions X-X' and Y-Y'. The light incident on the sample is the direction P-P'. The intensity of the light incident in the two crystal directions is X_i and Y_i. The light reflected from the two crystal directions, reduced by the absorption in each direction, is X_r and Y_r. The two reflected components may be resolved onto another plane at an angle ε to the polarizer direction. This plane when resolved onto the analyzer has a component A_r, indicating that light is transmitted to the eye.

Any transparent material when examined in incident, cross polarized light behaves as an isotropic material, i.e., the surface appears dark. This effect can be used to advantage in studying subsurface structure which often produces polarization effects that can be seen.

SET-UP OF AN INCIDENT LIGHT, POLARIZING MICROSCOPE FOR CONTRAST ENHANCEMENT

1. Remove the analyzer from the light path in the body tube.
2. If the polarizer is in the form of a circular filter, insert it into the filter carrier and set the orientation of the polarizer to the zero mark. If the polarizer is mounted in a fixed carrier, slide the carrier into the light path. If high illumination levels are used, place a heat absorbing filter in the illumination path prior to the polarizer (see Chapter 7, Section VI C).
3. Adjust the microscope for correct brightfield illumination per Chapter 3, Section IV C. The microscope is now set up for observations in plane polarized light.
4. For examinations in cross polarized illumination, insert the analyzer in the light path. For maximum contrast the angular position of the analyzer should be carefully adjusted so that it completely crossed with respect to the polarizer, i.e., the field of view should be completely dark with an isotropic sample, e.g., a first surface aluminized mirror, on the stage.
5. Contrast may be altered by rotation of the stage, especially if the sample has preferred orientation. The contrast may also be changed by the insertion of a compensating accessory inserted in the tube slot, see Chapter 7, Section VI F.

E. Problem Correction For Polarized Light Microscopy

Most, if not all of the problems noted in Chapter 2, Table 28 for brightfield, transmitted light or in Chapter 3, Table 3-1 for brightfield, incident light will be applicable as well to polarizing microscopes. Problems unique to polarizing microscopes are listed in Table 4-11.

VII. MODULATION CONTRAST MICROSCOPY

A. General

The modulation contrast technique, described by Hoffman and Gross (1975) is one of the most recent of the optical techniques for enhancement of contrast. Modulation contrast permits the examination of subtle sample features in either transmitted or incident illumination with a sensitivity that rivals the more involved techniques described above.

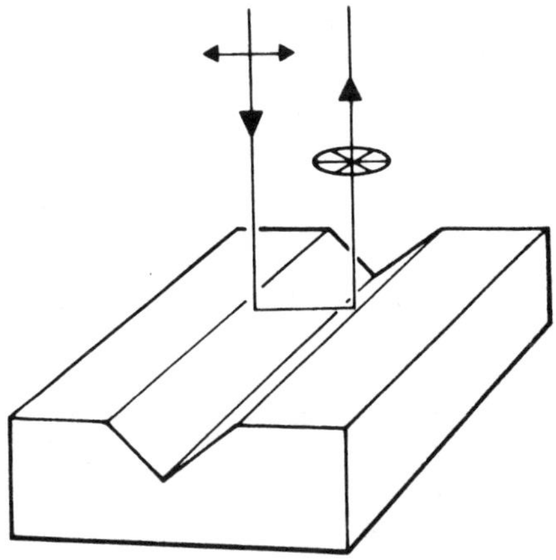

Figure 4-30. Plane polarized light incident on a groove, scratch or a wedge shaped etch pit may undergo multiple reflections before reentering the objective lens. The multiple reflections introduce some elliptical polarization which will be observed as light when examined with crossed polarized illumination.

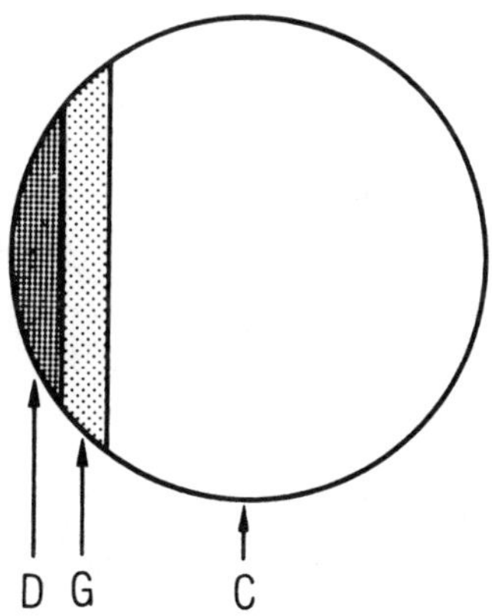

Figure 4-31. Arrangement of the modulation plate for modulation contrast. C is the clear area, G is a light gray neutral density filter area and D is a dark gray neutral density filter area. Redrawn courtesy of modulation Optics Inc.

TABLE 4-11
POLARIZED LIGHT MICROSCOPY-PROBLEMS AND THEIR CORRECTION

PROBLEM	CAUSE	CORRECTION
Field unevenly illuminated in plane polarized light.	Accessory partially blocking field of view.	Remove accessory from slot.
	Analyzer partially blocking field of view.	Completely remove analyzer from optical path.
No light from sample in cross polarized light.	Completely isotropic sample.	Chose another contrasting technique.
Field not completely dark in cross polarized light.	Analyzer not crossed completely.	For transmitted light, remove sample; for incident light, use a first surface, mirror a sample; then carefully cross the analyzer.
	Dirt on objective front lens or on condenser final lens.	Clean lenses per Chapter 10, Section VI.
	Defective polarizer or analyzer.	Refer instrument to service organization.
	Strained objective lens.	Refer to Chapter 10, Section V for testing of objective lenses.
Center of field is yellow-orange color in cross polarized light. Field not completely dark.	Filter polarizer is defective.	Refer instrument to service organization.
Cannot focus on sample.	Bertrand lens in the optical path.	Remove Bertrand lens from optical path.
Field of view very small.	Grain isolating diaphragm closed.	Open grain isolating diaphragm completely.

The brightfield microscope, either transmitted or incident, can be easily modified to produce modulation contrast by the addition of two simple and modestly-priced components; a slit to be mounted in the condenser system and a modulator plate to be mounted after the objective lens in a plane conjugate to the slit. The modulator plate is made of thin glass with two neutral density filter areas, offset to one side; this is shown in Figure 4-31. The clear area, C, has a transmission of 100%, the gray area, G, has a transmission of 15% and the dark area, D, has <1% transmission. The arrangement of the slit and modulator plate to form the modulation system is shown in Figure 4-32.

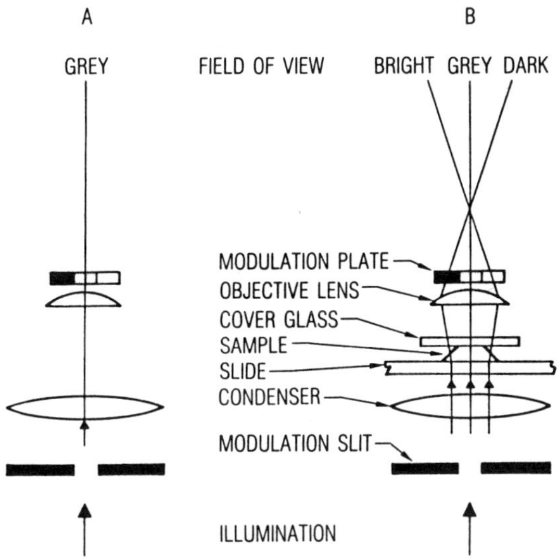

Figure 4-32. Schematic cross section and ray diagram of a transmitted light, modulation contrast microscope (A) without and (B) with a sample. Redrawn courtesy of modulation Optics Inc.

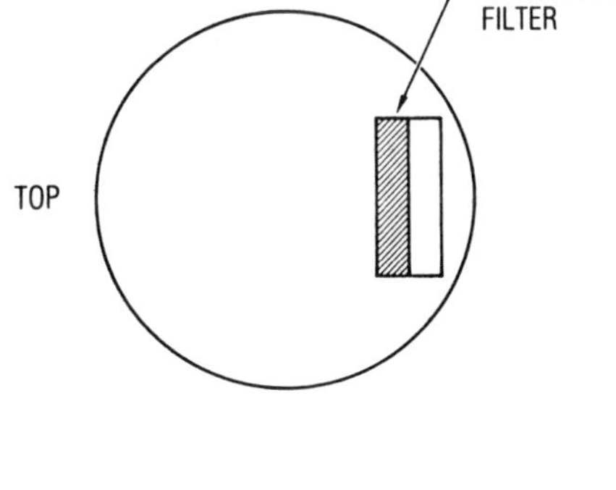

Figure 4-33. Arrangement of an advanced modulation contrast slit with a polarizing filter. If the microscope is equipped with a cap analyzer or a rotatable analyzer, the effective slit width can be changed from the open area only to the full area of the slit by uncrossing the analyzer to the polarizing filter. Redrawn courtesy of modulation Optics Inc.

The function of the modulation contrast system is illustrated for the transmitted light case in Figure 4-34. The light from the slit would be imaged, without a sample, on the gray area, this produces a uniform gray field of view. Any phase gradients may be considered to be a succession of tiny prisms. When a sample, for example consisting of a prism on either side of an area of zero gradient, is examined contrast is produced. The ray refracted from the right side of the sample is refracted into the clear area of the modulator plate and appears brighter than the ray from the zero gradient area. The ray refracted from the left side of the sample is almost completely absorbed by the dark area of the modulator plate, thus that area appears dark. In this technique the phase object is characterized by its gradient rather than by path difference.

A complete set of planachromatic objective lenses with installed modulator plates are available from Modulation Optics Inc. (100 Forest Drive, Greenvale, NY 11548). These objective lenses have the standard RMS thread to fit almost any instrument and range from 10x to 100x (oil immersion). Furthermore, two long working distance objectives with 20x and 40x magnifications are made.

Condenser systems have been developed specifically for certain models of Leitz, Nikon, Olympus and Zeiss transmitted light microscopes. Other instruments can be custom modified by Modulation Optics Inc. These condenser systems are equipped with a turret that will hold either two or four apertures to match specific modulation contrast objective lenses. The slits currently supplied have two distinct areas, as shown in Figure 4-33; area 1 is completely open, while area 2 is equipped with a polarizing filter. By rotating a polarizing, contrast control filter mounted on the condenser to a crossed position with the filter in area 2, it is possible to use area 1 only for illumination. Alternately, it is possible to illuminate the sample with area 1 plus any desired brightness of light from area 2 by rotating the contrast control to other positions. This adjustment produces varying degrees of contrast in the sample image.

B. Procedure For Modulation Contrast Microscopy

The set up of a microscope equipped with the commercially available modulation contrast objective lens and condenser system for transmitted light will be described. This procedure will be essentially the same for the incident light system. Also, the setup should vary only slightly for those instruments that have been custom modified by Modulation Optics, Inc.

SET-UP OF A MODULATION CONTRAST MICROSCOPE

1. If they are not already present, attach the modulation contrast objective lenses and condenser system to the microscope. If the microscope is a polarizing instrument, remove the analyzer from the optical path.
2. Place the sample on the stage, and correctly set-up the instrument for brightfield illumination per Chapter 2, Section IV C. For this step, the modulation contrast slit should be removed from the condenser path. If the turret condenser is provided, merely rotate it to the empty position. Note: a dark blue or green filter for the illumination is recommended by the manufacturer.
3. Observe the modulator plate in the rear focal plane of the objective lens. Three methods are given:
   a. Remove an eyepiece and look down the tube of the microscope. This is simple but not too effective because of the small image.
   b. Replace an eyepiece with an eyepiece telescope and focus the telescope to produce a sharp image of the modulator plate.
   c. If the instrument has a Bertrand lens, place the lens in the optical path. If the Bertrand lens is focusable, critically focus the modulation plate image.
4. Move the sample to a clear area of the slide. A first surface mirror may be used for the incident light case.
5. Select the slit aperture corresponding to the objective lens to be used. The slit is either on a separate insert or is mounted in the turret as a part of the condenser. The lower extension of the slit is used to manually rotate or translate the slit.
6. Place the contrast control filter in the light path before the condenser system. Note that rotating the polarizing filter influences the brightness of part 2 of the slit.
7. Locate the slit image so that part 1 of the slit is superimposed on the gray region of the modulator plate and part 2 of the slit is in the clear region of the modulator plate, as shown in Figure 4-32a. Maximum contrast of the sample image is obtained when the polarizer is rotated to the position that completely extinguishes part 2 of the slit.
8. Return the eyepiece to the microscope or remove the Bertrand lens from the optical path and move the sample into viewing position.
9. Readjust the condenser position:
   a. Close down the field diaphragm.
   b. Adjust the condenser focus to achieve sharp focus of the field diaphragm.
   c. Reopen field diaphragm until its image is just outside field of view.

10. The sample may now be observed in modulation contrast. A comparison may be made with the brightfield image by simply removing the slit from the optical path.
11. Repeat steps 3 through 9 each time the objective lens is changed.

C. Problem Correction For Modulation Contrast Microscopy

For general problems which can be observed in the brightfield image refer to Chapter 2, Section IV e. The problems unique to modulation contrast microscopy are given in Table 4-12.

TABLE 4-12
MODULATION CONTRAST MICROSCOPY-PROBLEMS AND THEIR CORRECTION

PROBLEM	CAUSE	CORRECTION
Low contrast image.	Incorrect alignment of slit.	Readjust slit location.
	Contrast control polarizer missing.	Place contrast control in position.
	Incorrect position of contrast control polarizer.	Rotate contrast control polarizer to increase contrast.
Illumination of image uneven.	Incorrect slit.	Replace with correct slit.
	Slit not superimposed on modulator plate.	Repeat set-up procedure.
	Illumination incorrect.	Correct illumination per Chapter 2, Section IV C and repeat Step 9 above..
Slit is not imaged onto the gray region of modulator plate.	Wrong slit in optical path.	Replace with correct slit.

VIII. AMPLITUDE CONTRAST MICROSCOPY

A. General

This technique, which appears quite similar to the phase contrast technique, was based on work by A. Wilska (1953, 1954). The annulus in the condenser system is identical however, the construction and function of the ring in the imaging system is however quite different. The ring, as shown in Figure 4-34, is composed of two layers of material immersed in the cement between two lens elements in the objective. The soot or carbon black layer is a nonreflecting medium that absorbs 75 to 80% of the incident light. Since the soot layer has a finite thickness and high refractive index of 2.32 it causes retardation in addition to absorption of the sample rays; this layer by itself would produce a phase contrast effect. The second layer is transparent cryolite with a refractive index lower than the cement. Thus, the phase of light rays passing through the cryolite is advanced more than that of rays passing through an equal thickness of cement. The thickness of the cryolite is adjusted so that the retardation suffered by the image ray in the soot layer is just matched by the advancement of the ray in the cryolite, thereby eliminating any phase effect due to the system. Preparation of these rings are described by Pluta (1962, 1967).

The amplitude or anoptral contrast technique described above was made available by Reichert for a period of time. They supplied individual objectives with amplitude rings and later included amplitude rings in their remote phase contrast assembly, see Section IV B.

B. Applications

According to Pluta, amplitude contrast provided greater resolving power, image contrast and image sharpness when used with a middle quality achromatic objective lens than he obtained with a brightfield microscope with a good quality apochromatic objective lens at the same magnification and N.A. The technique is particularly suited for the investigation of tiny amplitude details and

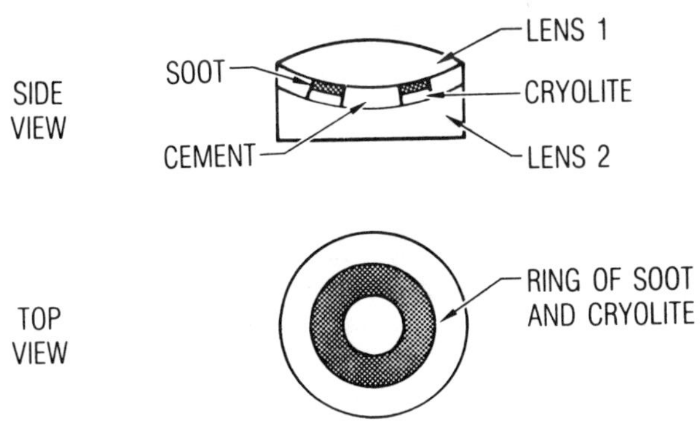

Figure 4-34. Details of an amplitude contrast ring. This ring differs from that of the bright phase contrast ring in Figure 4-16 in that the cryolite layer exactly compensates for the phase retardation of the soot layer, thus there are no phase effects with the amplitude contrast ring.

fine structures in stained biological cells and tissues. It should also be valuable for the observation and identification of small particles in other branches of microscopy as well.

C. Procedure For Amplitude Contrast Illumination

The set-up of the amplitude contrast microscope as developed by Reichert is identical to the procedure described earlier for phase contrast, see Section IV D,E.

Leitz (1972) has also reported that a form of amplitude contrast can be achieved using a DIC system of the Nomarski or Smith type, as described in Section VI B. In this technique the analyzer is rotated somewhat out of the crossed position. However, this rotation should be less than $\pm 45^{\circ}$ because this angle yields a brightfield image.

IX. FLUORESCENCE MICROSCOPY

A. General

This method of contrast enhancement depends on the property of fluorescence that is exhibited by certain materials. Fluorescence is that physical process wherein electromagnetic energy of a certain wavelength is absorbed by a material (excitation) and a portion of that energy is reemitted as electromagnetic energy of a somewhat longer wavelength (emission). For optical microscopy, the excitation energy is usually blue to ultraviolet light and the emission energy is a longer wavelength light that may be almost any color of the visible spectrum. In the case of simple fluorescence, the emission light decays exponentially when the excitation light is extinguished.

Materials that exhibit fluorescence may be divided into two categories. The first is a primary fluorescent material in which the fluorescence is due to "doping" of the structure with trace elements and the other is secondary fluorescence that results when an area of a nonfluorescing material is selectively "tagged" with a fluorescent compound called a fluorochrome. It is worthwhile to note that as little as 10^{-18} g. of a fluorochrome can be observed with a properly configured and aligned fluorescence microscope.

B. Applications

Fluorescence microscopy has proven to be a very important technique in the qualitative and quantitative correlation of chemical and physicochemical properties of materials with their structures. In the fields of biology and medicine it has become an indispensable tool. For example, it has important applications in fundamental and applied research as well as in routine investigations in medical diagnosis such as histology, cell biology, haematology and genetic karyotyping studies. Particularly notable techniques are the immunofluorescence method for the detection of antigens and chromosome investigations that are made possible by the observation of band structures that are illuminated by adsorption of certain fluorochromes.

Applications for fluorescence microscopy in the materials sciences include sedimentary petrography, coal petrography, microchemistry, forensics, environmental analysis and microelectronic fabrication. For the latter field, Kern, et al. (1982) have developed a technique using fluorochromes to decorate areas of corrosion and defects in the glassivation of integrated circuits.

The fluorescence microscope combined with a photometer yields an extremely powerful analytical instrument. An important application of this combination in the biomedical field is cytofluorometry, i.e., the quantitative determination of nucleic acids, amino acids, proteins, and biogenic amines as well as the fluorochromes added to cells and cell organelles for test purposes.

C. Fluorescence Instrumentation

Fluorescence microscopy can be accomplished using either a transmitted or an incident light instrument. The early instruments were generally brightfield, transmitted light microscopes with two special filters and a strong source of blue to ultraviolet light substituted for the standard visible light source. Such an instrument is shown in Figure 4-35a. The excitation filter is placed between the light source and the sample to filter out all light except that which causes the sample to fluoresce. The barrier, or suppression, filter is placed at some point between the sample and the eye. The purpose of the barrier filter is to filter out all of the light from the source; at the worst it may be dangerous to the eye and at the least it reduces the contrast of the fluorescent image.

The proper selection of the two filters is necessary to optimize contrast since the energy relationship between the exciting light and the emitted fluorescent light ranges from 100:1 to 1000:1. This ratio is further increased because only a fraction of the total field of view may emit fluorescence.

Transmitted light, fluorescence microscopy was greatly improved by using darkfield illumination as described in Section III. Using this type of illumination affords a background that is essentially black with only the fluorescent radiation and the light scattered by the sample entering the objective. Darkfield illumination does have some limitation at higher magnifications because the numerical aperture of the objective lens must be less than that of the condenser. As a result, this reduces the resolving power of the microscope. To increase the resolution an oil immersion darkfield condenser may be used but these are tedious to align and must be used with nonfluorescing oils.

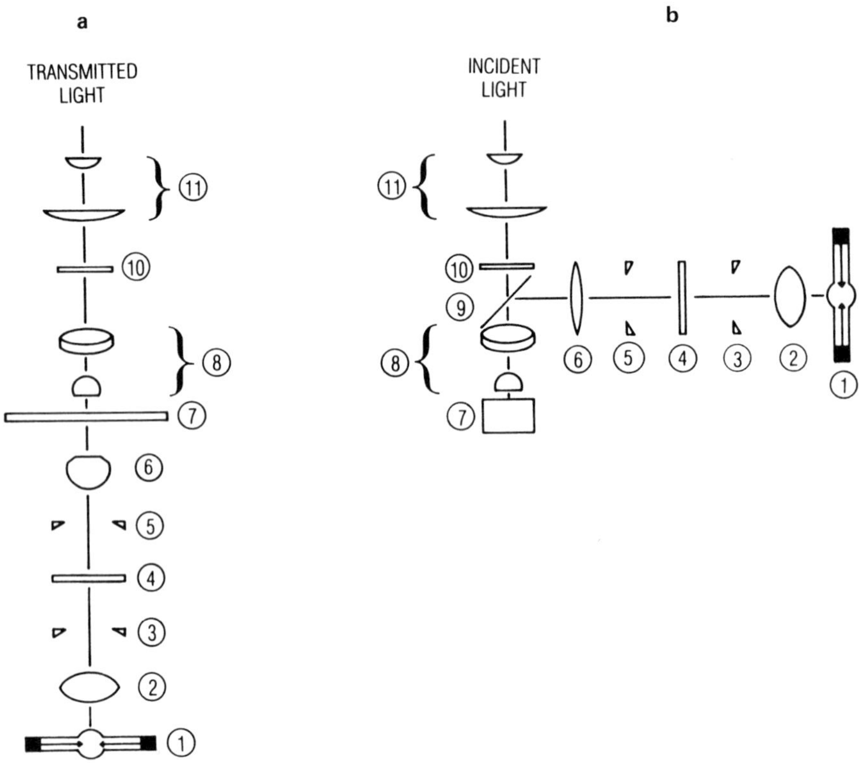

Figure 4-35. Cross sections of transmitted and incident light, fluorescence microscopes.

1. Illumination lamp
2. Lamp condenser
3. Field diaphragm
4. Excitation filter
5. Aperture diaphragm
6. Condenser lens
7. Sample
8. Objective lens
9. Partially reflecting mirror
10. Barrier Filter
11. Eyepiece

Another important feature of darkfield illumination for fluorescence microscopy is that the sample is irradiated by a hollow cone of illumination rather than the full cone of illumination. This reduced flux of radiation minimizes the effect of "fading", i.e., an exponential decay in the intensity of the fluorescence that results from exposure to the excitation radiation.

As mentioned above, fluorescence microscopy may also be achieved with the incident light instrument. For some time a standard instrument was used with the usual excitation and barrier filters, this is shown in Figure 4-35b. This arrangement was roughly equivalent in sensitivity to the transmitted light, brightfield instrument. However, a major improvement in sensitivity was made when dichroic mirrors were substituted for the usual plane glass reflector in the illuminator, this is described in Chapter 7, Section VI E. For example, a typical dichroic mirror will reflect over 90% of the excitation illumination to the sample and then transmit over 90% of the fluorescent radiation to the eye. By way of comparison, a standard plane glass reflector reflects only 50% of the excitation radiation to the sample and transmits only 50% of the resulting fluorescent radiation to the eye. In summary, the dichroic filters, in combination with the appropriate excitation and barrier filters, give an intense excitation to the sample and a very low background of nonfluorescent light in the image.

Optimum fluorescence microscopy requires careful selection of all components of the system, including the light source, each of the filters and the dichroic filter, if used. A dual illuminator system that will give either brightfield or fluorescent illumination is desirable for alignment purposes. When the system is in the brightfield position the excitation filter should not be in the light path. If short wavelength, ultraviolet light, is the excitation radiation, the condenser system should have quartz lenses, first surface collector mirror and first surface beam directing mirrors. The most common source of illumination for fluorescence microscopy is the mercury vapor arc lamp because it has a spectrum made up of a number of discrete wavelength emissions. The emissions in the ultraviolet, blue and green regions of the spectrum are particularly useful, easily isolated with filters, and relatively intense. A xenon or a tungsten halogen lamp can be used a blue source with appropriate filtration.

A word of caution is in order with regard to the use of mercury vapor arc lamps especially with quartz optics. The very strong 253.6 nm ultraviolet radiation in the illumination is damaging to the eyes, thus any leakage of the illumination prior to the excitation filter is a potential hazard. The presence and location of this ultraviolet radiation can be established by using a fluorescent card as a detector in a darkened room. Use laboratory safety glasses (with glass lenses) during this test.

The ideal fluorescence objective lens is designed with optical elements that are free from any fluorescence. Some transmitted light objective lenses have a preliminary barrier filter as its first element, thereby minimizing fluorescence within the lens. Generally, apochromatic and semiapochromatic objective lenses should not be used for this purpose due to the fluorescence of the fluorite optical elements. The objective lenses for incident light instruments as well as the condenser systems for transmitted light must be capable of transmitting the excitation radiation. If ultraviolet radiation is used, specially designed lenses with quartz elements and nonfluorescing cements must be used.

If the barrier filter is used as a cap on the eyepiece, even the eyepiece must be free of fluorescence. For very weakly fluorescent samples, a monocular instrument is preferred since the image brightness is 3 to 5 times greater than with the binocular instruments.

The excitation filter should have sufficiently high transmission in the region of the desired excitation light but sufficiently low transmission in all other areas of the visible spectrum. In this way it is possible to eliminate undesired fluorescence and falsification of the fluorescent colors. The barrier filter must absorb the excitation light but should not narrow the spectrum of the fluorescent radiation. Both absorption filter systems and the more sharply selective interference filters are available. The dichroic mirror should be chosen for the specific excitation and fluorescent wavelengths. For a listing of various filters see Chapter 7, Section VI. Refer to microscope manufacturers for dichroic mirror systems.

Mixed contrasting techniques have be used in which one is fluorescence illumination. The simplest of these includes a partially reflecting mirror arranged so that brightfield, white light illumination can be added to the fluorescent image. Any ratio of the two illumination types can be achieved by adjusting the brightness of the white light source. In this way, the fluorescing and the otherwise dark areas of the sample can be quickly compared.

Leitz and Reichert have marketed an interesting mixed contrasting system using both fluorescent and phase techniques. The instrument, shown schematically in Figure 4-36 is a transmitted light instrument with nonfluorescing, phase contrast, objective lenses, See Section IV, and a special phase annulus. The annulus is transparent to visible light just as the standard annulus however, in this case the normally opaque areas of the annulus transmit ultraviolet light. In this way the white light forms a phase image in the usual fashion that is superimposed on the fluorescent image. Other mixed forms of illumination have been described by Koch (1972); these are listed in Table 4-13.

Three important accessory items for fluorescence microscopy are: quartz slides and cover glasses from Walter C. McCrone Associates, Inc., 2820 S. Michigan Ave., Chicago, Il 60616; for increased transmission of ultraviolet excitation radiation and nonfluorescing immersion oil from R.P. Cargille Laboratories Inc., 55 Commerce Road, Cedar Grove, NJ 07009.

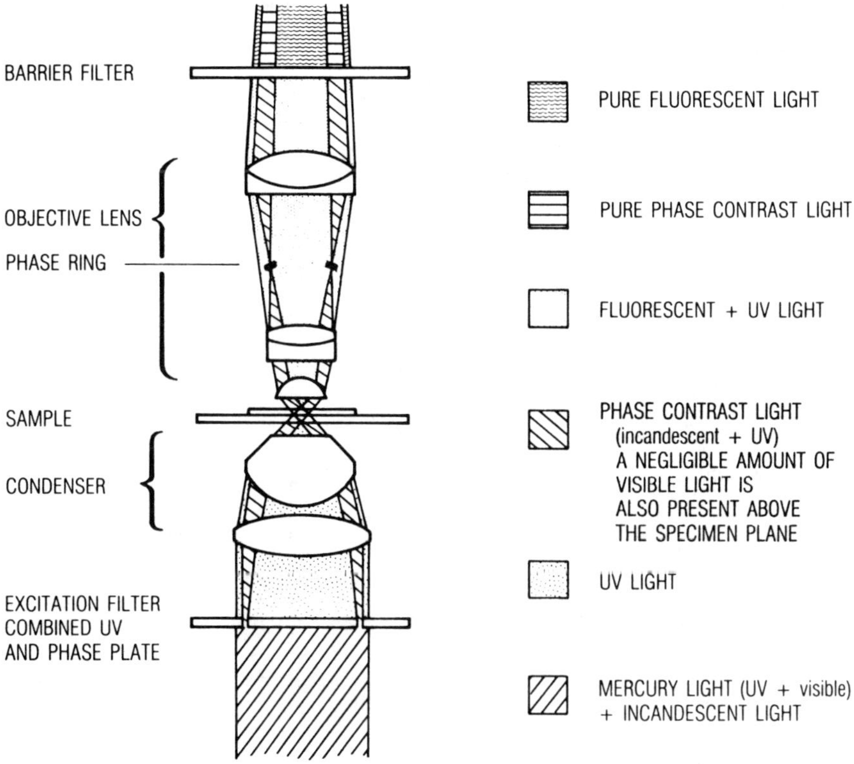

Figure 4-36. An example of mixed fluorescent illumination; in this case the fluorescent illumination is augmented with phase contrast illumination. Redrawn courtesy of Wild Leitz USA, Inc.

TABLE 4-13

MIXED ILLUMINATION METHODS USING FLUORESCENCE MICROSCOPY

ILLUMINATION #1	ILLUMINATION #2	APPLICATION
TRANSMITTED LIGHT DARKFIELD	INCIDENT LIGHT FLUORESCENCE	BIOMEDICAL
TRANSMITTED LIGHT DARKFIELD	TRANSMITTED LIGHT DARKFIELD FLUOR.	BIOMEDICAL
TRANSMITTED LIGHT PHASE CONTRAST	INCIDENT LIGHT FLUORESCENCE	BIOMEDICAL
TRANSMITTED LIGHT AMPLITUDE CONTRAST	INCIDENT LIGHT	BIOMEDICAL
TRANSMITTED POLARIZED LIGHT	INCIDENT LIGHT FLUORESCENCE	CRYSTALLOGRAPHIC STUDIES
TRANSMITTED LIGHT FLUORESCENCE	INCIDENT LIGHT FLUORESCENCE	EXAMINING SAMPLES WITH TWO DIFFERENT FLUOCHROMES

D. Alignment Of The Mercury Vapor Arc Illuminator

A typical arc illuminator, shown in Figure 2-7, consists of the arc lamp, reflector, focusable lamp condenser lens, filter slot(s) and a ventilated, light-tight housing. For optimum illumination of the fluorescent sample the illuminator must be correctly aligned and focused. The items requiring alignment to the axis of the illuminator are the lamp and the reflector; the lamp condenser is fixed and defines the illuminator axis. Since this alignment is required only when the lamp is replaced, it is described in Chapter 10, Section VI.

The illuminator must also be carefully focused to provide the even illumination of the sample. Since this procedure requires the projection of the illumination onto a large, unprotected surface, care must be exercised to avoid damage to the eyes by short wavelength ultraviolet radiation. This may be assured by using an appropriate UV absorbing filter in the illuminator filter slot, See Chapter 7, Section VI. The use of laboratory safety glasses (with glass lenses) is also recommended as an additional UV filter. This procedure is accomplished as follows:

PRELIMINARY FOCUSING OF MERCURY VAPOR ARC ILLUMINATOR

1. Place a UV absorbing filter in the illuminator filter slot, See Chapter 7, Section VI.
2. Turn on the arc lamp and allow to stabilize for about 2-3 minutes.
3. Place a sheet of white paper on the stage of the microscope.
4. For a transmitted light instrument remove the condenser system and obtain an even illumination of the paper using the lamp condenser adjustment. For an incident light instrument, remove the objective lens and obtain an even illumination of the paper using the lamp condenser adjustment. If the sheet of paper cannot be evenly illuminated the illuminator must be aligned, see Chapter 10.
5. Replace the condenser system or the objective lens and proceed with the set-up of the microscope as described below.

E. Procedure For Transmitted, Fluorescence Illumination

The set-up of a transmitted light fluorescence microscope is relatively straightforward, however, successful examination of the sample requires attention to a number of details such as selection of the filters to match the subject fluorochromes and use of the appropriate slide and cover glass materials. Cleanliness is very important for both the sample and the microscope optics. For example, dirt or oils on the image side of the sample may produce anomalous fluorescent effects while dirt or oils on the source side of the sample may absorb excitation light, thus reducing or eliminating sample fluorescence.

SET-UP FOR TRANSMITTED LIGHT, FLUORESCENCE MICROSCOPY

1. Adjust the microscope for correct brightfield (Chapter 2 Section IV C) or darkfield (Chapter 4, Section III D) illumination. If a dual illuminator is used, perform this step using the visible light source. Note: Do not use a mercury vapor lamp source for this procedure unless an ultraviolet absorbing filter is in place in the excitation filter slot (see Chapter 7, Section 6).
2. Install the appropriate barrier filter (see Chapter 7, Section 6). This is inserted in a slot above the objective or is contained in a revolving turret.
3. Place a fluorescent sample on the stage.
4. Replace the ultraviolet absorbing filter in the excitation filter slot with the appropriate excitation filter (see Chapter 7, Section 6).
5. Illuminate the sample with the fluorescent source.

6. Perform minor adjustments of the lamp condenser, if used, or the condenser system focus to improve uniformity and intensity of the image.
7. The sample may now be examined in transmitted light, fluorescence illumination.
8. For mixed illumination methods consult the appropriate chapter and section, align the instrument for that method, then follow with step 2.

F. Procedure For Incident, Fluorescence Illumination

The set-up of an incident light fluorescence microscope, like the transmitted light instrument, is quite straightforward. However, a successful examination of the sample requires attention to a number of details such as selection of the filters to match the subject fluorochromes and cleanliness of both the sample and the microscope optics.

SET-UP FOR INCIDENT LIGHT, FLUORESCENCE MICROSCOPY

1. Adjust the microscope for correct brightfield illumination, see Chapter 3, Section IV C. If a dual illuminator is used, perform this step using an incandescent lamp. Note: Do not use a mercury vapor lamp source for this procedure unless an ultraviolet absorbing filter is in place in the excitation filter slot (see Chapter 7, Section 6).
2. Install the appropriate barrier filter (see Chapter 7, Section 6). This is inserted in a slot above the objective or is contained in a revolving turret.
3. Place a fluorescent sample on the stage.
4. Replace the ultraviolet absorbing filter in the excitation filter slot with the appropriate excitation filter (see Chapter 7, Section 6). Also, if used, install the appropriate dichroic mirror in the illuminator.
5. Illuminate the sample with the fluorescent source.
6. Perform minor adjustments of the lamp condenser, if used, or the condenser system focus to improve uniformity and intensity of the image.
7. The sample may now be examined in incident light, fluorescence illumination.
8. For mixed illumination methods consult the appropriate chapter and section, align the instrument for that method, then follow with step 2.

G. Problem Correction For Fluorescence Microscopy

The first step in correcting any problems in fluorescence microscopy is to consult the Problem Correction Section for the corresponding brightfield illumination technique. If the problem is not corrected using that table then it is safe to assume that the problem is related to the fluorescence accessories; these problems are addressed in Table 4-14.

TABLE 4-14
FLUORESCENCE MICROSCOPY-
PROBLEMS AND THEIR CORRECTION

PROBLEM	CAUSE	CORRECTION
Fluorescence source lighted but no image.	Excitation shutter closed.	Open shutter.
	Wrong combination of excitation and barrier filters.	Use corect filter combination, see Chapter 7, Section VI.
	Dichroic mirror not installed.	Install correct dichroic filter, see Chapter 7, Section VI.
	Non-fluorescing sample.	Prepare proper fluorescent sample.
	UV absorbing filter in excitation filter position.	Replace UV absorbing filter with correct excitation filter.
	Contamination on excitation side of sample	Clean condenser lens or reverse side of glass slide.
Insufficient contrast.	The objective lens is fluorescing.	Use objective lenses designed for fluorescence work.
	The immersion oil fluoresces.	Use non-fluorescing immersion oil.
	Fluorescing contamination on filters, objective, or condenser lenses.	Clean lenses, reverse side of glass slide, and filters per Chapter 10, Section VI.
Irregular illumination.	The mercury arc lamp not centered to condenser system.	Recenter lamp per Chapter 10, Section VI.
	The condenser system not properly aligned.	Realign the condenser per Chapter 10, Section VI.

X. VIDEO ENHANCED CONTRAST

All of the preceding contrast enhancement techniques operate on or modify the optical system of the microscope to provide increased contrast for the image. In this day of computerization we have the opportunity to operate on a video image presented to an electronic system thereby producing an image that has been enhanced in any one of a number of ways. This process has come to be referred to as video enhanced contrast (VEC). Needless to say, that primary image can be a brightfield image or one that is the product of the various optical contrast enhancing techniques described in this chapter.

One of the early uses of VEC was the preparation of the image for quantitative analysis; this aspect of electronic image analysis is discussed in Chapter 5, Section IV. These instruments are generally described as "Image analyzers". Recently, however, turnkey systems have become available that are primarily designed for image enhancement.

The general arrangement of the image enhancement type, image analyzer is shown in Figure 4-37. The detector, generally some form of a video camera, is affixed to the microscope in the same location as a photographic camera. The output of the camera is usually operated on by an analog contrast enhancement system. The signal then may be directed to a digital system which provides numerous additional contrast enhancing routines. The output of the image analyzer is usually a video monitor. A permanent image may be obtained at this point by video recording or photographing the monitor image. A direct image with reasonable resolution and gray level may be produced with a video quality printer.

A typical enhancement of an image first involves the manipulation of the gain and black level of the signal from the video camera as shown in Figure 4-38. In this case, the signal gain or amplification and black level are both reduced so that the image contrast range fills the entire gray scale of the system. As a result of the redefined signal gain and black level, the areas of interest will be enhanced.

Other features or artifact usually contribute to some extent to the image. Typical of these are dust particles in the optical system, lens defects, video detector faults and contributions from uneven illumination. These artifacts contribute a fixed background superimposed on the image of the sample that can be removed by defocusing the microscope or moving to a blank area of the sample and acquiring an image. By digitizing this image and subtracting it from the similarly digitized sample image, a final highly refined image of the sample may be produced. Further improvement of the image may also be accomplished in reducing the random noise in the image with signal averaging. A typical sequence of video enhancement is shown in Figure 4-39.

Each image analyzer system is unique, therefore it is impossible to give anything beyond a general description of the various functions available on a given instrument. A brief listing of some of the various contrast enhancing functions is given in Table 4-15. The operating instructions are necessarily specific to each system and are beyond the scope of this handbook.

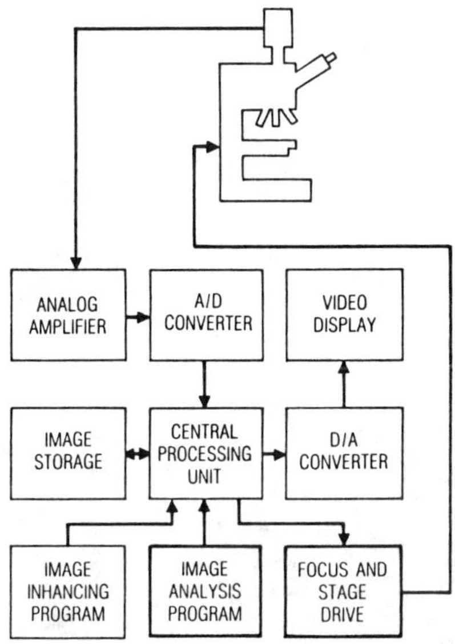

Figure 4-37. Electronic image enhancement is one of the newest techniques for improving contrast in light microscopy. This block diagram contains the basic elements of a typical electronic image analyzer. Such a system is primarily used to evaluate some microstructural parameter such as size or concentration, however it can also be effectively used without the image analysis program to electronically enhance the contrast of a micrographical image.

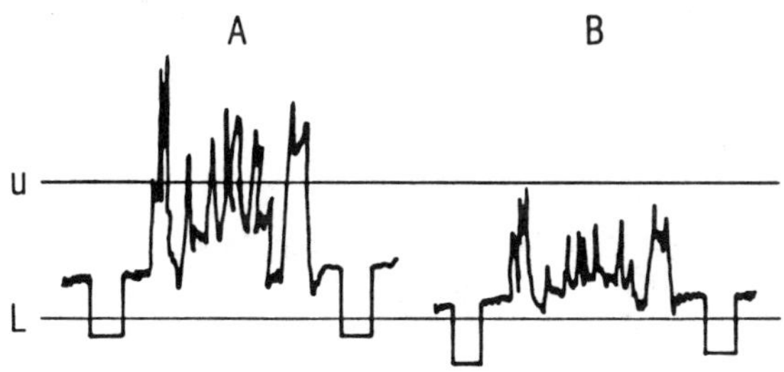

Figure 4-38. The effect of gain and black level on the contrast of a video image. The lines, U and L, represent the limits of contrast available for the video image. In (A), much of the raw video image is saturating the system, i.e., much of the image signal is above the upper level and is imaged at full brightness, whereas in (B), both the gain and the black level has been reduced so that the image falls within the usable contrast range of the system.

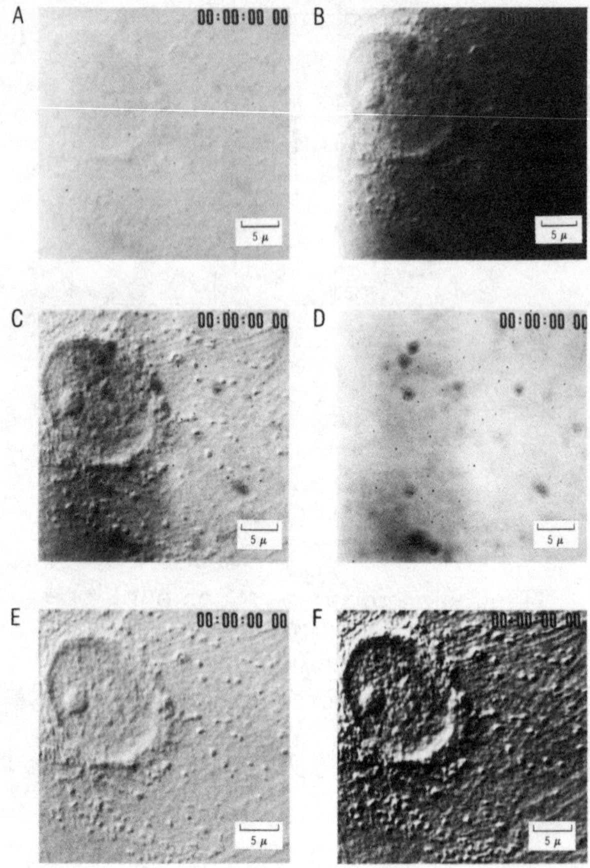

Figure 4-39. A series of photomicrographs illustrating the usefulness of electronic image enhancement are reprinted courtesy of Photonic Microscopy Inc. and by permission of American Laboratory.

A. Original DIC image of a Swiss 3T3 fibroblast in matched refractive index medium
B. Analog contrast enhanced image of A
C. Analog shading corrected image of B
D. Out of focus background or "mottle" image
E. Image C after subtracting mottle image, D.
F. Image F after digital contrast enhancement using exponential rolling average.

TABLE 4-15
TYPICAL IMAGE ANALYZER ENHANCEMENT FUNCTIONS

FUNCTION	APPLICATION
Analog contrast enhancement.	Amplify video signal, readjust black level.
Image accumulation.	Integrate (sum) a number of images to improve "noisy" dark images.
Background subtraction.	Remove background "mottle" from each image.
Pseudo color display.	Transform gray levels into user designated colors.
Digital contrast enhancement.	Enhance, suppress, or isolate areas of the gray level, etc.
Differential edge enhancement.	Enhancement of vertical and/or enhance-ment.horizontal edges by various methods.
Spacial filtration.	Suppression or enhancement of specific high or low frequency spatial information.
Arithmetic operations.	Interframe or intraframe memory addition, subtraction, multiplication or division

---

Another area in which video and computer techniques are finding new applications is that of video intensified microscopy (VIM). In VIM a very weak image, such as that in fluorescence microscopy, is made visible through the use of various techniques such as frame blanking, image intensifying cameras and photon counting cameras. These techniques are described by Reynolds (1964), Hayakawa, et.al.(1986) Wick (1987) and Commare (1988).

REFERENCES

Allen, R. D., G. B. David, and G. Nomarski, Zeit. Wiss. Mikroskopie und Mickroskop. Tech., 69/4, 193, (1969).

Bradbury, S., Proceedings, R.M.S., 15/4, 270, (1980).

Bluhm, J. W. and M. N. Haller, American Laboratory, p. 22, April (1976).

Commare, D. L., American Laboratory, p56, Dec. (1988).

Hayakawa, T., et.al., Photochemistry and Photobiology, 43, (1), 95, (1986).

Hoffman, R. and L Gross, Applied Optics, 14/5, 1169, (1975).

Kern, W., R. B. Comizzoli, and G. L. Schnable, Industrial Res. and Develop., p. 151, May, (1982).

Koch, K. F., LEITZ-Mitt. Wiss. u. Techn., 7, 206, (1972).

Lang, W., American Laboratory, p.46, April (1970).

Leitz Technical Information Bulletin, XII/2, 14, (1972).

Mott, B. W. and H. R. Haines, Research, 4/1, 24, (1951).

Nicastro, A. J., American Laboratory, p.12, Dec. (1982).

Pluta, M., The Microscope, 16, 211, (1968).

Reynolds, G. T., IEEE Transactions. Nucl. Sci., NS-11, 147, (1964).

Richardson, J. H., "Polarimetry", Chapter 36, P.205, in "Systematic Materials Analysis", Vol IV, Edited by Richardson, J. H. and R. V. Peterson, Academic Press, Inc. New York, NY, (1978).

Smithson, F., Min. Mag., 28, 384, (1948).

Wick, R. A., Applied Optics, 26, (16), 3210, (1987).

Wilska, A., Nature, 171, 353 and 698, (1953).

Zernike, F., Z. Tech. Phys., 16, 454, (1935).

5

Quantitative Microscopy

I. INTRODUCTION

The first question the microscopist often seeks to answer about a sample is what is the appearance or nature of its structure. This initial information is qualitative in nature and includes such properties as shape, color, and texture. Following these preliminary studies other questions following very close behind are sure to be, how large is a certain portion of the sample or how many items are present or what volume fraction is a given phase of the whole. Questions of a quantitative nature such as these latter ones are the basis for this chapter.

Measurements of dimensions require definitive starting and ending points; this assumes that the sample has sufficient contrast to visually identify these points. Therefore, if the sample lacks intrinsic contrast, the microscopist should resort to one of the techniques described in Chapter 4 to provide induced contrast for the structure.

Successful quantitative microscopy of a sample requires extended periods of considerable concentration on the part of the microscopist, therefore it is imperative that the correct adjustment of the microscope be undertaken for each observation to minimize eye strain. Likewise, every aspect of ergonomics should be considered to minimize general operator fatigue; this is discussed in Chapter 10, Section, IV.

II. TRANSVERSE MEASUREMENTS

A. Length Measurements

Determination of transverse lengths, i.e., in the dimensions parallel to the stage of the microscope, is the most frequently performed quantitative measurement. Two methods are used. The first method is suitable for large samples requiring only low magnifications to see the structures of interest. this method requires a calibrated mechanical stage and an eyepiece with a cross-wire. The sample is carefully focused and aligned by means of the mechanical stage so that on edge of the structure to be measured is exactly under the cross-wire. The value of the mechanical stage calibration for that point is recorded. Then the sample is traversed using the mechanical stage until the other edge of the structure is exactly under the cross-wire. The reading of the stage calibration is noted once again; the difference between the two values is the desired dimension of the structure. The magnification used should be high enough to accurately set the edges of the desired structure on the cross hair but low enough that the structure remains recognizable as it travels across the field of view. This procedure is used for the measuring microscope, the toolmaker's microscope, the traveling microscope, the cathetometer and the optical comparator.

The second method for the measurement of transverse lengths consists of comparing the length of some desired structure on the sample with an accurately ruled stage micrometer, see Chapter 7, Section IV I. To make this comparison a secondary scale or reticle is used; this is located in the eyepiece. These eyepieces are available in two different types, the fixed reticle eyepiece and the interchangeable reticle eyepiece. The former eyepiece comes equipped with a reticle that is permanently positioned and cannot be removed; a true cross-wire eyepiece is of this type.

The interchangeable reticle eyepiece, shown in cross section in Figure 5-1, has a threaded reticle cell in which any of a variety of reticles may be placed, see Chapter 2, Section III E. A reticle should be installed or changed only in a dust-free, dry area. Carefully unscrew the reticle cell, A, from the eyepiece then unscrew the reticle retainer, B, and remove the existing reticle, storing it in a dust free container. Clean the desired reticle with lens tissue and lens cleaner as described in Chapter 10, Section VI. Place that reticle in the cell, replace the retainer on the reticle cell and then return the reticle cell to the eyepiece. If the alpha-numerics on the reticle are inverted, repeat the above steps, inverting the reticle in the process.

The Zeiss Axiomat also has provisions for using reticles at some intermediate focal plane that is conjugate to the eyepiece reticle; this is particularly useful if a photomicrograph of the sample

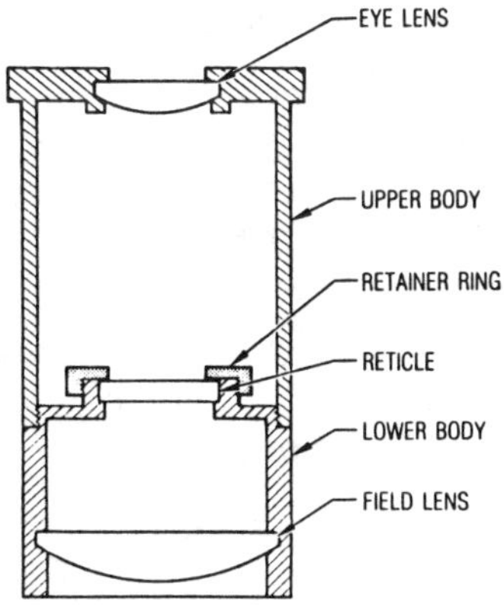

Figure 5-1. Cross section of an interchangeable reticle eyepiece.

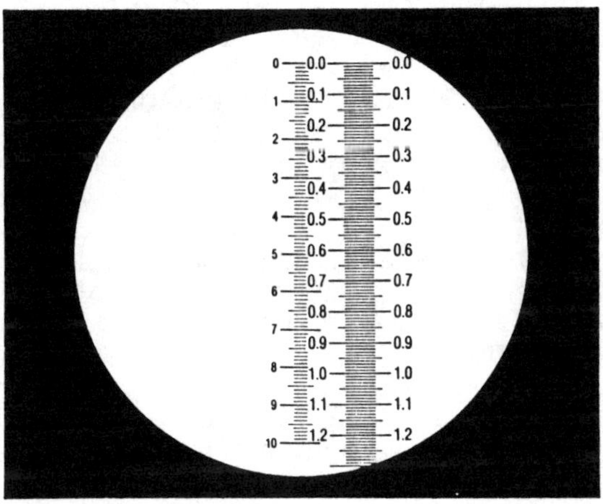

Figure 5-2. Calibration of the eyepiece reticle (center) with a stage micrometer (right). The origin of stage micrometer is lined up with the origin of the reticle and the value of the stage micrometer corresponding to the 100 line on the reticle is recorded; in this case the stage micrometer value is 1.22 mm. Therefore, each individual reticle division corresponds to 1.22 mm / 100 or 0.012 mm (12 μm). This process must be repeated for each objective lens used.Redrawn courtesy of Wild Leitz USA, Inc.

is desired with the reticle superimposed. This procedure can also be accomplished with other camera systems by using a reticle eyepiece as a camera lens, see Chapter 8, Section II.

It must be emphasized that all reticles have arbitrary calibrations since the calibration process incorporates the combined effect of the objective magnification, zoom lens setting (if any), draw tube setting (if present), and even the interpupillary setting of non-compensating binocular heads.

The procedure for calibration of the reticle with the stage micrometer is given in Table 5-1. If the reticle eyepiece is not equipped with a ruled reticle, one of the latter type should be

TABLE 5-1
CALIBRATION OF AN EYEPIECE RETICLE

1. Set up the microscope per the instructions in Chapter 2, Section IV C or Chapter 3, Section IV C. Particular attention should be given to the correct focusing of the ruled reticle eyepiece. If the microscope has a binocular head with compensating adjustments, it is very important that these be properly set to the value of the microscopist's interpupillary spacing. If the binocular head has a compensating adjustment on only one eyepiece, the specific interpupillary spacing used for the calibration must be recorded and used for all subsequent measurements. If the microscope has compensating adjustments on both eyepieces and they are correctly set for each observer's eyes, the reticle calibration values will be correct for any interpupillary spacing.
2. Select an objective lens and record its magnification as well as that for the ruled reticle eyepiece and any zoom system, Chapter 2, Section III C. If the microscope has a graduated draw tube, Chapter 2, Section II A, record that value.
3. Focus carefully on the stage micrometer using the mechanical stage to align a reference point on the stage micrometer with the zero point of the ruled reticle.
4. Determine the number of stage micrometer units that correspond to a single reticle unit, see Figure 5-2. This value, the calibration value for that specific lens system, should be recorded with the information listed in step 2.
5. Repeat steps 2-4 for each objective lens/eyepiece combination. If a zoom system is to be used, measurements should be made at several zoom values for each objective lens/eyepiece combination.
6. Prepare a table listing the calibration values for each reticle eyepiece/objective lens combination for that specific microscope. Indicate any zoom or graduated draw tube values.

installed as described in Chapter 7, Section VII. A stage micrometer with the most frequently used units should be selected since they are available with either metric or English units.

If a specific length value, e.g., 10 μ per division, is desired for the reticle units, this may be achieved by adjusting the zoom system or the graduated draw tube, however the latter procedure will cause some degradation of the image quality.

Once the calibration values have been determined, the dimension of interest on the sample may be determined using the procedure in Table 5-2. The highest magnification objective lens should be used that will allow the structure of interest to be within the total reticle scale; this assures that the measurement will be made with maximum precision and minimum eyestrain. The sample is positioned so that the structure of interest is coincident with the reticle scale and the number of reticle units corresponding to the

TABLE 5-2
LENGTH MEASUREMENT WITH A RETICLE EYEPIECE

1. Set up the microscope per the instructions in Chapter 2, Section IV C or Chapter 3, Section IV C. Particular attention should be given to the correct focusing of the reticle eyepiece. If the binocular head has a compensating adjustment on only one eyepiece, the specific interpupillary spacing used in the calibration process of Table 5-1 must be used. If the microscope has a binocular head with compensating adjustments for both eyepieces, they must be set to those values used in the calibration process of Table 5-1; this arrangement will be correct for any interpupillary spacing.
2. Place the sample on the stage and focus it critically.
3. Select an objective lens and ruled reticle eyepiece such that the image of the dimension of interest on the sample is less than the length of the reticle image.
4. Record the magnification for the objective lens and the eyepiece as well as that for any zoom system. If the microscope has a graduated draw tube, set it to the value recorded in Table 5-1.
5. Position the sample by translation and/or rotation until the dimension of interest is spanned by the reticle ruling.
6. Determine the number of reticle units that correspond to the dimension of interest.
7. Obtain the number of length units per reticle division from the tabulated data of Table 5-1 for this microscope arrangement recorded in Step 4.
8. Multiply the number of reticle units obtained in Step 6 by the number of length units per reticle units from Step 7. This is the length of the dimension of interest.

dimension of interest is determined. The number of reticle units times the calibration value for that objective lens/zoom lens (if any)/eyepiece combination equals the length of the structure of interest. A mechanical stage (see Chapter 7, Section IV) is an ideal device for accurately positioning the sample. The correct measurement direction for the dimension of interest can be achieved by rotating the sample on the stage, rotating the revolving stage (if available) or by rotating the reticle eyepiece in the body tube.

Transmitted light samples have a finite thickness that, depending on the setting of the aperture diaphragm, is all more or less in focus; this situation produces some uncertainty in length measurements of structures that do not have vertical edges. Figure 5-3a illustrates this problem.. The problem can be minimized by using the aperture in its full open position thereby producing the shallowest depth of field possible.

Since the incident light sample is essentially flat and usually opaque the blurring of structural edges as described for the transmitted light case is not a problem; this is shown in Figure 5-3b.

If the sample lacks contrast, one of the contrasting techniques described in Chapter 4 may be used, however, a word of caution should be given about the phase contrast technique since it does produce halos that may obscure the real edge of structural elements.

Measurements of length may be made from photomicrographs, however the magnification must be accurately known. Ideally the negative of each photomicrograph. should be marked with a line that corresponds to a given length at the sample plane; this may be easily achieved with the reticles described above. In this way any enlargement of the photomicrograph will also exhibit the appropriate length relationship. A second method for determining lengths without the reticle in the photomicrographic image, requires the preparation of a photomicrograph of the stage micrometer using the same set-up as for the sample. Since distortion can affect the magnification in different areas of the image, it is important for critical measurements to check the entire image of the stage micrometer for linearity.

These procedures are suitable for the length measurement of a wide variety of samples ranging from individual particles to organisms and biological structures. If the structures are equidimensional, i.e., round or regular polygons, one measurement per object is sufficient. If the objects are elongate or irregular in shape, more than one measurement is required per object. In this case it is necessary to rotate the sample, the stage or the eyepiece to effect the measurements in the various directions. Dimensional measurements of the structures on the surfaces of integrated circuit electronic chips or wafers usually require high precision and accuracy; the following section addresses this concern.

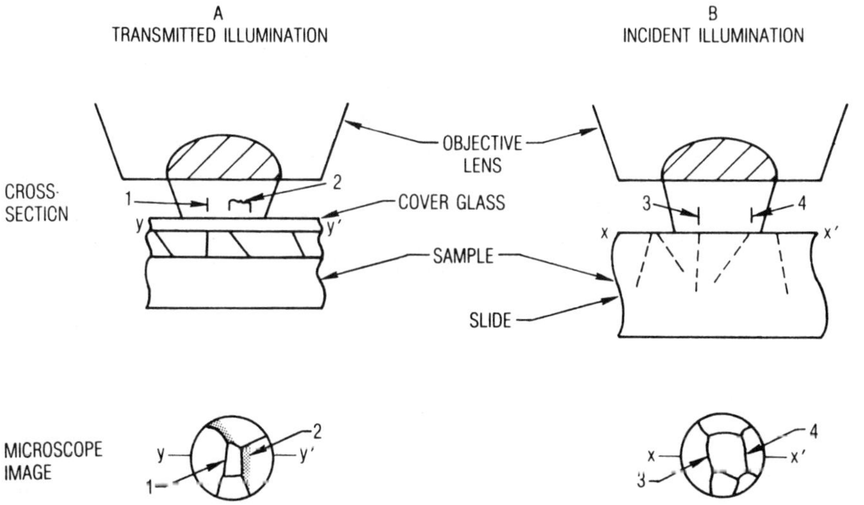

Figure 5-3. A comparison of quantitative measurements with brightfield, transmitted and incident light microscopes. In the transmitted case, the transparency of the sample combined with the depth of field of the objective lens gives a wide or diffuse image of a grain boundary unless the boundary is vertical; compare boundaries 1 and 2. This diffuse nature makes accurate location of some strain boundaries difficult. However, since the brightfield incident light microscope images just the surface of the sample, all properly etched boundaries are of similar sharpness.

B. Precision and Accuracy of Length Measurements

Measurements involving the calibrations on the mechanical stage of a standard microscope are limited in accuracy by the quality of the rulings; these are generally accurate to about ±0.1 mm. Precision in this case depends on the ability to correctly set edges of the image coincident with the cross-hair and to correctly read the scale; this is on the order of about ±0.2 mm.

The typical stage micrometer is uncertified and may have errors as large as 1%. These stage micrometers may be certified by a competent metrology laboratory. Precision stage micrometers may also be purchased that are certified accurate to ±0.05 micron for each 0.01 mm division and to ±0.1 micron for the entire scale when used at a specified temperature of 20° C. If a certified stage micrometer is carefully handled and maintained at ambient temperature, the accuracy should be retained. Since the reticle is a secondary standard, the accuracy of measurements using the ruled reticle is dependent on both the care used in the calibration of the reticle and the care used in making the actual measurement.

The requirement to know the values of numerous small spacings in the fabrication of semiconductor devices and other high technology fields has spurred considerable research in the factors and techniques involved in performing precise and accurate measurements of line widths. Notable research in this area has been conducted by Nyyssonen (1980) of NIST (National Institute of Standards and Technology, formerly the National Bureau of Standards). The major difference between their set-up of the microscope for brightfield microscopy and that described in Chapter 3 Section IV C and Chapter 4 Section IV C is the setting of the aperture diaphragm at 67% of the full objective lens aperture rather than at than at the 80 to 90% recommended in Chapters 2 and 3. They also recommend the use a Wratten No.60, green filter in the illumination for all measurements. The micrometer eyepiece or the image shearing eyepiece are necessary to provide the required precision and accuracy. TV type imaging systems with image analysis will also provide speed as well as additional accuracy and precision; these are discussed in Section IV D below.

One very important result of the line width work at NIST was the generation of two Optical Microscope Linewidth measurement standards, SRM 474 and SRM 475, see Figure 5-4. These standards are designed for the measurement of opaque and clear lines in the range of 0.5 to 10 μ on integrated-circuit photomasks in transmitted light illumination. The patterns of these SRMs include both opaque lines on a clear background and clear spaces on an opaque background. The accuracy of the linewidths is ±0.05 μ or better. A series of other SRMs are planned for

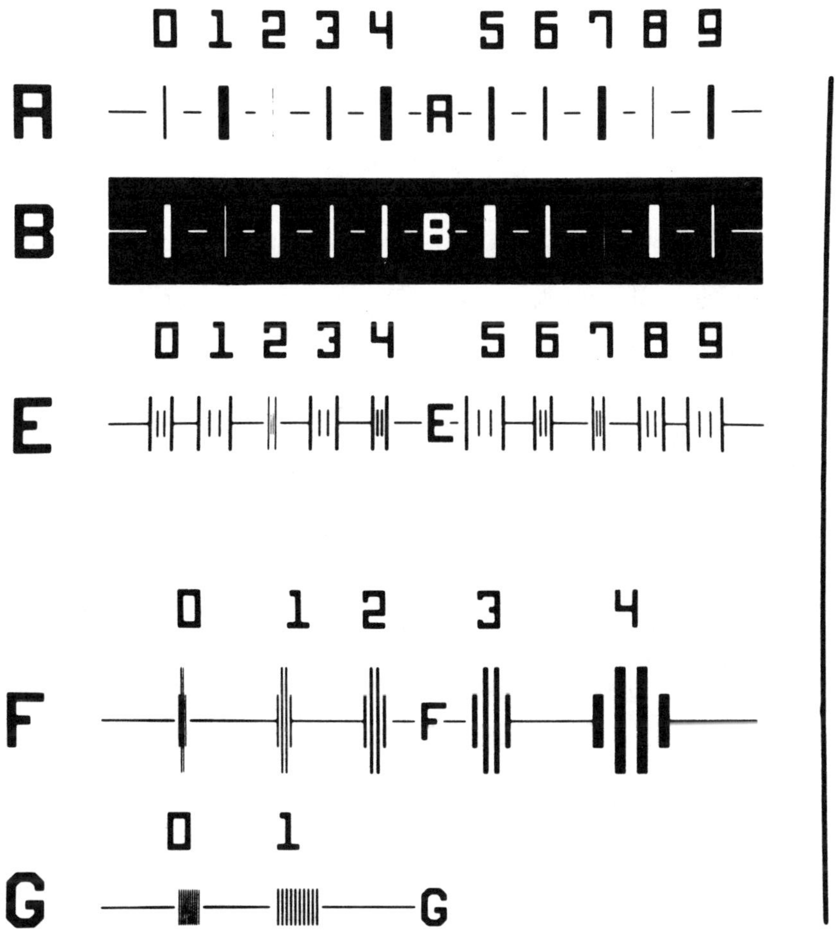

Figure 5-4. Features of the basic pattern on the Standard Reference Material, SRM 475 photorepeater linewidth standard. This SRM is specifically developed for the measurement of opaque and clear lines on electronic, integrated circuit photomasks with a transmitted light microscope. Linewidths range from 0.5 to 10 µm with an accuracy of 0.05 µm or better. Line A consists of black lines of known width on a clear background, Line B is the reverse of Line A, Line E consists of line pairs for calibrating line spacings, Line F is pairs of lines and spaces for determining line-to-space ratios on video-micrometer systems, and Line G is series of lines for checking mechanical and optical errors in a measurement system. For optimum image quality the microscope should be equipped with objective lenses that are designed for use without cover glass. Approximate scale: 1 mm = 4.4 µm. Courtesy of The National Institute of Standards and Technology (NIST).

other types of photomasks and for use with the incident light microscope.

These linewidth standards, SRM 474 and SRM 475 are available from the National Institute of Standards and Technology, Washington, D.C. 20234.

C. Angular Measurements

The measurement of angles in the plane of the stage is an important parameter in the fields of optical crystallography, petrography and metallography. These measurements are easily made if the microscope is equipped with a calibrated revolving stage. The calibrations are usually readable to $\pm 0.1^{\circ}$ with the aid of a vernier. In this procedure one limb of the angle is aligned with a cross-wire of the eyepiece and the angular value of the stage recorded. The stage is then rotated in that direction which will bring the other limb of the angle in coincidence with the same cross-wire with the least rotation required. The value of the stage calibration is once again recorded. The angular relationship of interest is then the absolute value of the difference between the two recorded readings.

A second method for the measurement of angles makes use of the protractor or goniometer eyepiece (see Chapter 7, Section VIII) in place of the regular eyepiece. This eyepiece has a full 360° scale with a vernier. Angular measurements are made in the same manner as above, except in this case the cross-wire is rotated rather than the sample.

D. Particle Number

Another directly measurable property of a microscopical sample is number; the number of red blood cells, the number of bacteria, the number of dirt particles, etc. Since these samples are usually suspensions, the value that is usually sought is the number per unit volume, therefore a container of fixed volume is used to hold the sample. A reticle in the eyepiece or rulings engraved directly on the container are used to establish the confines of the area in which the counting takes place. The procedures for the use of these systems are beyond the scope of this book; for these the reader is directed to the references.

The Whipple disk, shown in Figure 5-5, is one such measuring reticle. It consists of a square divided into 100 equal areas (10 X 10) with one of these areas further divided into 25 equal areas. The dimensions of this reticle are such that with a 10X objective lens and a 10 X eyepiece an area of about 1 mm^2 is covered by the gridwork. This reticle is generally used with a special sample cell for counting bacteria, plankton, other organisms, dust and miscellaneous particles. A typical application is given in "Standard Methods For The Examination of Water And Wastewater" (1976).

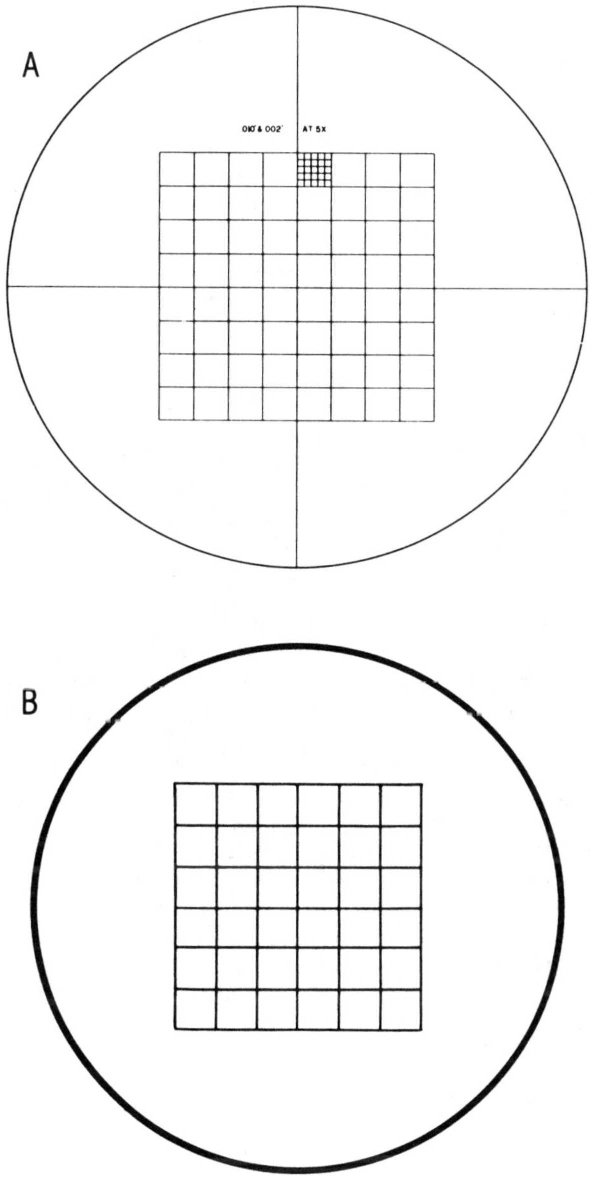

Figure 5-5. (A) is a Whipple disc reticle used for counting of various particles, including bacteria, dispersed in a liquid and (B) is a Howard disc reticle used for counting filament of mold in food products. Courtesy of Cambridge Instruments Inc., Optical Systems Division.

The Sedgewick-Rafter cell is normally used with the Whipple disk, it has a rectangular opening 50 X 20 X 1 mm that is enclosed with a special rectangular cover glass. The volume of the cell is 1 ml.

The Whipple disk is calibrated for each objective lens is exactly the same manner as the eyepiece reticle, see Table 5-1. When used with the Sedgewick-Rafter counting cell only those objective lenses 20X and lower are used. The shallow working distance of the standard higher magnification objective lenses will result in breakage of the cover glass when attempting to focus at the bottom of the cell. Long working distance objective could be used to advantage if higher magnifications are required, see Appendix 1.

The net micrometer is similar in construction to the Whipple disk and is used for counting large particles. The net micrometer is also calibrated as described in Table 5-1 for the eyepiece reticle.

The Howard disk, Figure 5-5, is used with the Howard cell in the determination of the mold content of foods, and the concentrations of suspensions and emulsions. The field of the Howard disk is divided into rectangles which are 1/6 the diameter of the field. The Howard cell has two calibration lines scribed on the bottom surface of the cell that are spaced 1.382 mm apart. When the disk is used in a 10 X Huygens focusing eyepiece and a 10 X objective lens, the diameter of the reticle is usually somewhat larger than the calibration spacing. The magnification of the calibration markings is increased with a zoom system or increasing the length of the graduated draw tube until the reticle diameter is exactly 1.382 mm. If the microscope is not equipped with either a zoom system or a graduated draw tube, the eyepiece should be slowly drawn out of the body tube while noting the image (some refocusing may be required). When the diameter of the disk matches the spacing of the calibration lines, note the distance that the eyepiece has been withdrawn. Make a cardboard collar that fits around the body of the eyepiece with that exact length and use it as a spacer when these measurements are made.

The Howard cell consists of a chamber that is 0.1 mm deep when confined with the appropriate cover glass. Thus the volume of fluid examined in each field is:

$$V = (1.382\ \text{mm}/2)^2 \times 3.14 \times 0.1\ \text{mm} = 0.15\ \text{cu mm} \qquad (23)$$

In addition to the obvious convenience of a reference frame, the squares that make up the Howard disk field are also used to determine the length of various mold filaments and thereby provide further classification of the sample. A typical procedure using the Howard disk is given in Horowitz (1975).

The hemacytometer is a counting chamber which has a grid ruled on the floor of the counting area. A number of different rulings are available, among these are the improved Neubauer grid (see Figure 5-6), the Burker grid, the Malassez grid, the Thoma grid and the Turk grid. The better quality hemacytometers have rulings which meet the specifications of the National Institute of Standards and Technology (NIST) or other standards organizations.

The primary use of the hemacytometer is in the determination of the number of white and red blood cells per cubic millimeter of whole blood or spinal fluid. The measurement procedure is described in Levinson and MacFate (1969). A method for the examination of plankton in water using a hemacytometer is described in "Standard Methods For The Examination Of Water and Wastewater" (1976). Other suspended particulates such as yeasts, various microorganisms and dust have been measured using the hemacytometer.

Another form of a calibrated counting chamber is the milk smear slide. This slide consists of a frosted glass slide with clear 1 cm^2 circular areas delineated on the top surface. In use, a 0.01 ml sample of milk is dispersed evenly over one of the circular areas and allowed to dry. After staining, the slide is examined for bacterial or somatic cells using an oil immersion objective lens and an eyepiece with a simple cross hair. In this application, the diameter of the field of view in the eyepiece is measured using the stage micrometer; see Table 5-1, for the procedure; in this case, substitute "diameter of the field of view" for "reticle unit". The measured diameter the area of the field of view is calculated from the diameter and from this the number of fields-of-view that must be counted to obtain the number of cells per milliliter of milk. This procedure is described in detail in Marth (1978).

III. AXIAL MEASUREMENTS

It is easy to forget the third dimension of the sample, especially when engrossed in a detailed study of the structure in the plane of the stage. However, every sample does have dimensionality in the direction of the microscope axis, even in metallurgical samples where it is only the slight differences in height of adjacent grains due to selective etching. It is the measurement of this axial dimension, sometimes known as "geometric thickness", that is the subject of this section.

The methods that will be discussed are the fine focus method, the optical section method, and the interferometric method. The first method can be accomplished on most incident and transmitted light microscopes but the remaining techniques require special instruments or at least special accessories. The relative useful range of the methods as well as their accuracy and precision are given in Table 5-3.

A

B

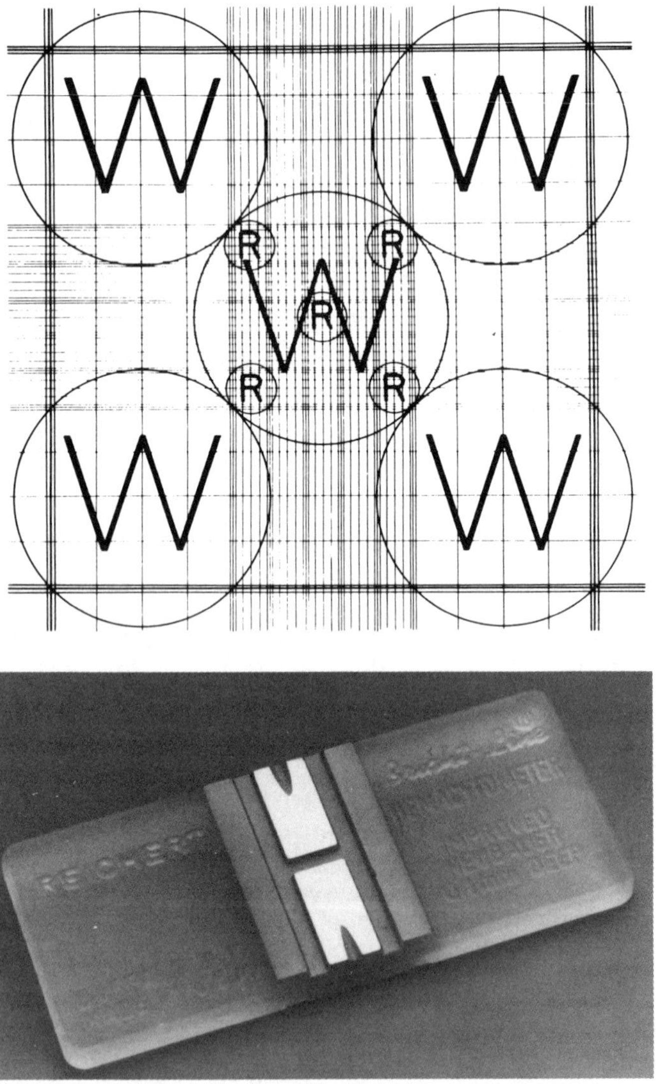

Figure 5-6. One form of haemacytometer ruling is shown in (A). The haemacytometer counting chamber, (B), is primarily used for counting blood cells but is also useful for bacteria, spores, yeasts, etc. Courtesy of Cambridge Instruments Inc., Optical Systems Division.

A. Fine Focus Axial Measurements

Most of the more advanced microscopes are equipped with a calibrated micrometer drum on the fine focus knob. Since this scale indicates distance along the microscope axis, it would seem to be the basis for a rather simple method for measuring axial dimensions on a sample, however, most microscopists do not rely on this technique. This hesitancy arises because the technique couples the quantitative reading of a micrometer scale with the subjective determinations of best focus. Nevertheless, reasonably accurate measurements can be made with care.

Before any axial measurements are made using the micrometer drum of the fine focus it is mandatory that the value for each micrometer division be known. The fine focus knob is usually calibrated in 1, 2 or possibly, 4 micron increments; this value can be obtained from the manufacturer's instruction manual for the instrument or can be determined experimentally as described in Table 5-4.

This procedure is suitable for both transmitted and incident illumination instruments with either a standard or inverted stand. England (1957) has described a very accurate calibration procedure for the fine focus, based on interferometry.

The measurement of the axial dimension of an opaque material in normal incident illumination, as shown in Figure 5-7, is described in Table 5-5. Transparent materials may also be measured in this way if no mounting medium is used and the measurements are made at the side of the material, i.e., not through the transparent material.

American Optical has produced an Optical Depth Gauge which is basically an upright stand instrument with a very large graduated drum which serves as the fine focus knob; this is used to make the axial measurements. In operation the lower surface of interest is located with the 10X objective lens then carefully focused with the 45X objective lens. At that point the vernier of the fine focus is set to the zero of the fine focus drum. The fine focus knob is then adjusted until the upper surface of interest is sharply focused. The height difference is then read directly from the graduated drum. The instrument is not equipped with a stage, thus the sample rests on the bench or in the case of a large sample, the microscope rests directly on the sample. Normal incident illumination is built into the microscope. Unitron and Zeiss also make depth measuring microscopes similar to this American Optical instrument.

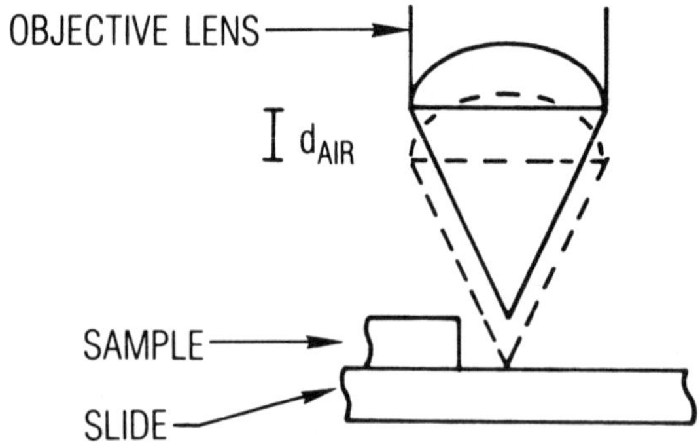

Figure 5-7. Procedure for measuring sample height (Z axis dimension) using the fine focus. Focus critically on the upper surface of the sample with a higher power objective lens and note the reading of the fine focus knob. Move the sample to one side and focus on the top surface of the slide supporting the sample. The total number of fine focus divisions times the length per division gives the sample thickness.

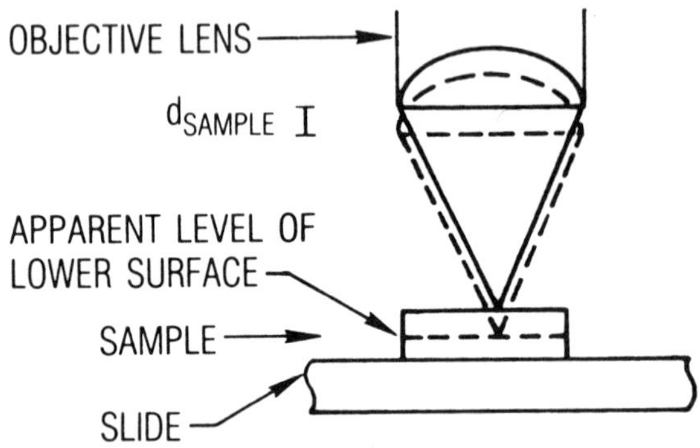

Figure 5-8. Procedure for measuring the thickness of a transparent sample with polished and parallel sides. Focus critically on the upper surface of the sample with a higher power objective lens and note the reading of the fine focus knob. Then focus on the lower side of the sample. The total number of fine focus divisions times the length per division times the refractive index of the sample gives the sample thickness. If an oil immersion objective lens is used, the latter value must be further divided by the refractive index of the immersion oil.

TABLE 5-3
AXIAL MEASUREMENT METHODS

METHODS	RANGE	ACCURACY	PRECISION
Fine Focus	1 - 200 μm	± 1 μm(1)	± 1 μm
	0.04 - 8 mils	± 0.04 mils	± 0.04 mils
Optical Section	1 μ - 20 mm(2)	± 1 μm	± 1 μm
	0.04 - 800 mils	± 0.04 mils	± 0.04 mils
Interferometry			
a. dual beam	300 - 2×10^4 Å(3)	± 300 Å	± 300 Å
	1.2 - 80 μin.	± 1.2 μin.	± 1.2 μin.
b. Multiple beam	5 - 10^5 Å(4)	± 2 Å	± 2 Å
	0.02 - 400 μin.	± 0.01 μin.	± 0.01 μin.

(1) Quoted by Zeiss for their Depth Measuring Microscope.
(2) Total range not achievable on the same instrument.
(3) Lower value is for visual measurements; measurements from photographs have a lower value of about 150 Å (0.6 min).
(4) Entire range may not be achievable on the same sample. Tolansky (1970).

The Unitron instruments, the TM Universal measuring microscope and the RMM Rollscope, however are equipped with a dial micrometer that indicates the axial measurements. The measurement procedure with these instruments is the same as described above.

For special applications Gaertner Scientific Corporation supplies a wide selection of building blocks for the construction of measuring microscopes; arrangements of these may be made to fit a wide variety of applications.

It will be recalled that the procedure described above allows the measurement of thickness of a transparent material if the measurement is made at the side of the material. In addition to this, a transmitted light microscope may also be used to measure the thickness of a transparent material directly through the material. In this case the refractive index of the material must be known.

The measurement of the axial dimension of a transparent material in transmitted light illumination, as shown in Figure 5-8, is described in Table 5-6.

B. Light Sectioning Axial Measurements

The light sectioning microscope is a specially constructed incident light instrument that permits a rapid and accurate measurement of axial dimensions. In this instrument the objective lens is mounted at an angle to the sample surface ranging from 30 to 75°. The sample is viewed at that angle or a prism is used to redirect the image from the objective lens to a vertically mounted body tube. The illumination is imaged as a slit and directed at the sample either vertically or at the same angle as the objective, this latter arrangement is shown in Figure 5-9.

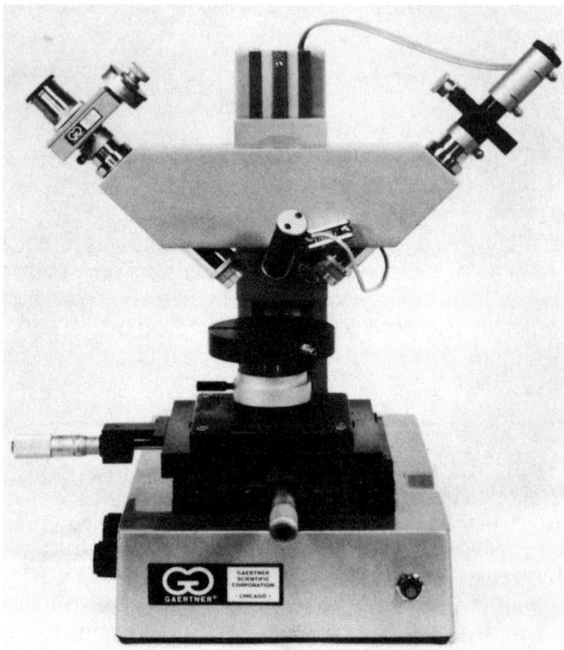

Figure 5-9. A light sectioning microscope is used to measure differences in the height of various areas on a sample surface. Light in the form of thin line is imaged on the sample by the right 45° tube. The effect of the sample surface irregularities on the illumination is measured in the microscope that makes up the left 45° tube. Courtesy of Gaertner Scientific Corp.

TABLE 5-4
CALIBRATION OF THE FINE FOCUS MICROMETER DRUM

1. Record the total number of divisions for a complete revolution of the micrometer drum.
2. Carefully measure and record the thickness of a cover glass using a machinist's micrometer (either metric or English).
3. A microscope slide is prepared by lightly marking it in a small central area with a felt tip pen; this provides for easy identification of that surface.
4. Using a very small drop of cyanoacrylate cement glue the measured cover glass to the microscope slide so that one edge of the cover glass partially covers the marked area on the microscope slide. The cover glass should lie flat against the glass slide. Also mark the upper surface of the cover glass with the felt tip pen.
5. Set up the microscope for either brightfield, transmitted or incident light illumination. Use a high magnification dry objective lens and a full aperture diaphragm setting. Set the fine focus near the center of its travel.
6. Very carefully focus on the upper surface of the cover glass using the coarse focus, then locate the edge of the cover glass that is over the marked area of the microscope slide in the center of the field of view and finally, critically focus on the marked area of the glass slide with the fine focus knob. If the objective lens hits the cover glass a lower power objective must be used. Record the value of the fine focus micrometer drum for the marked area of the glass slide.
7. Focus upward until the marking on the top of the cover glass is sharply focused and record the value of the fine focus micrometer drum. If a full revolution or more of the fine focus knob is required, that number should also be recorded. If the travel of the fine focus mechanism is exceeded, set the fine focus knob nearer the lower end of its range and repeat steps 6 and 7.
8. The focus then should be returned to the marked area of the glass slide; the micrometer scale value should agree with the initially recorded value. If good agreement is not obtained, the measurements should be repeated.
9. The axial distance corresponding to each division of the micrometer drum is equal to the thickness of the cover glass in microns (μm) or thousandths of an inch (mil) divided by the total number of micrometer drum divisions traversed from bottom to top of the cover glass (include the number of full rotations of the fine focus knob multiplied by the number of divisions per complete rotation).

TABLE 5-5
DETERMINATION OF AXIAL DIMENSION USING FINE FOCUS IN NORMAL INCIDENT ILLUMINATION

1. Set up the microscope for incident light illumination, see Chapter 3, Section IV C.
2. To minimize the depth of field and thereby optimize the sharpness of focus, use the highest magnification objective lens that also has sufficient working distance to encompass the axial dimension of interest. However, the magnification should be low enough that the field of view includes both the high and low points to be measured. To further minimize the depth of field, the aperture diaphragm should be fully open.
3. The fine focus knob is rotated until it is near the bottom of its travel.
4. The lower of the points of interest on the sample is focused very carefully with the coarse focus knob. Optimum focus on that lower point of interest is obtained with the fine focus knob and the value of the micrometer scale recorded.
5. The microscope is then refocused upward using the fine focus knob until the upper point of interest is sharply focused; the value of the micrometer fine focus at that point is then recorded. If a full revolution or more of the fine focus knob is required, that number should also be recorded (It must be emphasized that this technique cannot be used if the axial dimension exceeds the total travel of the fine focus mechanism).
6. The focus then should be returned to that of the original lower point and the micrometer scale value should agree with the initially recorded value. If good agreement is not obtained the measurements should be repeated.
7. The axial dimension of interest is then the total number of micrometer drum divisions between the lower and the upper points of interest (include the number of full rotations of the fine focus knob multiplied by the number of divisions per complete rotation) times the number of units per division, as determined in Table 5-4.

TABLE 5-6
DETERMINATION OF AXIAL DIMENSION FOR TRANSPARENT MATERIALS USING FINE FOCUS

1. Set up the microscope for brightfield transmitted light, see Chapter 2, Section IV C.
2. To minimize the depth of field and thereby optimize the sharpness of focus, use at least a 40X objective lens. To further minimize the depth of field, the aperture diaphragm should be fully open.
3. The fine focus knob is rotated until it is near the bottom of its travel.
4. The lower surface of the sample is focused very carefully with the coarse focus knob. Optimum focus on that lower surface is then obtained with the fine focus knob and the value of the micrometer scale recorded.
5. The microscope is then refocused upward until the upper most surface of the sample is sharply in focus; the value of the micrometer fine focus at that point is then recorded. If a full revolution or more of the fine focus knob is required, that number should also be recorded (It must be emphasized that this technique cannot be used if the axial dimension exceeds the total travel of the fine focus mechanism).
6. The focus then should be returned to the lower surface; the micrometer scale value should agree with the initially recorded value. If good agreement is not obtained the measurements should be repeated.
7. The total number of micrometer drum divisions between the lower and the upper surfaces of the sample is recorded (include the number of full rotations of the fine focus knob multiplied by the number of divisions per complete rotation).
8. The thickness of the sample is then obtained by multiplying the total number of micrometer drum divisions, M_t, by the number of microns per division (μm/d), as obtained in Table 5-4, and then multiplying that product by the refractive index, η_m of the sample, as expressed below:

$$\text{Thickness} = M_t \times (\mu m/d) \times n_m \qquad (24)$$

If an oil immersion lens is used for this measurement the thickness is given by:

$$\text{Thickness} = M_t \times (\mu m/d) \times n_m/m_i \qquad (25)$$

where η_i is the refractive index of the immersion oil. A permanently mounted thin section may also be measured, in this case the effect of the mounting medium may be neglected and procedure in Table 5-5 used.

If the sample is perfectly planar, the slit produces a straight bright line on the surface of the sample when viewed through the eyepiece. If the sample has surface structure in the path of the line, the line deviates toward the light source for elevations and away from the light source for depressions. The amount of the deviation from the straight line in the image of the sample is a measure of the axial dimension of the structure that produced it. Thus, a reticle or micrometer eyepiece can be used to measure the axial dimension. This measurement is generally easier to perform and has less potential subjectivity than the fine focus measurements described above. On the negative side, the light sectioning must be performed on a special instrument.

As the angle between the sample surface and the objective lens decreases there is an effective magnification of the effect of a given structure on the light from the slit. Table 5-7 lists the magnification ratio for various angles of light section.

TABLE 5-7
MAGNIFICATION RATIOS FOR VARIOUS ANGLES OF LIGHT SECTION

ANGLE BETWEEN SAMPLE AND OBJECTIVE LENS AXIS	MAGNIFICATION RATIO
30°	1.73
45°	1.41
60°	1.00
75°	0.52

Light sectioning microscopes are available from Gaertner Scientific, Leitz, Nissho (Tokyo) and Zeiss. The Nissho is supplied in the USA by Vickers instruments, Inc. Nikon had supplied a surface finish microscope that had as accessories two special objective lenses with permanently mounted prisms that project the slit image onto the sample at a 45° angle. Due to the diversity of instruments available, a general procedure is not given.

C. Two Beam Interferometry

In Chapter 4, Section VI the principles of the interference microscope were discussed with emphasis on the contrasting technique. To produce brightness or color contrast only the low orders of interference are needed. Another equally important application of two beam interferometry is in the measurement of sample dimensions along the axis of the microscope. The transmitted light techniques are important in the biomedical field while the incident light techniques find much application in the metallurgical and microelectronics fields.

The transmitted light, two beam interferometer permits the measurement of either the thickness (T) or the refractive index (η_s) of a sample, as given below:

$$T = \frac{OPD}{(\eta_s - \eta_m)} \tag{26}$$

$$\eta_s = \frac{OPD}{T} + \eta_m \tag{27}$$

where OPD is the observed path difference and η_m is the refractive index of the surrounding medium. The thickness of the entire section is an important parameter in quantitative cytochemical calculations. For further information in this area the reader is referred to Barer and Dick (1957) and Hale (1958).

The wide variety and the uniqueness of the various transmitted light, two beam interferometers precludes the inclusion of a general procedure for setup of these instruments; the reader is referred to the manuals for the specific instruments.

In the incident light, two beam interferometer the reference surface is adjusted so that there exists a virtual wedge between between the sample and reference wavefronts. Each point on the sample surface is then compared with other point on the reference surface. As many orders of interference (or fringes) are used as necessary to define the sample surface irregularities; the number of fringes is largely limited by our ability to count them. The practical number of fringes range from 10 to 50 depending upon the contrast of the image. As might be expected, a photomicrograph will improve our ability to make these measurements.

The surface of the sample must be reflective and for maximum contrast the reference surface should have the same reflectivity. The interference image can be compared to a

Figure 5-10. Two beam interferometer image of a beryllium metal surface. Wavelength of the illuminating light is 546 μM. 200X.

topographic map on which a given fringe passes through areas of similar elevation, as shown in Figure 5-10. The fringes are produced as the sample elevation changes by lambda/2, where lambda is the wavelength of the illuminating light, as shown below:

$$D = n \times \lambda/2 \tag{28}$$

where n is the number of fringe spacings involved and λ is the wavelength of the light. Thus, if a sodium vapor arc lamp with a wavelength of 0.589 μ is used, the difference in elevation from one fringe to the next is 0.295 μ. It is important to note at this point, that some error is introduced in this value if a high numerical aperture objective lens is used; this may be as great as 10%.

The interference measurements are usually made first with white light and then with monochromatic light. Due to the wide range of wavelengths represented, the white light produces broad fringes however its use does produce strongly colored low order fringes that can be easily identified. This latter aspect is very important when a large step is encountered on the sample. Once the fringes are identified then the illumination can be changed to monochromatic light for more accurate measurements of the dimensions.

The two general types of instruments used for two beam interferometry are the self referencing systems and the independent referencing systems, as described in Chapter 4, Section VI C. It is important to note that the independent referencing system, although not easy to adjust for DIC, can be used for both DIC and interference fringe measurements. The same is not true for the self referencing systems; different Wollaston prisms must be used for DIC and for interference fringe measurements. The characteristics of the two systems are given Table 5-8.

The Reichert polarizing, or Nomarski, interferometer, shown schematically in Figure 5-11a is the one example of the self referencing instrument. The virtual wedge between the sample and the reference surface, which in this case is also the sample, is designed into the Wollaston prism. Since the sample is also the reference, critical adjustment of the sample surface relative to the axis of the microscope is not necessary.

In the Nomarski interferometer image the fringes are in a fixed direction, northwest to southeast. In order to make measurements on a step or other structure, it must be aligned at right angles to the fringe direction or in the northeast to southwest direction. This is most easily accomplished with a revolving stage.

As mentioned in Chapter 4, Section VI, the self referencing system is free from "flutter"; that is, the interference detail in the image does not shift due to flexing of the microscope frame since the sample and reference beams have essentially identical

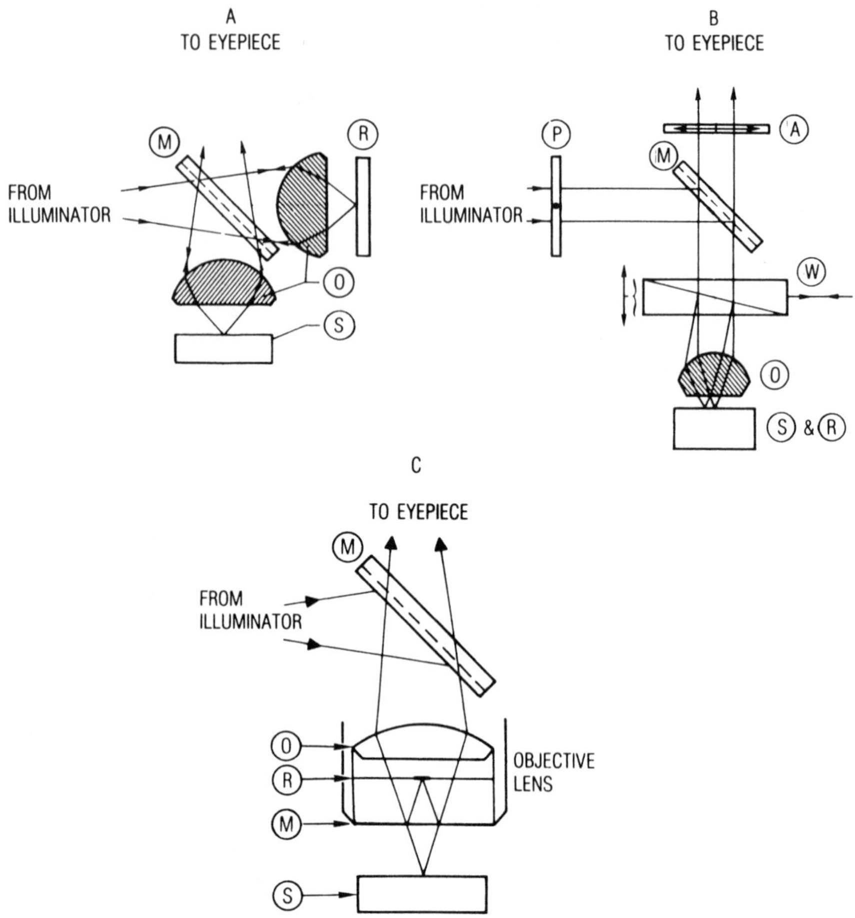

Figure 5-11. Schematic diagrams of three two-beam interferometers: (A) Michelson / Linnik, (B) Nomarski and (C) Mirau.

A. Analyzer
M. Partially reflecting mirror
O. Objective lens
P. Polarizer
R. Reference surface
S. Sample
W. Wollaston prism

TABLE 5-8
CHARACTERISTICS OF THE TWO TYPES OF TWO BEAM INSTRUMENTS FOR INTERFERENCE FRINGE MEASUREMENTS

SELF REFERENCING TYPE	INDEPENDENT REFERENCING TYPE
Requires a polarizing microscope.	Interference fringe accessories may be set up on any microscope.
Instrument very easy to set up up and use.	Set-up and use requires care and some experience.
Sample requires only the usual leveling necessary for incident light microscopy	Leveling critical to obtain the desired number and direction of fringes; use leveling stage.
Image is very stable - has no "flutter" due to vibration.	Requires very rigid instrument and vibration causes "flutter".
Limited range of fringe spacings.	Fringe spacings infinitely variable.
Structural elements must have a certain minimum free distance to provide reference beam area.	Separate reference surface does not limit the structural density of the sample.
Sample must be greater than a given size for each objective lens to produce fringes.	No lower limit to sample size to produce fringes.
Interpretation can be difficult for complex samples.	Interpretation is comparatively straightforward.

paths. This feature is ideal for photomicrography. However this feature is not without its disadvantage since one cannot distinguish an elevation from a depression on the surface of the sample by lightly depressing the microscope stage as described below.

Since the sample surface is also the reference surface there will be two images of each structure on the sample, i.e. every detail of the sample will have an identical image located along a SW-NE direction. In other words, each image serves as the reference for the other image. A direct consequence of this is that the displacement of a fringe in one image is opposite to the displacement in the other image, as shown in Figure 5-12. The

Figure 5-12. A typical Nomarski fringe image. The displacement of the fringes are in opposite directions in the two images of each object.

symmetrical nature of these Nomarski fringes makes quantitative measurements difficult even for very simple structures.

The spacing between the Nomarski double images is inversely proportional to the magnification of the objective lens used. If the sample viewed is very small it may happen that the visible separation is larger than the sample. In that case there is no overlap and no fringes are produced. The minimum sample size, S_S, to achieve overlap is given by the following inequality:

$$S_S > 1.35\ \mathrm{mm}/M_{Obj} \tag{29}$$

where M_{Obj} is the magnification of the objective lens used. If such a small sample must be examined, a replica can be made on a large planar surface then that replica can be examined using the interferometric technique.

The self referencing system is best suited for samples on which the surface structure of interest is not too dense, since it is necessary to have some nonstructured area to serve as a reference surface.

A number of the independent referencing, two beam interferometers have been produced and some of these are still available. The most popular are the Michelson, the Mirau, and the Michelson/Linnik arrangements; these are shown in Figure 5-11b, c & d. Leitz and Zeiss offer specific microscope stands specifically designed for their Michelson/Linnik systems. The Leitz instrument is a very open design, upright stand that will accept relatively large samples. It has provisions for matching the sample with four different reference mirrors that have 5%, 25%, 50% and 90% reflectivities, respectively. This system is also available as an attachment for their upright stand metallographic microscopes. The Zeiss instrument is configured as an inverted stand that will also accept relatively large samples that are only limited by their weight and access to the instrument. Wild produced an attachment for their M12 and M20 microscopes that was also equipped with four different reference mirrors with reflectivities of 4%, 30%, 60%, and 90%.

Nikon produces an accessory set for their metallurgical microscopes consisting of two Mirau interferometer objective lenses, a leveling stage and a micrometer eyepiece for measuring the fringe displacements. Leitz also produces a Mirau attachment that may be used with their multiple beam interference attachment and their 10X, H20X, and H32X objective lenses.

Watson offers 10X and 20X Michelson interferometer objective lenses that may be attached to any relatively rigid microscope stand. Each Watson objective lens has a built-in condenser system that must be externally illuminated from the side; a fiber optic illuminator is ideal for the purpose.

When an independent reference interferometer is used with an upright stand microscope it is imperative to have a leveling stage or its equivalent. The author has used modeling clay as a substitute for a leveling stage for this application, this will be described in the set-up procedure.

Typical applications of two beam interferometry are measurements of step and layer thicknesses on semiconductor devices, individual and total thicknesses of multilayer optical coatings, as well as the examination of polished and etched metal, glass, plastic, and ceramics samples.

D. Procedures for The Incident Light, Two Beam Interferometric Microscopes

Since three distinct types of two beam interferometers, the Nomarski, the Michelson and the Mirau are represented in this section, set-up procedures will be given for each.

SET-UP OF THE REICHERT NOMARSKI INTERFEROMETRIC MICROSCOPE

1. Set up the incident light microscope for crossed polarized light operation, see Chapter 4 Section VII. The aperture diaphragm should be set between 0.3 and 0.7 of the full disk diameter, see Chapter 3, Section IV C.
2. For the MeF, MeF2 and Zetopan models remove the objective lens and install the interferometer between the condenser assembly and the objective lens. For the MeF3 the interferometer is installed in a slot at the right and beneath the stage. For the POLYVAR-MET the interferometer is installed in the front of the instrument just above the revolving nosepiece. For the latter two instruments, the interferometer may be removed from the light path by sliding it away from the center of the instrument.
3. Set the rotatable analyzer at 45°, this will yield a brightfield type image. Focus the microscope on the sample. NOTE: For the later interferometer accessories that have built-in polarizing filters, the interferometer must be slid out of the optical path and the brightfield module slid into place.
4. Focus the sample critically, then return the microscope to the interferometer condition by reversing the procedure of step 3. NOTE: Since the sample is also the reference surface there will be a double image of the sample, i.e., every detail of the sample will have an identical image located along a SW-NE direction. The spacing between the double images is inversely proportional to the magnification of the objective lens used.

5. Since the fringe direction is fixed, the sample must be rotated and/or translated until the strongly colored fringes cross over the structure of interest at right angles. Document the location of the zero order black fringe as it traverses the sample; this gives the approximate number of fringes for the displacement of interest. NOTE: The interferometer for the MeF, MeF2 and the Zetopan has two adjustment levers; (1) translates the fringe pattern across the image and (2) changes the fringe spacing over a limited range. Lever 1 can also be used for this adjustment.
6. Change to monochromatic light either by using an absorption filter and a line spectrum lamp or by using an interference filter with the white light source. Do not make any changes to the interferometer! Identify the zero order fringe and measure the exact fringe displacement in the area of interest. The difference in elevation, D, is given by equation (28). If the entire surface exhibits deviations from flatness then the equation is at best an approximation. Note: It is not possible to determine whether the difference in elevation is due to an elevation or a depression by this measurement.

For problems encountered in Nomarski two beam interferometry refer to Table 4-9.

There are several versions of the independent reference, two beam interferometers either as a complete instrument or as an attachment. The Michelson instruments are very similar in operation, therefore the following instructions are specific to the Watson interferometer attachment but are also suitable for the Leitz reflected light interference microscope/attachment. The several Mirau instruments vary widely in arrangement; therefore, only a very generalized procedure is included in the following set-up with appropriate steps indicated by an asterisk.

SET-UP OF AN INDEPENDENT REFERENCING TWO BEAM INTERFERENCE MICROSCOPE

1. (For attachments only) affix the interference system, Figure 5-13, to an upright stand microscope. For the Watson attachment, secure the system by tightening the knurled ring, A, until it is tight against the tube of the microscope.*
2. Place a sample on the stage. Ideally a leveling stage should be used. Otherwise, mount the sample on a glass slide using modeling clay and level it with a metallurgical sample leveling press, See Chapter 3, Section II C.*
3. Move the objective toward the sample until the spacing between them is about 2 mm. Turn on illumination and focus the sample.*

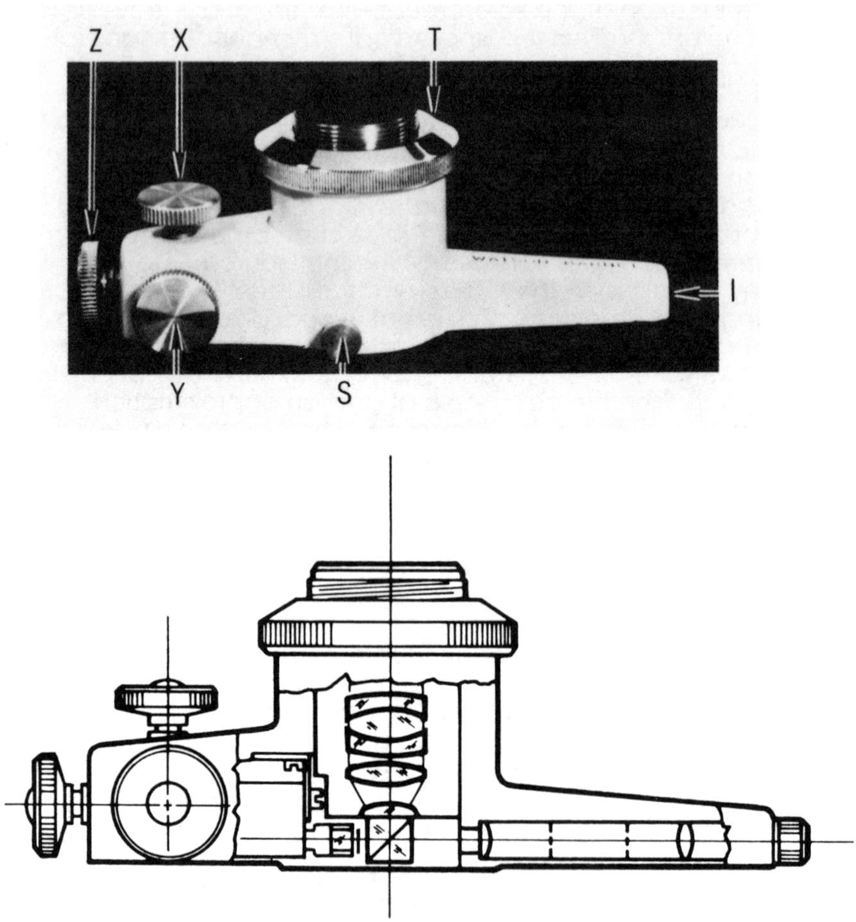

Figure 5-13. The Watson two beam interferometer system. The system is attached to the microscope body in the place of an objective lens and locked into position with the knurled knob T. Illumination is admitted to the system through the opening I. The reference beam may be shut off by inserting knob S. The reference surface is tilted with knobs X and Y and focused with knob Z. The internal arrangement is shown in the cutaway view. Courtesy of Hacker Instruments, Inc.

4. Remove the filter, D, and illuminate the system with an incandescent source through the filter opening.
5. Remove the shutter, F, from the optical path to illuminate the reference surface.
6. Focus the reference surface carefully with the path difference screw, K; this surface is in focus when any dust particles on the reference may be seen sharply focused.
7. Now focus the sample critically. If the black, zero order fringe surrounded by brightly colored fringes can be seen in the field of view when the sample is sharply focused, proceed to step 9. If no fringes are seen, continue with step 8.*
8. Examine the rear focal plane of the objective lens. This may be accomplished by: (a) swinging the setting lens, if available, into position over the eyepiece, (b) using the Bertrand lens, if available, or (c) removing the eyepiece and looking down the tube of the microscope. If the interferometer is out of adjustment two bright circles of light will be seen.*
9. Make the two circles of light coincident by adjusting the leveling stage or if the sample is mounted in modeling clay by manually leveling the sample in the clay. Reverse the procedure in step 8 and refocus the sample if necessary.*
10. Fringes should now be seen, if not repeat Step 8. Adjust path difference screw until the black, zero order fringe surrounded by brightly colored fringes can be seen in the field of view.
11. Spread the fringes to the desired spacing and rotate them so that they are at right angles to the structure to be measured by adjusting the tilting screws, G and H. Concurrently retain the zero and low order fringes in the field of view by careful adjustment of the path difference screw.
12. Document the location of the zero order black fringe as it traverses the sample; this gives the approximate number of fringes for the displacement of interest.*
13. Change to monochromatic light either by using an absorption filter and a line spectrum lamp or by using an interference filter with the white light source. Do not make any changes to the interferometer! Identify the zero order fringe and measure the exact fringe displacement in the area of interest. The difference in elevation, D, is given by equation (28).*
14. To distinguish between an elevated structure and a depressed structure, press lightly on the microscope stage or focus away from the sample using the fine focus knob; under these conditions the fringes will appear to move toward high areas and away from low areas.*

E. Multiple Beam Interferometry

Multiple beam interferometry is the fourth and most sensitive method for measuring sample dimensions in the axial direction. Tolansky (1970) has observed as small as 5 A steps, this corresponds to less than a four atom thickness of silver or gold. The major reason that this technique is more sensitive than the two beam technique is the sharply defined, narrow fringes produced by the multiple beam technique. The conditions to achieve these sharply defined fringes are: (1) highly monochromatic light must be used; (2) the sample must be normal incidence illuminated with a collimated beam of illumination (consisting of parallel rays); (3) The sample and the reference flat should have high and matching reflectivities; (4) contact is necessary between the reference flat and the sample surface; and (5) the sample must be of adequate size (>4μ).

The fringes have no identity in multiple beam interferometry since monochromatic light is used. Therefore, if a fringe cannot be followed across a step, the magnitude of the step should be determined using a two beam interferometer then once that has been established, a more precise measurement may be made using the multiple beam interferometer.

Other limitations of the multiple beam interferometer are: (1) direct contact with the sample is required, therefore the sample and/or the reference surfaces may be scratched; (2) only isolated structures may be observed on the sample since the adjacent sample surface interacts with the reference beam; and (3) Flatness of the reference mirror is typically good only to about 20 A, therefore fringe displacements in this range cannot be attributed, without great care, to the sample.

Accessories are available from both Leitz and Nikon for achieving multiple beam interferometry with their incident light microscopes. These accessories include the objective lenses, the interferometer system which mounts directly on the objective lens and several semi-transparent reference mirrors with various reflectivities designed to match a wide variety of sample reflectivities. However for optimum results with samples of lower reflectivity, the sample should be carefully coated with a thin, vacuum deposited layer of silver and observed using the highest reflectivity reference mirror. Tolansky (1970) has demonstrated that the evaporated coating increases the sample's reflectivity without masking its surface structure.

Multiple beam interferometry is routinely used to measure the thickness of highly reflective thin films and coatings; it is particularly valuable if the coatings are less than a fringe in thickness.

Multiple beam interferometry can be accomplished using an upright stand, normal incidence light microscope, however

Tolansky (1970) has outlined the changes to the instrument that must be accomplished for successful operation.

The procedure outlined below is based on the commercial instrument manufactured by Leitz. This particular microscope has an adjustable assembly to hold and orient the reference mirrors that affixes directly to the objective lenses, this arrangement is shown in Figure 5-14. The objectives suitable for these studies are listed in Table 5-9.

TABLE 5-9
RECOMMENDED OBJECTIVE LENSES FOR USE WITH THE LEITZ MULTIBEAM INTERFERENCE ATTACHMENT

Pl	3.2x/0.06
	5x/0.09
	10x/0.18
H	20x/0.40
H	32x/0.60

SET-UP OF A MULTIPLE BEAM INTERFEROMETRIC MICROSCOPE

1. Replace the existing illuminator lamp with a sodium lamp, and allow lamp to operate for about 2 minutes to achieve full brightness. A weak, stationary system of fringes may be noted without a sample present, this should be ignored.
2. Lower the stage of the microscope and attach the basic unit of the interferometer to the desired objective lens; it should rest firmly against the stop. The unit is held in place with the fixing screw.
3. Using the two adjustment screws, align the lower part of the basic unit so that it is approximately parallel to the knurled ring immediately below the fixing screw.
4. Select a reference mirror with a reflectivity that closely matches that of the sample. Handle the reference mirrors only with the suction device provided. Install the mirror in the multibeam head by first unlocking the lever on the face of the mechanism, see Figure 5-13, inserting the mirror, then relocking the lever. NOTE: If the mirror is dirty it may be cleaned with a 1:1 ethanol diethyl ether mixture and a clean soft rag (caution: this is a very flammable mixture; perform this cleaning in a fume hood)
5. Securely affix the multibeam head to the basic unit.
6. Rotate the knurled ring to bring the reference mirror to its uppermost position.
7. Level the sample as described in Chapter 3, Section II A and place it on the stage of the microscope.

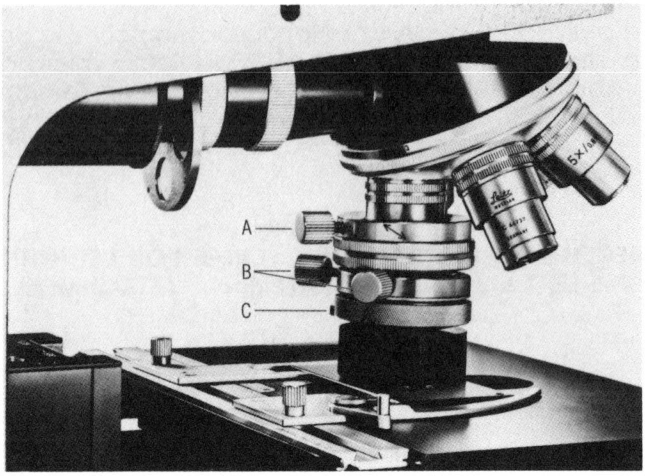

Figure 5-14. The Leitz multiple beam interferometer is clamped to any one of several special objective lenses with the upper knurled knob, A. The double knurled ring just below the clamping knob adjusts the height of the reference surface above the sample. The two lower knobs, B, provides for leveling of the reference surface within C. Courtesy of Wild Leitz USA, Inc.

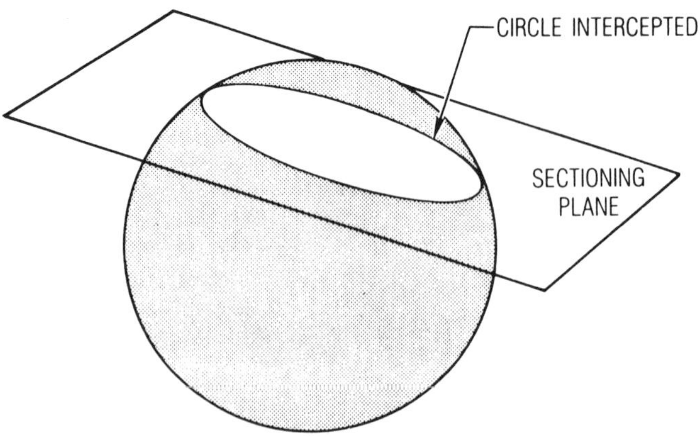

Figure 5-15. Cross sectioning of a grain in a sample; in this example, simplified to a sphere. The sectioning process produces a circular surface ranging from the full diameter of the sphere to a single point depending on the position of the sectioning plane.

8. Carefully focus on the sample. Recognition of the sample surface is difficult if a high reflectance mirror is used; in this case the approximate location of the sample surface may be determined by closing down the field diaphragm of the illuminator and focusing on the edge of the diaphragm. After the sample surface is located, return the field diaphragm to its correct position, just outside the field of view.
9. Close the aperture diaphragm of the normal incidence illuminator, then slowly move the reference mirror to within about 0.3 mm of the sample surface, by means of the knurled ring of the basic unit. The latter is accomplished by observing the process from the side so that the gap between the sample and the mirror may be clearly seen.
10. Continue to move the reference mirror closer to the sample, but at this point, observe the sample through the microscope. Stop as soon as interference fringes become visible in the sample image.
11. Using the two adjustment screws, align the interference fringes so that they are at right angles to the structure to be measured; the width of the fringes may also be adjusted in the same way. Note: the reference mirror should not yet be in contact with the sample!
12. Carefully observe the field of view in the microscope image while gently moving the mirror toward the sample. Contact with the sample will be indicated by a sudden change in the direction and/or the spacing of the fringes. When this occurs, turn the knurled ring back slightly until the fringes just begin to run uniformly through the field of view again. The sample may now be studied with multiple beam interferometry. The differences in elevation may be calculated using equation (28).

IV. DERIVED LINEAR MEASUREMENTS

The microscope has long been used to measure simple properties of samples such as isolated particle size and various surface dimensions. The real triumph in mensuration, however, resulted from an understanding of the structure of materials and the application of relatively simple mathematical expressions to obtain important but less visually obvious information about the sample. Some of the properties determined are grain size, volume fraction of various phases and particle spacings. It is important to realize that these measurements are all designed to give information about solid materials based on an examination of a randomly oriented polished surface.

A. Grain Size

Let us visualize a material of uniform grain size as it would appear in three dimensions; we will find it not unlike a mass of soap bubbles. Thus, if we examine the surface produced by making a random slice through a material of uniform grain size, we would not be surprised to find that the diameter of a specific grain in this plane would vary, depending on where we cut through the grain, this is illustrated in Figure 5-15. In fact, the grain diameter measured on a random, planar cut through the material would range from a point to the full diameter.

To convert the average planar grain diameters to the actual grain diameters for a material that has equant and uniform sized grains, a conversion factor must be applied. A conversion factor of 1.126 is given by the American Society for Testing and Materials, ASTM, (1985); 1.27 is given by Fryxell and Chandler (1964) and 1.5 is given by Coble (1961). Recognizing the range of conversion factors in use, Lennon (1966) developed a nomograph, shown in Figure 5-16, relating the planar and the actual grain diameters as a function of the correction factor. To obtain the actual grain diameter, a straight edge is placed so that it connects the average linear intercept diameter on the left scale with the assumed conversion factor on the right scale; the average grain size is then given at the intersection of the straight edge with the center scale.

Thus, the diameters of the grain surfaces in the random, planar cut can be related to the grain size of the material and this process of preparing the cut, whether it be a thin section, microtomed section or a ground, polished and etched surface, can be useful in characterizing the material. It should be emphasized that thin sections may be difficult to analyze unless the grain size is significantly larger than the thickness of the section, i.e., 0.001 inch; this is illustrated in Figure 5-3. For this reason, most quantitative analyses are performed on polished and etched samples.

A comparative method for grain size determination was developed by the (ASTM). This method is useful for uniform grains size materials and is based on the comparison of the grain structure of the sample at 100X magnification with a series of graded hexagonal patterns. A typical ASTM pattern is shown in Figure 5-17.

Each ASTM grain size pattern is given a number that is related to the size of the hexagonal pattern. This number, N, is often used alone to indicate the grain size of the sample. N may also be used in the following expressions to give actual grain dimensions:

$$n(100X) = 2^{N-1} \text{ per sq.in.} \qquad (30)$$

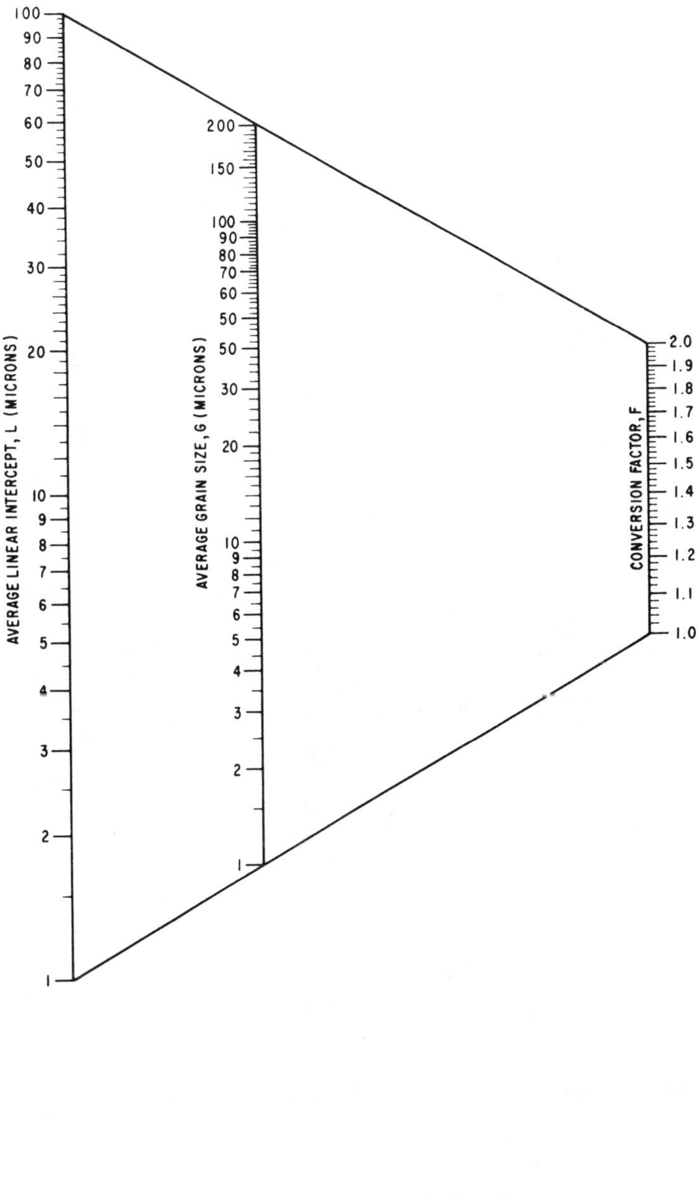

Figure 5-16. A nomograph for determining the average grain size from the average linear intercept distance for a range of conversionfactors. Line up the value of the average grain size on the left side of the chart with the desired conversion factor on the right and the intersection of that line with the center scale gives the average grain size. Courtesy of the American Ceramic Society.

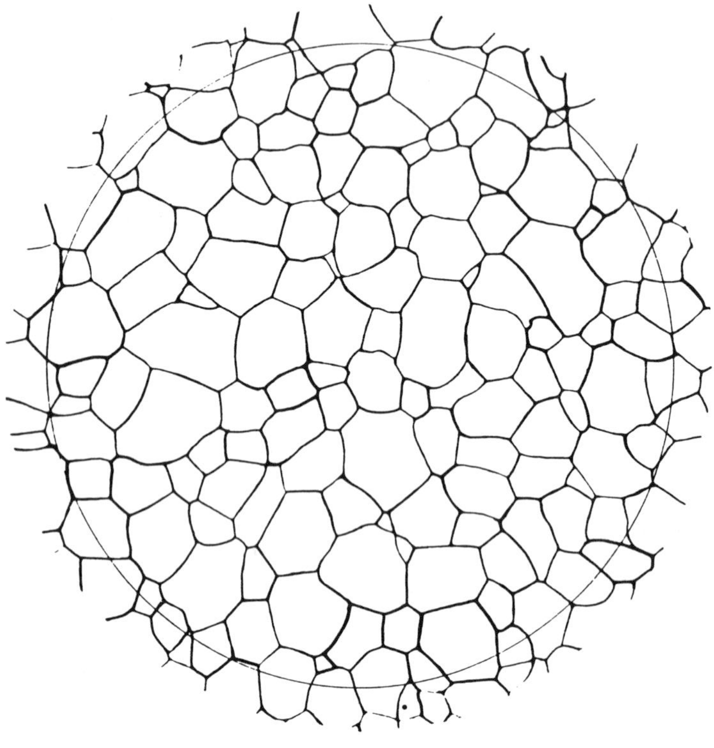

Figure 5-17. Typical drawing from the ASTM Grain Size Comparison Chart for lightly etched, untwinned grains. Copyright ASTM. Reprinted with permission.

or

$$a(100x) = 2^{1-N} \text{ sq.in} \quad (31)$$

where n = the number of planar grains per square inch, a = the mean planar grain area in square inches and N = the number of the ASTM grain size chart that corresponds to the observed structure at 100x. In practice, a photomicrograph of the structure is prepared of the sample structure at exactly 100x and visually compared with the ASTM chart. Some manufacturers also supply an eye-piece with reticles that exhibit the ASTM chart patterns, see Chapter 7, Section VII C. When these eyepieces are used with an objective lens of suitable magnification, the ASTM number corresponding to the sample grain size may be determined directly.

A second approach to the determination of grain size is the lineal intercept method. This method is free from subjective bias, amenable to statistical evaluation and is relatively easy to perform. Four manual procedures will be described for the lineal intercept method; these are described in Table 5-10.

TABLE 5-10
TECHNIQUES FOR LINEAL INTERCEPT DETERMINATION OF GRAIN SIZE

TECHNIQUE	APPLICATION	ACCESSORIES REQUIRED
Sample translation	Measure large grains.	Graduated mechanical stage and a cross-wire eyepiece.
Image traversing	Fine grains.	Screw micrometer eyepiece, mechanical stage optional.
Photomicrographic	All	Photomicrographic capability.
Hilliard	All	Photomicrographic capibility.

As described in the above table, the sample translation techique uses a microscope equipped with a cross-wire reticle and a graduated mechanical stage. A magnification is chosen that is just high enough to comfortably resolve the structure of the sample. However, with excessive magnification, nonuniform movements of the image resulting from the operation of the mechanical stage may interfere with the measurements. The sample is placed on the mechanical stage in a random orientation, carefully focused and then translated with the mechanical stage. The number of grain boundaries (P) that intercept the cross-wire of the eyepiece and the corresponding length of the traverse (L) are then recorded. Since the number of grain boundaries intersected and the number of corresponding grains (N) are approximately equal for a large sampling, the number of grains per unit length (N_L) are given by:

$$N_L = \frac{N}{L} \approx \frac{P}{L} \tag{32}$$

The average grain diameter (d_a) is given by:

$$d_a = \frac{L}{N} \approx \frac{L}{P} \tag{33}$$

To randomize the analysis, the sample should be removed from the mechanical stage after a statistically valid number of readings have been made and replaced at another random orientation on the mechanical stage for further lineal analysis.

The second technique for lineal analysis, the image traversing technique uses the moveable cross-wire of a screw micrometer eyepiece to traverse the fixed sample image. This technique is to be preferred over the sample translation technique for fine grained materials since the stage micrometer is difficult to move with sufficient precision to delineate the structure. A magnification is chosen for this technique that is high enough to resolve the boundaries but low enough to provide a reasonable number of grains to be measured in each field. It is very convenient to use a mechanical stage to translate the sample if several fields of view are to be examined. Likewise, a revolving stage will permit the easy rotation of the sample to provide random counting directions.

To determine the grain size with the image traversing technique, both the screw micrometer eyepiece and the microscope must be critically focused as described in Chapters 2 or 3, Section IV C. Furthermore, the screw micrometer must be calibrated for the specific microscope arrangement used, as described in Chapter 7, Section VIII. The initial and final micrometer readings and the number of grain boundaries traversed by the moveable cross-wire are tabulated for each field. The actual distance traversed on the sample for each field is given by the total number of micrometer divisions times the calibration value determined above. N_L and d_a are calculated using equations

(32) and (33). Since the number of grains in a single field may be relatively few, it may be necessary to sum the results of several fields to satisfy statistical requirements.

The photomicrographic technique for lineal analysis uses one or more photomicrographs of the sample. These photomicrographs must have sufficient resolution, magnification and flatness of field that the grain boundaries are clearly delineated. Likewise, the total magnification of the photomicrographs must accurately determined, as described in Chapter 8, Section V B. Lines of random direction are drawn or otherwise superimposed on the photomicrographs. The length of each line and the number of grain boundaries intersected is tabulated for each line. The actual length of the lines on the sample is obtained by dividing their length on the photomicrograph by the magnification that was obtained above. The actual lengths and the number of grain boundaries may each be summed, if necessary, to achieve statistical significance. N_L and d_a are calculated using equations (32) and (33).

A variation of the photomicrographic technique was developed by Hilliard (1964) to improve the precision of the measurement. He employs a test circle drawn on a transparent plastic sheet that overlays the photomicrograph, or is inscribed on the ground glass screen of the microscope camera. The circle may also be incorporated in an eyepiece reticle for direct observation. To correctly apply the test circle, its size must be related to the magnification such that the average number of intersections for each application of the circle is greater than six. In operation, the circle is applied randomly to the photomicrograph or other image and the number of grain boundaries intersecting the circle recorded. If the perimeter of the circle is so close to a grain boundary that the intersection cannot be resolved, one count is recorded; if the perimeter passes through a triple junction of grains, one and one-half counts are recorded. If the circle is applied more than once to an image, it should not overlap a previous circle.

The tabulated grain boundary intersections are converted to an equivalent ASTM grain size number using the nomograph in Figure 5-18. If the readings are made on a photomicrograph, a projection microscope or on the ground glass screen of the camera, the tabulated is used to compute the number of intersections per 10 cm of circle perimeter. This value is located on the right hand scale and the magnification of the image used is located on the middle scale. A straight edge joining these two values also intersects the left scale which gives both the equivalent ASTM number (G) and the average linear intercept distance L (bar) in microns. For example, if a total of 180 intersections were counted in 5 applications of a 20 cm perimeter circle to a 100x photomicrograph, the number of intersections per 10 cm of perimeter is 18. A straight line joining 18 on the right scale and 100 on the

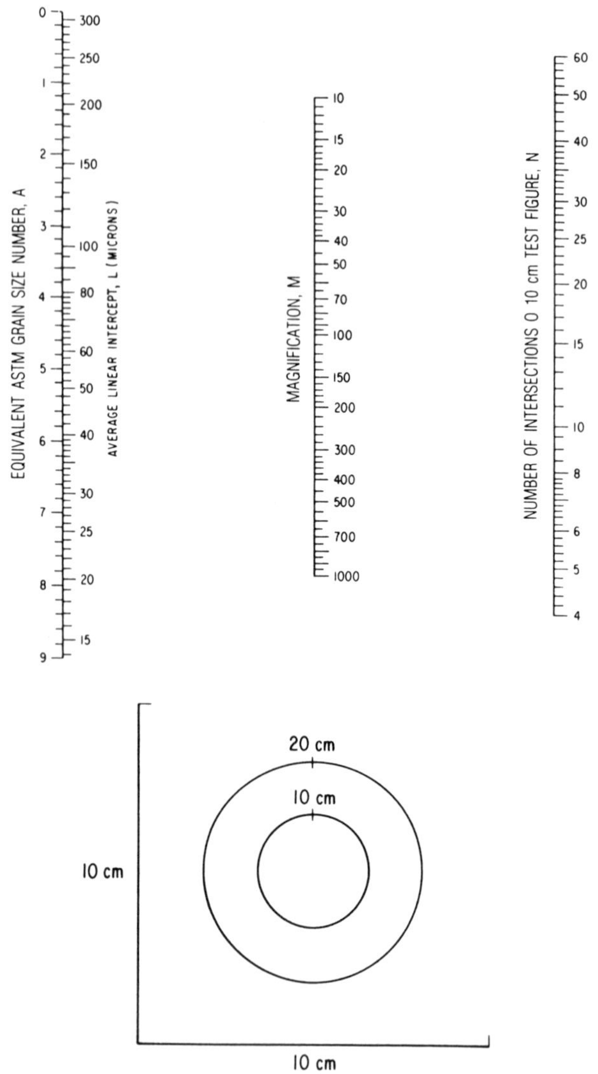

Figure 5-18. Nomograph and test circles for determination of grain size using the Hilliard method. Above is the nomograph for converting the measurements made with the circles, below, to either ASTM grain size or average linear intercept size. The circles, (10 and 20 cm in circumference) are to be drawn on transparent plastic sheet for use with photomicrographs or directly on the viewing or focusing screen of the microscope for direct measurement of the image. Reprinted from Metals Progress, May 1964, courtesy of ASM International.

center scale intersects the left scale at an ASTM number of 5 and an average linear intercept distance of 55 μ.

If a reticle eyepiece with an inscribed circle is used the apparent circumference, C, in cm is determined for each objective lens using a stage micrometer. An equivalent magnification is calculated for each objective lens (and zoom setting, if used), from the expression:

$$M(\text{equivalent}) = 10/C. \tag{34}$$

The number, M(equivalent) is then used as the value for the center scale of the nomograph. Measurements in which values of M or N are not represented on their respective scales may be used by doubling or halving one of the two quantities. For example, if N or M is halved, then G is increased by two units and L (bar) is halved. For a material with reasonably equant and uniform sized grains a standard deviation of 0.3 in the ASTM value or 10% in L (bar) may be expected if a minimum of 35 intersections are counted.

An excellent historical review of the various aspects of grain size measurements is given by Vander Voort (1984).

B. Volume Fraction

The concentration of various phases in a material has great importance in both fundamental and applied research. Likewise, the quality of most items of commerce depends on the composition of one or more components. This composition is easily monitored in a multiphased material by microscopically determining the volume fraction of the various phases. From this information it is an easy matter to obtain the weight fraction or percentage of the various phases.

The volume fraction, as with grain size, is normally determined on polished and etched samples of opaque and transparent materials alike. However, it is possible to use thin sections for this purpose if the grain size is relatively coarse.

There are several methods used to determine volume fraction; among these are areal analysis, lineal analysis and point counting. Success with any of these methods requires some means of distinguishing the various phases. A beginning requirement is that all grain boundaries on the sample surface must be adequately etched. Grain shape, color, surface relief,and/or differential staining or etching then will permit the identification of the various phases. If thin sections are used, polarized light illumination often provides sufficient contrasting.

The basic relationship between the area fraction of a phase and its volume fraction was determined by Delesse (1948); this may be expressed as:

$$\frac{A_p}{A_t} = \frac{V_p}{V_t} = VF_p \tag{35}$$

where A_p, V_p and VF_p are the surface area, volume and volume fraction of phase P, respectively. A_t and V_t are the total area and volume.

Manual determination of areas is difficult because it involves the measurement of areas with a planimeter or by tracing areas and weighing the sections corresponding to the various phases. However, areal analysis or its equivalent is now easily accomplished using one of the various image analysis systems described later.

The mathematical basis for the determination of volume fraction using lineal analysis is attributed to Rosiwal (1903) who published for the benefit of petrographers in their analysis of rock compositions. This relationship is given by:

$$\frac{L_p}{L_t} = \frac{V_p}{V_t} = VF_p \tag{36}$$

where L_p, is the length of a traverse in phase P and L_t the length of the total traverse.

The simplest technique for lineal analysis requires a photomicrograph on which the various phases are clearly distinguishable. Several random lines are drawn or otherwise superimposed on the photomicrograph. The lengths of individual segments of each line corresponding to a given phase are tabulated; the sum of these is the L_p for that phase and the sum of the L_p values for all phases is L_t.

Lineal analysis can be performed using a mechanical stage and cross-wire eyepiece if the structure is large grained. In this procedure the sample is critically focused and the value of the mechanical stage graduation noted. The stage is then translated until a boundary of that first phase is reached. A reading of the mechanical stage graduation is noted and the difference from the last reading tabulated for that phase. This process is repeated for each grain of each phase in the traverse. The volume fraction is then computed for each phase as above. In the same way a screw micrometer eyepiece can be used for fine structures. While requiring little in the way of accessories these procedures are very tiring and not recommended for large studies or complex structures.

Two of the more impressive light microscope accessories were developed for the purpose of simplifying the tedious procedures of lineal analysis; these are the integrating stage and the integrating eyepiece. These devices, described in Chapter 7, Sections IV B and VII, had six micrometer spindles which were used to sum the individual L_p values; the sum of all of these spindle values yielding L_t. Hurlbut also developed a mechanical counting device which provided for the motorized translation of a mechanical stage, see Chapter 7, Section IV C. Individual counters were simultaneously activated by their respective switches to record the translation through each phase separately. Again, the sum of the various counters gave L_t.

The efficacy of lineal analysis was investigated by Chayes (1956) who suggested that the quality of results to be expected depend on the measuring process used, on the instrument and on the operator. He states that a lineal analysis of a structure over a length of 1 mm has the same precision as a point count involving about 1000 points. An oft-repeated rule-of-thumb is given by Hennig (1958), which states that the total length of a lineal analysis traverse should be at least one hundred (100) times the average grain size.

Thompson (1930) described point counting, the third procedure for determination of volume fraction. In this procedure a grid network is superimposed over the image of the sample. This may be accomplished for the direct image using a specially designed reticle, see Chapter 2, Section III E. A photomicrograph may be so analyzed using a transparent sheet ruled with the grid pattern. The actual type of grid arrangement is not important as long as the points are evenly distributed. This method is particularly suited for the analysis of materials with a finely divided phase.

In this technique the number of points in each phase are noted and summed for the total number of points; these are used to express the volume fractions of the various phases as determined by:

$$\frac{I_p}{I_t} = VF_p \tag{37}$$

where I_p is the number of points in phase p and I_t is the total number of points measured.

The magnification of either the eyepiece image, if an eyepiece reticle is used or the photomicrograph, if the transparent sheet is used should be adjusted so that the average grain size is about equal to the distance between the grid points.

Swift had developed a mechanized point counting accessory, See Chapter 7, Section IV B. It consisted of a controller connected to a stepping mechanical stage. A number of switches on the controller advanced the stage one step while registering a count on their respective 4 digit counters. In operation, the microscope is focused on the sample and then the switch is depressed that is associated with that particular phase; this records a count for that phase and steps the stage to the next measurement point. The process is then repeated. The points accumulated for each individual phase and the total of all points are used with equation (37) to calculate the various volume fractions.

A statistical evaluation of point counting is given by Neyer (1966) and by Hennig, loc. cit.

C. Particle Spacings

The spacing between particles is a very important parameter in such fields as dispersion hardening of materials. Fullman (1953) has given a general expression for the mean free distance between particles that is independent of the size, shape or degree of dispersion of a particulate second phase:

$$MFD_p = \frac{1 - VF_p}{N_p} \qquad (38)$$

where MFDp is the mean free path between particles of phase P, VF_p is the volume fraction of phase P and N_p is the number of particles of phase P per unit length of a random straight line traverse. The volume fraction can most easily be determined using point counting and the number of particles per unit length can be determined using a process similar to lineal analysis except only the number of particles of phase P are recorded per unit length. MFD_p is considered to be the mean edge-to-edge distance between particles of phase P.

The mean center-to-center distance between particles or the mean particle spacing, S_p, for phase P is given by:

$$S_p = \frac{1}{N_p} \qquad (39)$$

When the particles are very small or the distance between them is very large, S_p approaches MFD_p. In this case it is preferable to measure the easily determined S_p.

D. Computerized Image Analysis

The various mechanical systems that have been developed to aid in the quantitative analysis of microscopical samples do simplify operator effort to some extent, however the real gains in productivity and accuracy in this field must be attributed to computerization. A number of these computerized image analyzers with a wide range of capabilities are currently available.

A typical image analyzer consists of a video camera which attaches to the microscope in the same location as a photographic camera, an image enhancing system and an image analyzing computer. Figure 4-37 shows a block diagram of a typical image analyzer.

The video camera used for this purpose is usually a high resolution monochrome unit, often operating at TV scan rates. Various image tubes are available for special applications..

The image information from the video camera is then processed in a video enhancement system as described in Chapter 4, Section X. The purpose of the image enhancement is to increase the contrast between the various structures or phases present so that each one can be uniquely characterized in the subsequent computer analysis.

The enhanced information is then prepared for the computer by an analog-to-digital (a/d) converter. The computer is programmed to operate on the resulting data using various algorithms that define a variety of object parameters. Certain outputs of the computer may be fed directly to a printer to give tabular alpha/numeric information while other outputs may be used to produce a computer modified image of the sample on a video monitor.

Interactive devices such as a light pen may be used with the monitor image to define certain areas of interest that may be further acted upon by the computer. Other outputs of the computer may be used to fine focus the image or translate the sample on the stage.

Morton (1972) has listed several factors that should be considered to assure that meaningful results are obtained in computer analysis: (1) the proper magnification must be selected, (2) the image must accurately represent the structure to be measured, (3) the features to be measured should have an unique range of gray level, (4) the image analyzer must be correctly operated to get valid data, (5) the proper algorithm must be chosen for the specific application. Table 5-11 illustrates the wide range of measurements that may be made on a structure.

TABLE 5-11
TYPICAL SAMPLE PARAMETERS MEASURED WITH AN IMAGE ANALYZER

Particle count	Particle size
Intercept count	Total area of features
Area % of features	Projected Length
Oversize count	End count
Interior (hole) count	Anisotropy
Closest approach	Area fill
Largest dimension	area measure
Length measure	perimeter
Convex perimeter	Aspect ratio
Mean chord	Maximum horizontal chord
Martin's radii	Feret's diameter
ASTM grain size	Porosity
Number of particles/area	hole area/total area
Average area/particle	Average diameter/particle
Average perimeter/particle	Equivalent circle diameter
Equivalent sphere volume	Surface of equivalent spheres
Volume %	Weight %
Plating thickness profile	Mean free path
Specific surfaces	Perimeter/unit area

V. CRYSTAL PROPERTY MEASUREMENTS

A. Optical Crystallography

Without a doubt, the most sophisticated use of the optical microscope is in the field of optical crystallography. The polarizing instrument used for these studies is referred to as a "petrographic" microscope. The sample for these studies ideally consists of crushed fragments of a translucent material in the size range of 100 to 250 mesh. However, with care, it is possible to identify a single sand sized grain. It is beyond the scope of this book to describe the procedures involved in optical crystallography however, sufficient information will be given to illustrate the correct set-up of the microscope.

Optical crystallographic techniques were originally developed for the identification of translucent minerals but has found use in chemistry, ceramics, drug identification, and forensics. Selected reading at the end of the chapter gives a number of texts on the various uses of the method.

Single phase transparent materials may be characterized as either isotropic (with a single refractive index) or anisotropic (with 2 or 3 unique refractive indices). Isotropic materials include cubic crystalline materials, glasses and amorphous materials. These materials are usually identified by their refractive index and cleavage shapes, if present. Dispersion, which is the difference in refractive as a function of the wavelength of light may also be used in the identification process.

The anisotropic materials that include all other crystal systems may be further subdivided. The uniaxial materials include the tetragonal, the hexagonal crystal systems, and the biaxial materials including the orthorhombic, monoclinic and the triclinic crystal systems. Strained isometric materials may also appear to be anisotropic in character. It is the location and identification of the 2 or 3 unique refractive indices of the anisotropic materials that contribute to the complexity of the optical crystallographic technique. A number of other optical crystallographic parameters, such as optical sign, axial angle, and cleavage angle may be measured. The anisotropic materials are further divided into the uniaxial (+ and -) and the biaxial (+ and -) classes.

The refractive index(s) of the grains are determined by comparing them with a suite of liquids with known refractive indices. These refractive index(s) plus the various other optical parameters of the sample material may be compared with tabulated data to identify the material.

The use of the polarizing microscope as a contrasting instrument was described in Chapter 4, Section VI. However, the critical use of this instrument is to be found in optical crystallography and petrography. For example, in the contrasting mode, the polarizer and analyzer need only be accurately crossed. In optical crystallography and petrography the polarizer and analyzer must not only be accurately crossed but their polarization directions must be parallel to the cross wires in the eyepiece. A procedure for setting the polarization directions of the polarizer and analyzer parallel to the eyepiece cross wires will be given later in this section. Furthermore, the orientation of the polarizer must be known. In the past, the orientation of the polarizer has been at the discretion of the manufacturer. If the microscopist assumes the wrong orientation of the polarizer, confusion and misinterpretation of the image will result. To eliminate this problem, the German industry adopted in 1972 the standard, DIN 58 879 (1975), which establishes the polarizer direction as "east-west"; this assumes that the direction the microscopist is facing while using the instrument is "north". A procedure will be given later in this section to verify the direction of the polarizer.

Examinations made with plane polarized illumination (without the analyzer in the light path) or with crossed polarized illumination are referred to as orthoscopic observations. Another method of examination critical to optical crystallography and petrography is conoscopic observation or simply, conoscopy. Conoscopic observation of a sample grain requires the use of an objective lens with a numerical aperture of at least 0.65 and the Bertrand lens, see Chapter 2, Section III C. The Bertrand lens, when inserted as shown in Figure 5-19, forms a telescope in combination

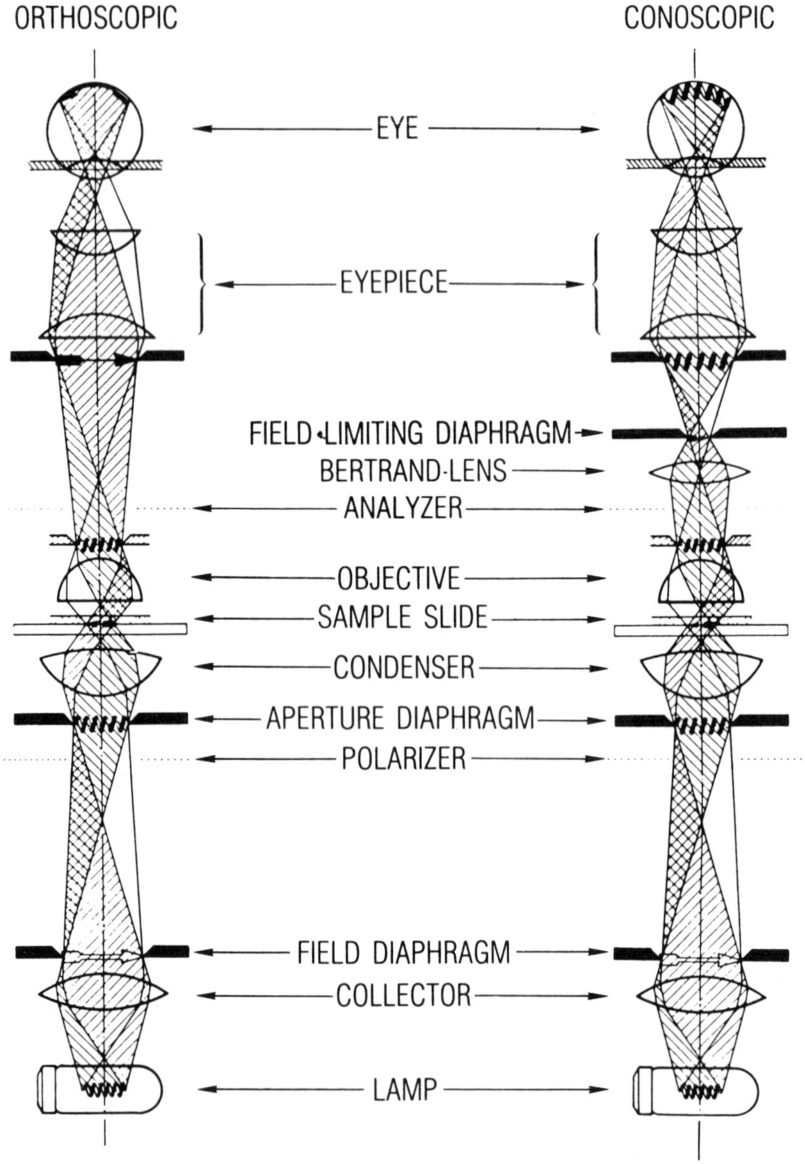

Figure 5-19. Ray diagram of a transmitted light, polarizing microscope in both the orthoscopic and conoscopic modes. In the orthoscopic mode the sample is imaged on the retina of the eye, whereas, in the conoscopic mode, the image at the rear focal plane of the objective lens, i.e., the lamp filament is imaged on the retina of the eye.

with the eyepiece that can be focused on the rear focal plane of the objective lens. The image that is observed in conoscopy is referred to as an interference image or interference figure; this image may be thought of as the information that is visible at the level of the rear focal plane of the objective lens. The contrast of the conoscopic image may be increased by closing down the grain isolating diaphragm so that the image is produce only by light from the desired crystal.

To obtain a good image for conoscopy it is necessary to have high aperture illumination, therefore it is common for the condenser to be equipped with an insertable, upper, high-aperture element; this element is inserted during conoscopy.

An evenly illuminated conoscopic image also requires a different type of illumination than that required for orthoscopy. True Koehler illumination which provides a bright evenly illuminated field of view for orthoscopic illumination, produces an image of the lamp filament at the rear focal plane of the objective lens, i.e., superimposed on the conoscopic image. Several manufacturers provide a "modified" Koehler illumination which makes use of a ground glass diffuser in the light path before the condenser. In this way the image of the lamp filament is eliminated. However, for optimum illumination of the interference image, e.g., in photomicrography of an interference image, Nelson or critical illumination is ideal. Alignment of the illuminator for Nelson illumination will be described in the set-up procedure.

Tho charactcrization or idcntification of a transparent crystalline phase is beyond the scope of this book therefore, the reader is referred to appropriate texts, such as Winchell, A. N. and Winchell, H. (1951), Winchell, A. N. and Winchell, H. (1964), Winchell, A. N. (1954) and Ehlers (1987). The various measurements that can be made using the three modes of observation are listed in Table 5-11.

The Hartshorne Rotation Apparatus, described in Chapter 7, Section IV C, is an accessory to the petrographic microscope that is very useful for orienting a single grain or grain fragments to aid in determining the optical properties.

Compensator accessories that fit into the tube slot, usually just above the objective lens are crucial to the interpretation of the images in both orthoscopic and conoscopic observations. These devices are discussed in Chapter 7, Section VI.

TABLE 5-12
OBSERVATION MODES FOR VARIOUS CRYSTAL PARAMETERS IN OPTICAL CRYSTALLOGRAPHY

ORTHOSCOPIC EXAMINATION

Plane Polarized Examination

a. Determination of refractive index, often after identification of crystallographic orientation by crossed polarized and/or conoscopic examination.
b. Determination of cleavage type(s) from fragment shapes.
c. Determination of dichroism or pleochroism after identification of crystallographic orientation by conoscopic observation.

Crossed Polarized Examination

a. Determination of sign (+ or -) of elongation.
b. Identification of crystallographic orientation, often after conoscopic examination.
c. Determination of crystal sign (+ or -) after identification of crystallographic direction by conoscopic observation.
d. Determination of extinction angle, often after identification of crystallographic direction by conoscopic observation.
e. Determination of anomalous interference colors after identification of crystallographic direction by conoscopic observation.

CONOSCOPIC EXAMINATION

a. Determination of axial angle.
b. Determination of axial dispersion
c. Determination of crystallographic orientation.

B. Optical Petrography

Petrography is the description and classification of rock types; optical petrography is the examination of rocks in prepared thin sections with a petrographic microscope. In this way, the various phases present and their arrangement within the rock may be determined. A number of newer "-ographies" are based on the methods of petrography; among these are ceramography and resinography.

The petrographic sample consists of a thin section of a rock or other material typically about 0.001 in.(0.025 mm) in thickness cemented to a glass slide and completed with a cover glass cemented in place. The cement used is a resin with a refractive index of about 1.54. A properly prepared sample is quite stable and will last for a long time without degradation.

The refractive index of grains in contact with the resin can be categorized as less than, equal to, or greater than that of the resin; these conditions are referred to as negative relief, zero relief and positive relief, respectively. Relief is further categorized in the first and last case as low, medium, and high depending on how distinct the boundary is between the grain and the resin. The index of a grain can also be compared with that of an adjacent grain of a known material. Other optical properties such as axial angle, optical sign, extinction angle, pleochroism, can be used if a grain of the unknown material is is the proper orientation with respect to the thin section plane. Table 5-13 lists the modes of observation required to make the various measurements.

TABLE 5-13
OBSERVATION MODES FOR VARIOUS CRYSTAL PARAMETER MEASUREMENTS IN THIN SECTION PETROGRAPHY

ORTHOSCOPIC EXAMINATION

Plane Polarized Examination

a. Determination of relief or relative refractive index, often after location of crystals with suitable orientation by conoscopic observation.
b. Determination of cleavage type(s) from fracture patterns.
c. Determination of dichroism or pleochroism after location of crystals with suitable orientation by conoscopic observation.

Crossed polarized Examination

a. Determination of sign (+ or -) of elongation.
b. Identification of crystallographic orientation, often after conoscopic examination.
c. Determination of crystal sign (+ or -) after location of crystals with suitable orientation by conoscopic observation.
d. Determination of extinction angle, often after location of crystals with suitable orientation by conoscopic observation.
e. Determination of anomalous interference colors after location of crystals with suitable orientation by conoscopic observation.
f. Determination of preferred orientation.

CONOSCOPIC EXAMINATION

a. Determination of axial angle.
b. Location of crystals with suitable orientation for subsequent measurements.

The problem of incorrect grain orientation with respect to the thin section plane can be minimized through the use of a universal stage. This accessory, shown in Figure 7-9, can provide from 3 to 5 degrees of rotation to the sample so that a given grain can be presented to the microscope in almost any orientation. The text by Emmons (1943) is an excellent detailed discussion on the use of the universal stage.

C. Procedure for Petrographic Microscopy

The set-up of the petrographic (or polarizing) microscope for orthoscopic observation has been described in Chapter 4, Section VI F. However, before the instrument is used for either optical crystallography or petrography, it is imperative that the orientation of the polarizer is known. If the instruction manual for the instrument is available that information should exist there, otherwise it may be obtained as described below.

DETERMINATION OF POLARIZER ORIENTATION ON A PETROGRAPHIC MICROSCOPE

1. View the light obliquely reflected from a horizontal surface (such as a tile floor) through a small piece of polarizing filter, as shown in Figure 5-20. A suitable filter is available from Edmund Scientific, 101 E. Glaucester Pike, Barrington, New Jersey 08007.
2. Rotate the filter around its axis until the light reflected from the surface is extinguished. Draw a vertical line on the filter while held in that position. That line represents the direction of polarization for that filter.
3. After the microscope has been set up for observation, place the filter on the stage. Use a low power objective lens and remove the analyzer from the optical path.
4. Observe the polarizing filter through the microscope as the stage is rotated and stop at an extinction position, i.e., when the observed light is at a minimum.
5. The polarizer orientation is located 90° to the line drawn on the polarizing filter. This direction should be recorded for use with all observations on that microscope.

The polarizer/analyzer orientation must be exactly parallel to the cross wires of the eyepiece if the extinction angle grain is to be accurately measured. To verify this condition, a suitable orientation test slide must be used. Such a test slide consists of a few elongate grains of a uniaxial or biaxial material that exhibit parallel extinction. Suitable materials for a slide are natrolite, or cancrinite crystals mounted in balsam and completed with a cover glass. Raw minerals and mounted slides suitable for this purpose are available from Ward's Natural Science Establishment, Inc., Box 92912, Rochester, NY 14692. Some synthetic materials also make excellent samples such as yttrium phosphate.

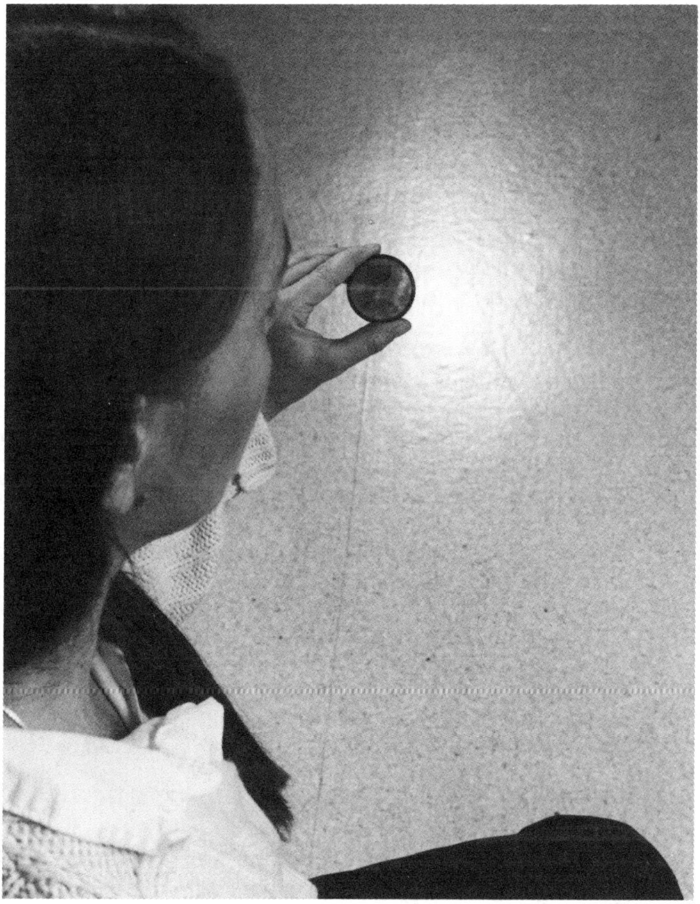

Figure 5-20. Illustration of a simple procedure for determining the approximate direction of polarization in a polarizing filter. Look through the filter at a distant reflection of light from a floor and rotate the filter until the reflected light seen through the filter is at a minimum. At that point, mark a line on the filter at right angles to the floor; this line is the approximate direction of polarization of the filter.

PROCEDURE TO VERIFY THAT THE POLARIZER/ANALYZER ORIENTATIONS ARE PARALLEL TO THE EYEPIECE CROSS WIRES

1. Set-up the polarizing microscope as described in Chapter 4, Section VI.
2. Place the orientation test slide (described above) on the stage and focus carefully with a 10X objective lens.
3. Insert the analyzer into the light path. If the instrument has rotatable polarizer and analyzer, adjust the analyzer for maximum extinction of the illumination background.
4. Move the slide so as to center a long grain on the test slide in the field of view and rotate the stage until it is fully extinguished.
5. Move the analyzer out of the light path. If the polarizer/analyzer orientation is correct, the edge of the grain should be exactly parallel to one of the eyepiece cross wires. It may be necessary to repeat steps 3 to 5 several times to assure a valid measurement.
6. If the polarizer/analyzer is misaligned with respect to the eyepiece cross wires:
   a. Instrument must be professionally serviced if polarizer and analyzer are not rotatable.
   b. The microscopist may make the adjustment as follows if both the polarizer and analyzer are rotatable.
7. Rotate the polarizer slightly in the direction opposite to the observed rotation and then repeat steps 3 through 5.
8. Repeat step 7 until the edge of the grain is exactly parallel to one of the eyepiece cross wires.

The measurement of refractive index is achieved in optical crystallography by comparing the desired refractive index of the crystal with that of an immersion fluid of know refractive index. In petrography that comparison is made with the permanent mounting resin or with an adjacent crystal.

The procedure generally used for the determination of refractive index uses axial illumination and the phenomenon known as the Becke line. Other methods for the determination of refractive index are the oblique illumination and the double-diaphragm methods; these are described by Saylor (1935). He has compared these three methods, as shown in Table 5-14 and has found that the double-diaphragm method is the most precise, followed closely by the Becke line method; the oblique illumination method was found to be the least precise.

TABLE 5-14
PROBABLE ERROR OF A SINGLE REFRACTIVE INDEX DETERMINATION RESULTING FROM AN UNCERTAINTY OF MATCH WITH FLUID*

OBJECTIVE LENS	SUBSTANCE	BECKE LINE	OBLIQUE	DOUBLE DIAPHRAGM
3.2X	ALUM**	0.00020	0.00023	0.00008
	KCl***	0.00023	0.00026	0.00014
10X	ALUM	0.00041	0.00021	0.00011
	KCl	0.00022	0.00027	0.00016
20X	ALUM	0.00042	0.00049	0.00014
	KCl	0.00017	0.00048	0.00020
45X	ALUM	0.00104	0.00088	0.00062
	KCl	0.00045	0.00125	0.00037

* Data from Saylor, loc. cit.

** $K_2O \cdot Al_2O_3 \cdot 4SO_3 \cdot 24H_2O$, $\eta = 1.45618$

*** $\eta = 1.49034$

Since the Becke line method is moderately precise and easily accomplished, this method will be described. The Becke line is a bright line which may be seen moving back and forth across the boundary of a grain as the focus of the microscope is adjusted above and below the correct focus for that boundary. This effect is shown in Figure 5-21. The procedure for the interpretation of the Becke line in axial illumination is given below for an isometric material. Saylor, loc.cit., has also given an indication of the effect of sample size on the accuracy of measurement; this is given in Table 5-15. The accuracy of the measurement will obviously decrease as the limit of resolution is approached.

TABLE 5-15
INFLUENCE OF SIZE UPON THE PROBABLE ERROR OF A SINGLE REFRACTIVE INDEX DETERMINATION*

OBJECTIVE LENS LENS	30 μm DIAM. GLASS FIBER	2 μm GLASS SPHERE
10X	0.00014	0.0006
45X	0.00034	0.0017

* Data from Saylor, loc. cit.

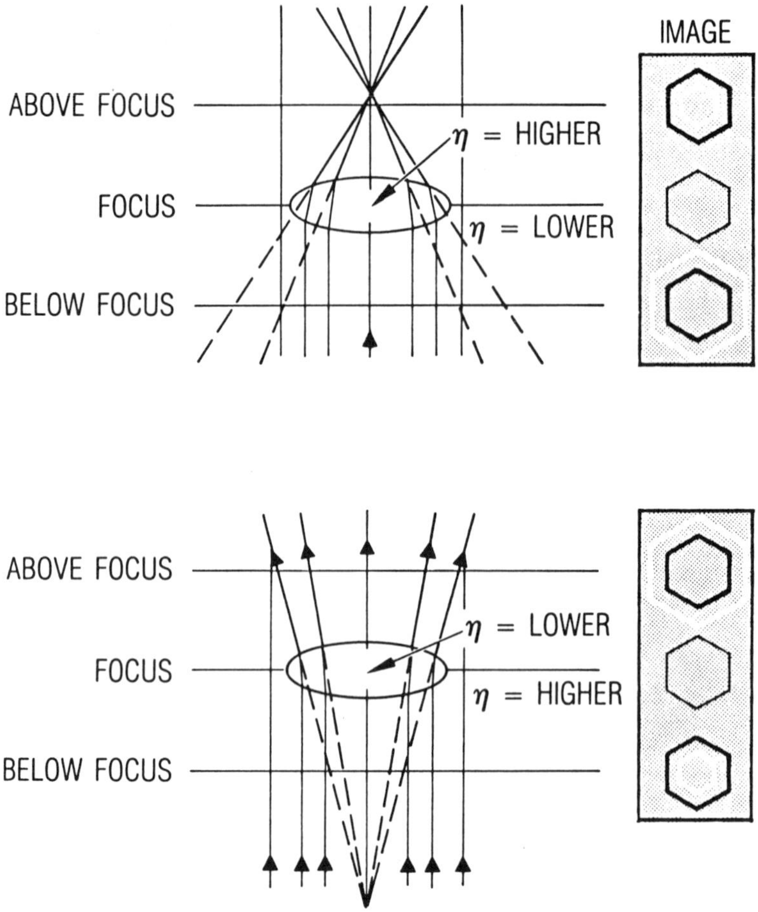

Figure 5-21. Illustration of the Becke effect. The ray diagrams on the left show the cross section of a grain mounted in an immersion fluid and the diagrams to the right show the position of the Becke line as seen in the eyepiece for each condition of focus. The Becke line is a weak to strong bright line which moves back-and-forth across the edge of an immersed grain as the focus of the microscope is moved above and below focus for the grain. At focus, the Becke line is coincident with the edge that produces it. The intensity of the line increases as the difference between the refractive indices of the grain and the immersion fluid increases. The direction of the line movement depends on the refractive index of the liquid relative to that of the grain as shown in the two drawings.

The measurement of the refractive index for uniaxial and biaxial materials uses the same procedure but each refractive index measurement must be preceded by the selection of a grain that is in the proper orientation to give the desired grain index for comparison with the immersion fluid. This interpretation of orientation is beyond the scope of this handbook and the reader is referred to Ehlers (1987).

MEASUREMENT OF REFRACTIVE INDEX USING THE BECKE LINE AND AXIAL ILLUMINATION

1. Set-up the microscope for orthoscopic observation as described in Chapter 4 Section VI.
2. Determine by viewing the microscope from the side the direction that the coarse and fine focus knobs must be rotated to move the objective lens away from the stage. Record this information for all subsequent observations.
3. Place a small amount of the crushed sample on the glass slide and add a large drop of refractive index fluid to the sample taking care not to contaminate the dropper with the sample and complete the mount with a cover glass. Verify that the sample grains are adequately surrounded by the fluid.
4. Place the sample on the stage and focus it carefully with a 10X objective lens. If the sample has very small particles a higher power objective lens may be necesary. Close down the aperture diaphragm to about 0.5 of its full aperture. The swing out final element of the condenser must be out of the light path.
5. Sharply focus the edge of a grain then move the fine focus in the direction that moves the objective lens away from the stage, as determined in step 2. A bright line that conforms to the shape of the grain edge will be seen to move away from the edge (either into the grain or into the fluid). Record the direction that it moves along with the refractive index of the immersion fluid. In this step the line moves toward the higher index medium.
6. Repeat steps 3 through 5 using a fresh portion of the sample material if possible, otherwise using small strips of blotter paper remove the existing fluid and retains the sample material. Add a new refractive index fluid. If the information from step 5 indicated the sample to have the higher index, use a higher index fluid. If, however, the fluid has the higher index use a lower index fluid for this test.

7. Repeat step 6 until the boundary between the grain and the fluid becomes very indistinct and the Becke line splits to form two lines one a pale orange and the other a light blue. These lines move in opposite directions across the boundary. The brighter of these is the one that should be used for the interpretation of that test. When these two lines are of equal brightness the index of the grain equals that of the fluid (in the yellow portion of the spectrum).

The accurate centration of the revolving stage to the axis of the petrographic microscope is crucial to the interpretation of conoscopic figures. The centering is accomplished in one of three ways: the stage may have centering knobs, the objective mount may have individual centering screws for each objective lens, or special polarizing objective lenses by Carl Zeiss have had two knurled rings near their lower end which provide centering for that lens.

Regardless of the centering mechanism, the procedure for centering the stage is the same. If the objective lenses or their mounts are centerable, all lenses should be centered for ease of locating a given area of the slide at different magnifications. However, with the centerable stage it is sufficient to align only the objective lens used for conoscopy. The centering procedure is given below.

CENTERING OF THE STAGE TO THE OPTICAL AXIS OF THE PETROGRAPHIC MICROSCOPE

1. Set-up the microscope as described in Chapter 2, Section VI C.
2. Select a small, circular particle, e.g., dust in the sample slide and carefully place it, either with a mechanical stage or manually, exactly at the intersection of the eyepiece cross wires.
3. Rotate the stage 180° and note the new location of that particle. (Note: do not allow the hands to rest on the stage as this will affect the position of the particle)
4. Mentally construct a line from the new location to the intersection of the cross wires, as shown in Figure 5-22.
5. Move the image of the particle along that line to its midpoint by moving the slide.
6. Move the image of the particle the remaining distance to the center of the cross wires using the centering adjustment of the stage, objective lens or objective lens mount.
7. Repeat steps 3 through 6 until the particle remains stationary when rotated.

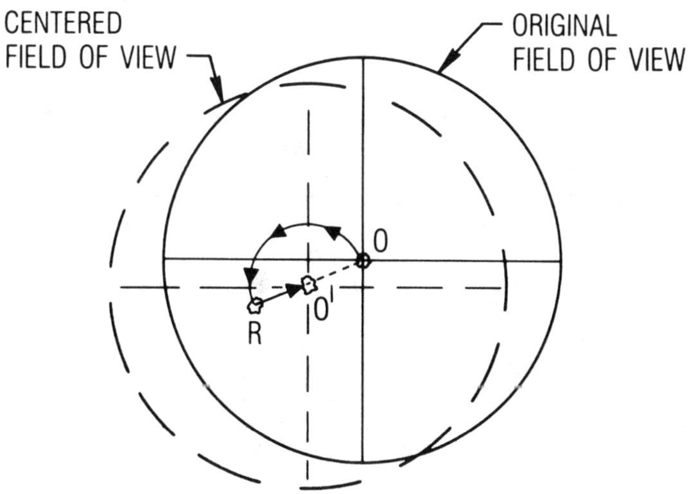

Figure 5-22. Procedure for centering either the stage or objective lenses. Place a slide on the stage and focus on dust particles on the slide surface. Move the slide, either manually or with a mechanical stage so that a recognizable, equant particle is centered under the cross wires of the eyepiece at position, O. Manually rotate the stage 180° and note the new position of the particle at R. Translate the image of the particle 1/2 of the way back to O on the line OR (point O') using the centering knobs for the stage or objective lens. Once again, move the slide, either manually or with the mechanical stage so that the particle is again centered under the cross wires of the eyepiece. The optical system should now be centered to the stage. If the grain is not stationary upon rotating the stage, repeat the above procedure. This process must be repeated for each objective lens.

The determination of such optical parameters as axial angle, optical sign and location of the unique axes for uniaxial and biaxial crystals requires the use of the petrographic microscope in the conoscopic mode. The set-up of the microscope for conoscopy is given below.

SET-UP OF THE PETROGRAPHIC MICROSCOPE FOR CONOSCOPIC OBSERVATION

1. Set up the polarizing microscope for orthoscopic observation as described in Chapter 4, Section VI F.
2. Locate a grain of interest and change to a high numerical aperture objective lens (0.65 N.A.). The aperture diaphragm is normally set at about 0.5 for measurement of refractive index.
3. Fully open the aperture diaphragm, and swing in the final element of the condenser lens.
4. Insert the Bertrand lens into the optical path and observe the resulting interference figure. If the Bertrand lens is centerable, center the interference figure with respect to the eyepiece crosswires. If the Bertrand lens is focusable, carefully focus the dark edge of the aperture at the rear focal plane of the objective lens.
5. If the microscope is equipped with a grain isolating diaphragm, close it as much as necessary to limit interfering images from any adjacent grains.
6. The microscope is now fully adjusted for conoscopic observation. To return to orthoscopic observation reverse the procedures of Steps 3 through 5.

D. PROBLEM CORRECTION FOR CONOSCOPIC OBSERVATION

The major problems connected with conoscopic observation are failing to make all the necessary adjustments to the microscope in transferring from orthoscopic to conoscopic observation and the reverse process. Familiarity with the specific microscope and the procedure above should minimize these problems. Unique problems are listed below.

CONOSCOPIC OBSERVATION, PROBLEMS AND THEIR CORRECTION

PROBLEMS	CAUSE	CORRECTION
Image of lamp filament in image.	Koehler illumination used.	Use a ground glass filter in condenser or critical illumination.
Indistinct bubble like object(s) in conoscopic image.	Decemented elements in objective lens.	Replace objective lens with undamdamaged one.

If true Koehler illumination is used for conoscopy the lamp filament will be seen sharply focused in the conoscopic image. To avoid this problem most manufacturers use a modified Koehler illumination which is achieved by inserting a ground glass plate in the condenser system. If the conoscopic image is to be critically photographed, even illumination of the conoscopic image may be achieved by the use of critical or Nelson illumination. The procedure for obtaining critical illumination is given below. It must be recognized, however that critical illumination is not useful for photomicrography of the orthoscopic image since the lamp filament will be observed, sharply focused, in the sample image.

SET-UP OF THE PETROGRAPHIC MICROSCOPE FOR CRITICAL (NELSON) ILLUMINATION

1. Set-up the petrographic microscope for orthoscopic observation as described in Chapter 4, Section VI F.
2. Remove any ground glass plates from the condenser system and focus the lamp filament image critically on the sample by adjustment of the condenser focus knob.
3. When this instrument is set-up for conoscopic observation that image will be uniformly illuminated.

E. DISPERSION STAINING

The concept of dispersion staining was first presented by Maschke (1872) but was developed into a full scale microscopical technique by Brown and McCrone (1963). This technique permits the identification of a large number of compounds by means of their optical dispersion. This technique has proven very valuable in the identification of a wide range of materials including dust particles, asbestos, glass fragments, and synthetic fibers.

All solids and liquids exhibit optical dispersion, that is, their refractive indices vary with wavelength of light. Liquids generally have a greater optical dispersion than solids, as shown in Figure 5-

23. In practice, fragments of a solid sample are mounted in a refractive index oil that matches the refractive index of the sample at some wavelength in the visible spectrum. When the sample is viewed in transmitted light using an annular or a central screen, as shown in Figure 5-24, the sample is seen in various colors. The identity of the sample may then be determined from tables in Volume 4 of McCrone, et al.(1973).

The systematic application of the dispersion staining technique requires a set of refractive index immersion oils with indices varying regularly in intervals of 0.01 or less over the range of 1.35 to 2.11. Fortunately, such a set is available from the R. P. Cargille Laboratories, Inc., Cedar Grove, NJ 07009.

F. PROCEDURE FOR DISPERSION STAINING MICROSCOPY

Any transmitted light microscope can be modified to perform dispersion staining. It, however, must be emphasized that only isometric materials can be determined, see Chapter 4 Section VI, unless a polarizing microscope with a revolving stage is used as the base instrument. The microscope must be fitted with objective lenses that incorporate both central and annular screens.

Central and annular screens that may be constructed and installed by the microscopist are described by Brown and McCrone (1963). Ideally, three identical objective lenses should be used in this case: (1) an unmodified objective lens for locating and orienting grains, (2) an objective lens equipped with a central screen and, (3) an objective lens fitted with an annular screen. The set-up has be simplified significantly by Walter C. McCrone Associates, Inc., (2820 S. Michigan Ave., Chicago, IL 60616) who offer 10X and 40X dispersion staining objectives. These objective lens are equipped with a three position, revolving disk that includes a clear opening, the central screen, and the annular screen, as shown in Figure 5-25.

In the following procedure, it is assumed that a polarizing microscope is the basic instrument and that a McCrone dispersion staining objective lens is used.

SET-UP FOR THE DISPERSION STAINING MICROSCOPE

1. If the microscope has a revolving nosepiece, remove one objective lens and install the dispersion staining objective lens. Note: it may be necessary to temporarily remove other objective lenses to allow clearance for installing the lens. Otherwise, install the dispersion staining objective lens on a quick change clutch.
2. Lock the dispersion staining objective lens in a position that will allow easy access to the revolving disk, using the knurled locking ring on the threads of the objective lens.

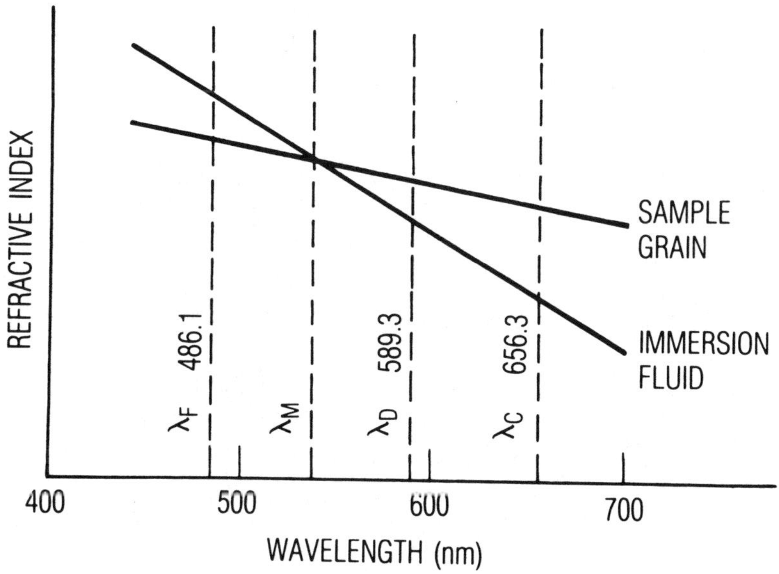

Figure 5-23. Graph of the refractive indices of a solid and an immersion liquid. The refractive indices of the two are identical at about 540 μm.

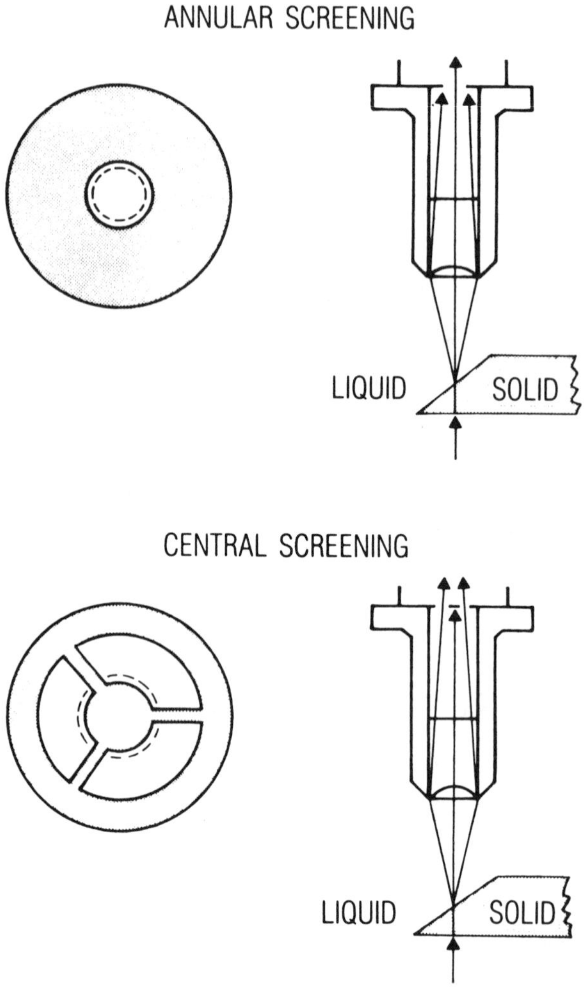

Figure 5-24. Illustration of the annular and central screening techniques for dispersion staining. The drawings on the left show the image of the two screens at the rear focal plane of the objective lens. The dotted lines indicate the position of the aperture diaphragm for optimum contrast. The screen and the position of the aperture diaphragm may be observed by removing the eyepiece, using an eyepiece telescope or inserting the Bertrand lens into the optical path of the microscope. The two cross sections on the right show the image rays used for the respective screens. Redrawn courtesy of McCrone Accessories and Components Division.

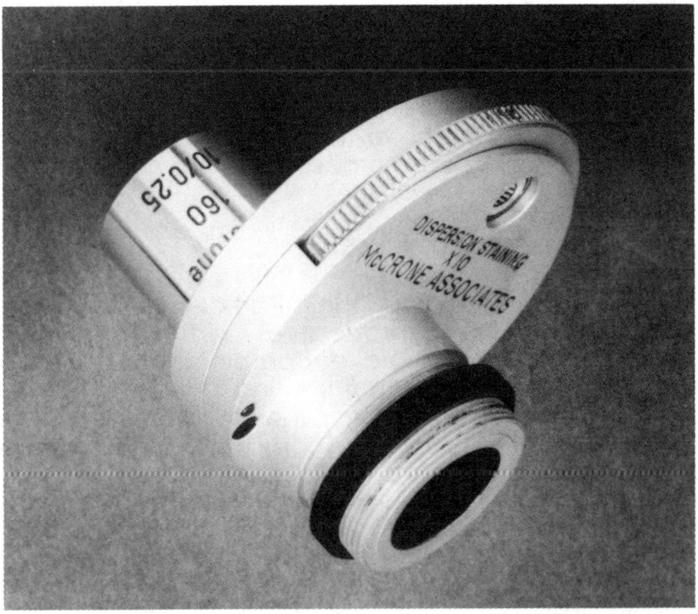

Figure 5-25. A McCrone dispersion staining objective lens. The turret provides an annular screen, a central screen, and an open position. The specific screen in position is shown in a window in the turret. Courtesy of McCrone Accessories and Components Division.

3. Rotate the revolving disk until the clear opening is in the optical path.
4. Place a sample in an appropriate immersion oil mount on the stage and focus carefully.
5. Set-up the instrument for plane polarized light observation as described in Chapter 4, Section VI. Carefully focus and center the aperture diaphragm using the condenser focusing and centering knobs. Observe this process with the Bertrand lens in place, then remove the Bertrand lens from the optical path.
6. The instrument is now set-up for dispersion staining studies.
7. Adjust the aperture diaphragm until it is slightly less in diameter than the opening in the annular stop for annular screening or slightly larger in diameter than the central opaque stop for central screening. This process may be examined with the aid of the Bertrand lens.

G. THE McCRONE PARTICLE ATLAS

McCrone and Delly in their Particle Atlas (1973) have developed a very useful procedure for the identification of particulate materials and have complemented this with an extensive pictorial catalog of natural and industrial materials. The Atlas includes a number of disciplines including optical microscopy, scanning electron microscopy, energy dispersive x-ray analysis and x-ray diffraction analysis. This discussion will cover only that portion of the Particle Atlas that deals with particle characterization by optical microscopy and the necessary instrumentation to perform these investigations.

The particles of various materials described in the optical microscopy portion of the Particle Atlas, Volume II are all characterized by means of a six-digit binary code, e.g., 110011. The significance of this code is given in Table 5-16. To characterize a particle, it is mounted on a glass slide using 1.660 refractive index mounting fluid and a cover glass. The six digit code for the particle is then determined by reference to the table. This code may then be used with tables and photomicrographs in Volume II of the Particle Atlas to identify the particle.

TABLE 5-16
DESCRIPTION OF PARTICLE CHARACTERISTICS FOR SIX-DIGIT "PARTICLE ATLAS"* CODE, $X_1X_2X_3X_4X_5X_6$

DIGIT	VALUE 1	VALUE 0
X_1	100% opaque in visible light spectrum.	Transparent, colored or colorless.
X_2	Colored.	Colorless or white if transparent, black if opaque.
X_3	Birefringent.	Isotropic or apparently so.
X_4	Refractive index, at least one R.I. > 1.660.	All apparent R.I.s $\leq$ 1.660.
X_5	One dimension of Particle is 1/4 that of either of the other two dimensions.	Particle is elongated or equant.
X_6	One dimension of particle is 4 times that of either of other two dimensions.	Particle is flat or equant.

*McCrone, et al., Loc. cit.

H. PROCEDURE FOR PARTICLE ATLAS EXAMINATIONS

The polarizing microscope is used for the determination of the six-digit code for a sample particle. The set-up of this instrument is given in Chapter 4, Section VI.

For comparison of the sample particle with the color photomicrographs in the Particle Atlas, the translucent sample is examined in plane polarized light or in slightly uncrossed, polarized light. The latter type of illumination is achieved by rotating the analyzer about 12° from the fully crossed position. Opaque samples are compared by using top, oblique illumination, preferably from two lamps, in addition to the slightly uncrossed, polarized light from below.

VI. MICROPHOTOMETRY

A. General

Photometry is the branch of science that deals with the measurement of light intensity. Microphotometry permits the

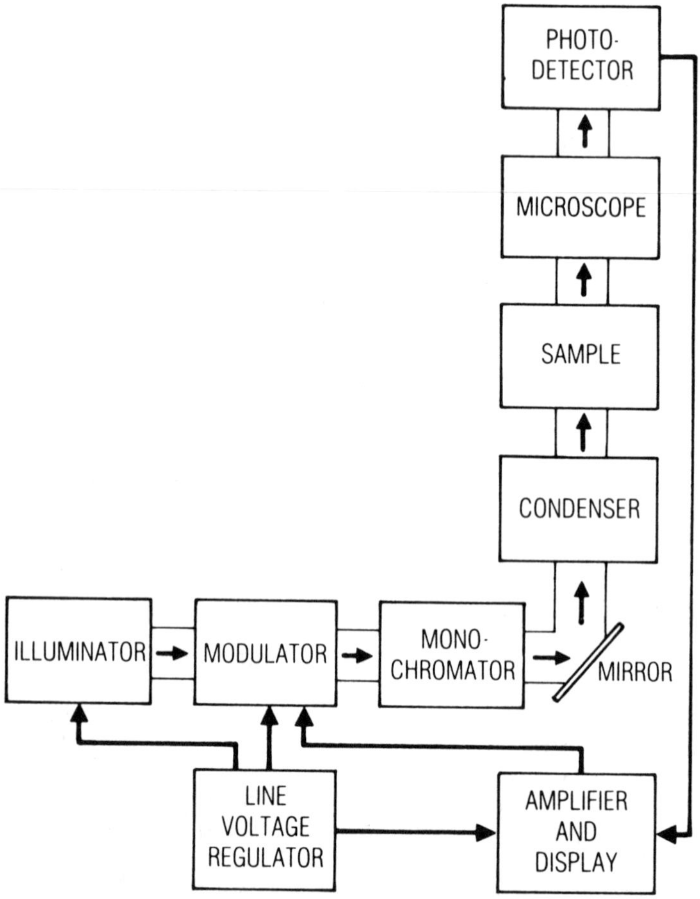

Figure 5-26 Block diagram of a single beam microphotometer.

measurement of intensity of light transmitted through or reflected from a sample as imaged in a microscope.

The photometric process may involve a dual beam or a single beam measurement. The dual beam systems compare the intensity from the sample with a reference beam from the light source and thereby compensate for variations in the illumination intensity. Berek developed an excellent dual beam system for the incident light microscope; this was subsequently marketed by E. Leitz. The single beam system is commonly used today since the light sources may be more easily stabilized. The single beam systems may be used either for incident or transmitted light. A typical single beam system is shown schematically in Figure 5-26.

The process of photometry can be applied to any type of sample, using any type of illumination or enhancement technique. Total absorption or reflectivity can be measured in white light. If a filter or a monochromator is included in the light path, the microphotometer can be used to measure intensity as a function of the wavelength of the light. Examples of this latter technique include measurement of: absorption spectra, transmitted or reflected color, and fluorescence spectra (microfluorimetry). One interesting use of this techniques coupled with a polarizing microscope is the measurement of preferred orientation of crystals in thin sections of rocks or ceramics. Larson (1969) has described the construction of a dual beam microphotometer for the quantitative measurement of reflectivity of bireflective materials in plane polarized light.

Microphotometry attachments are available from from both E. Leitz and Carl Zeiss for a number of their instruments. Continuous or step scanning stages are also available (see Chapter 7, Section V) that permit the measurement of intensity profiles across the sample.

B. Procedure for Microphotometry

As mentioned above the microphotometer attachment can be used with any type of illumination, therefore the basic set-up of the microscope proper can be obtained from previous chapters. The set-up of the microphotometer unit, however, is very specific for each system and is beyond the scope of this book.

REFERENCES

ASTM Standards, Section 3, p.117, American Society for Testing and Materials, 140 Race St., Phil. (1985).

Barer, R., and D. A. T. Dick, Exp. Cell Res., Suppl. 4, p.103, (1957).

Brown, K. M. and W. C. McCrone, The Microscope & Crystal Front, 13, 311, (1963), and 14, 39, (1963).

Chayes, F., "Petrographic Modal Analysis", John Wiley and Sons, Inc., N.Y. p.38, (1956).

Coble, R. L., J. Appl. Phys., 32, 793, (1961).

Delesse, M., Annales des Mines, 13, 379, (1948)

DIN 58 879, Bezugssystem der Polarisationsmikroskopie (Reference System of Polarized Light Microscopy), Beuth Verlag GmbH, Berlin and Cologne, Germany (1975).

Ehlers, E. G. "Optical Mineralogy", 2 Vol., Blackwell Scientific Publications, Inc. Palo Alto, CA (1987).

Emmons, R.C., "Universal Stage", Geol. Soc. Amer. Memoir 8, (1943).

England, J. B. A., J. Sci. Instruments, 34, 208, (1957).

Fracon, M., "Optical Interferometry", Academic Press, N.Y., (1966).

Fryxell, R. E. and B. A. Chandler, J. Am. Ceram. Soc., 47, 283, (1964).

Fullman, R. L., Trans. AIME, 197, 447, (1953).

Hale, A. J., "The Interference Microscope In Biological Research", E. & S. Livingston, Ltd, Edinburgh and London, (1958).

Hennig, A., Zeiss Werkzeitschrift, 30, 78, (1958).

Hilliard, J. E., Metals Progress, 99, May (1964).

Larson, L. T., "Equipment for the Quantitative Photometric Measurement of Reflectivity by Polarized Light Microscopy", Rev. Sci. Instrum., 40, #8, Aug.(1969).

Lennon. J. W., J. Am. Ceram. Soc., 49, 398, (1966).

Maschke, D., Pogg. Ann. Phys. Chem., 145, 549, (1872).

McCrone, et.al., "Particle Atlas", 2nd Ed., Ann Arbor Science Pub., Ann Arbor, FL (1973).

Morton, R. R. A., "The Design of Image Measurement Systems: Stereology and Quantitative Metallography", ASTM Special Technical Pub. 504, p.89, Phil. (1972).

Neyer, H., Zeiss Information, 60, 66, (1966).

Nyyssonen, D., Proc. Soc. Photo-Optical Instrum. Engrs., 221, 119, (1980).

Rosiwal, A., Bull. Geol. Soc. Amer., 14, 466, (1903).

Saylor, C. F., J. of Res. National Bureau of Standards, 15, 277, (1935).

The Hemacytometer: "Clinical Laboratory Diagnosis" Levinson, S.A. and MacFate, R.P., Lea & Febiger, Philadelphia, 7th Ed., (1969) p.773.

The Howard Cell: "Official Methods of Analysis of the Association of Official Analytical Chemists", Horowitz, W. , Editor. Association of Official Analytical Chemists, Washington D.C. (1975). p. 883.

The milk smear disc: "Standard Methods for the Examination of Dairy Products", Marth, E.H., 14th Ed., American Public Health Association, Wash. D.C., (1978), p. 125.

The Whipple Disc: "Standard Methods for the Examination of Water and Wastewater" published by the American Public Health Association, Washington, D.C., 14th Edition, (1976), p. 1024.

Thompson, E., J. of Geol., 38/3, 193, (1930).

Tolansky, S., "Multiple Beam Interference Microscopy of Metals", Academic Press, N.Y., (1970).

Vander Voort, G. F., "Practical Applications of Quantitative Metallography" ASTM Special Technical Pub. 839, Edited by McCall, J.L. and Steele Jr., J.H., p.85, Phil. (1984).

Winchell, A. N. "The Optical Properties of Organic Compounds", 2nd Ed., Academic Press Inc., NY, (1954).

Winchell, A. N. and H. Winchell, "Elements of Optical Mineralogy", Part III, Determination Tables, John Wiley and Sons, Inc., New York, (1951).

Winchell, A. N. and H. Winchell, "The Microscopical Characters of Artificial Inorganic Solid Substances", Academic Press, NY (1964).

6

Special Microscopes

I. INTRODUCTION

The microscope in either the upright or inverted stand configuration is to be found in nearly every scientific or engineering pursuit. However there are unique applications which require specialized instruments; it is these microscopes that are the subject of this chapter. In keeping with the general thrust of this handbook, the instruments discussed will be limited to those using visible light.

II. PORTABLE MICROSCOPES

The portable microscope can range from the simple hand lens to almost any microscope that can be carried in its case in one hand. An excellent example of the latter was produced by Wild, Heerbrugg Instruments, Inc. This full-sized, compound microscope was available with a wide range of accessories for transmitted light studies. It is easily transportable in a metal carrying hood that clamps over the stand and hermetically seals the instrument against the elements. Unfortunately, it is no longer available as a new instrument.

Some of the current, truly portable instruments are manufactured by Swift. Two of these instruments are shown in Figure 6-1. The transmitted light microscope may be used with sunlight as the illumination or with an integral battery powered illuminator. The three objectives mounted on a revolving nosepiece give total magnifications of 62X, 155X, and 620X. The weight of this instrument with its fitted cover in place is less than 2 lbs. Swift also offers an incident light microscope which also has an integral battery

A

B

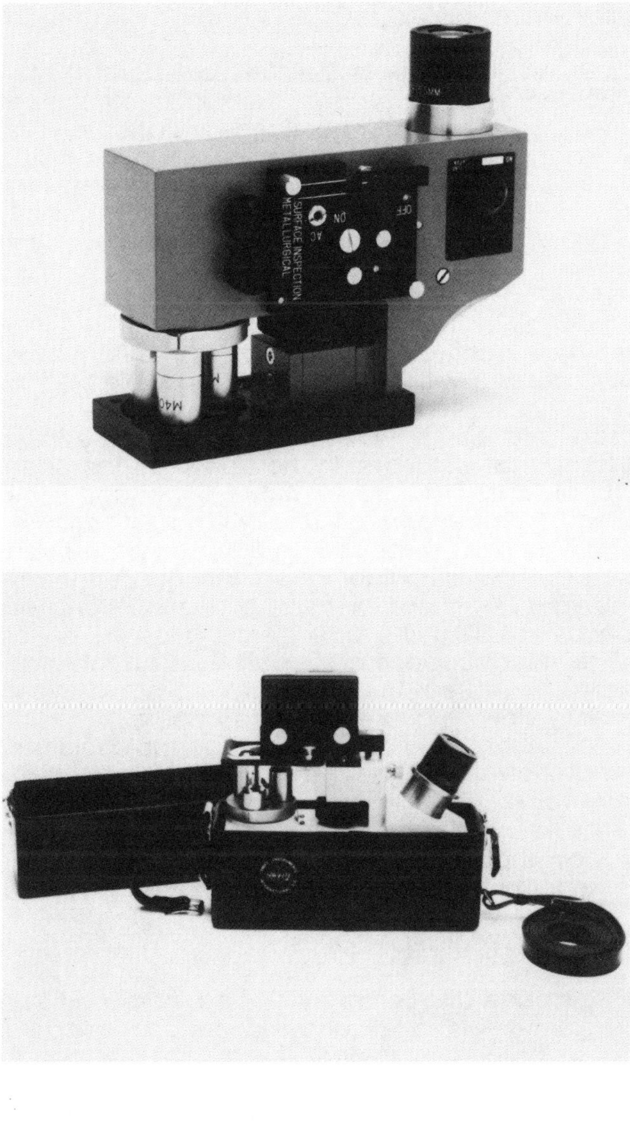

Figure 6-1. Two fully professional, portable microscopes; (a) an incident light microscope with built-in illuminator and (b) transmitted light microscope with attached illuminator. Courtesy of Swift Instruments, Inc.

powered illuminator and a revolving nosepiece. Magnifications are 100X, 200X, and 400X.

Struers, Inc., 26100 First St., Westlake, OH 44145 offers a portable, incident light microscope; this instrument, shown in Figure 6-2, has an integral illuminator powered by batteries in its carrying case. The standard version is equipped with a 10X objective lens and a 10X eyepiece. 20X and 40X objective lenses are available as options. The microscope is designed to rest directly on the surface to be examined. The weight of the microscope itself is 0.6 kg (1.3 lb).

The U.S. Department of Defense, recognizing the need for the high magnification (200X) examination of fiber optics terminations, has, accordingly, developed a military specification, MIL-M-83524, dealing with a hand held, portable microscope for this purpose.

Bausch and Lomb have developed a portable microscope similar to that described in the above military specification. This instrument, shown in Figure 6-3, will accept a number of objective lenses and eyepieces to yield magnifications of 40X to 800X. This instrument is equipped with a battery powered illuminator that permits: (1) normal incident illumination, (2) transmitted light through the fiber, (3) a combination of 1 and 2, and (4) oblique illumination by rotating the fiber termination to an off-normal position. The microscope is equipped with an integral, infrared absorbing filter to protect the viewer from possible eye injury due to inadvertently viewing a laser activated cable.

The light weight or small size of the portable microscope can pose a temptation to use these instruments in environments that are detrimental to them, hence care must be exercised. For example, the above instruments are not hermetically sealed, therefore they should not be used in wet environments or in situations where the temperature is less than the dew point.

III. LONG DISTANCE MICROSCOPES

Long distance microscopy is finding increased importance for scientists, engineers, medical personnel and law enforcement officers. These instruments are necessary for a number of reasons: the inability to be physically close to the object of interest, the hazardous nature of an event or experiment being viewed, or the importance of maintaining the isolation or privacy of living subjects.

The Gaertner Scientific Corp. model M353, shown in Figure 6-4, is a true microscope with a low power objective lens that has a working distance of 15.7 cm (6.20 inches) and magnification of 20X or 50X. Relay lens systems are also available which will increase the working distance of this microscope to 10 or 16 inches. A variety of mountings and illumination types make this instrument very versatile.

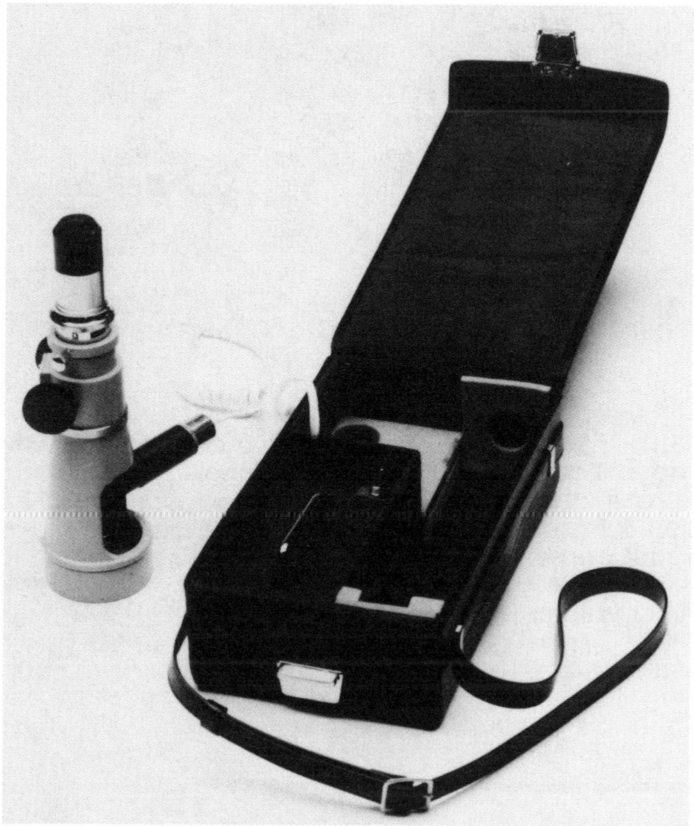

Figure 6-2. A fully professional, portable incident light microscope, PSM-2 and carrying case with battery supply for illuminator. Courtesy of Struers, Inc.

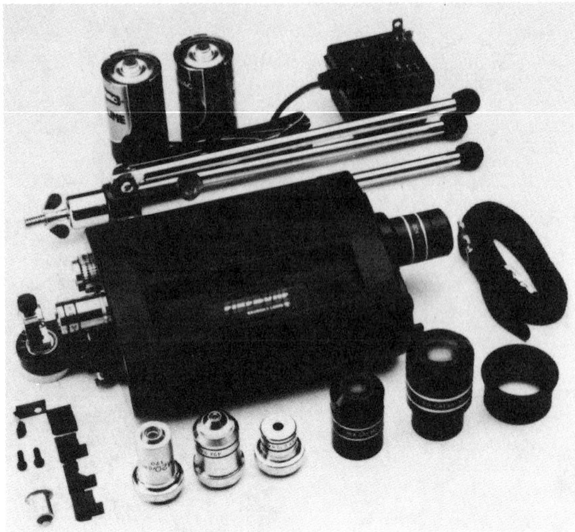

Figure 6-3. A portable hand held microscope for the inspection of fiber optical cable at magnifications ranging from 40X to 800X. Four integral methods of illumination are provided; normal incident illumination, light transmitted through the cable, simultaneous normal incident and transmitted illumination, and oblique illumination by rotation the cable support. Courtesy of Cambridge Instruments Inc., Optical Systems Division.

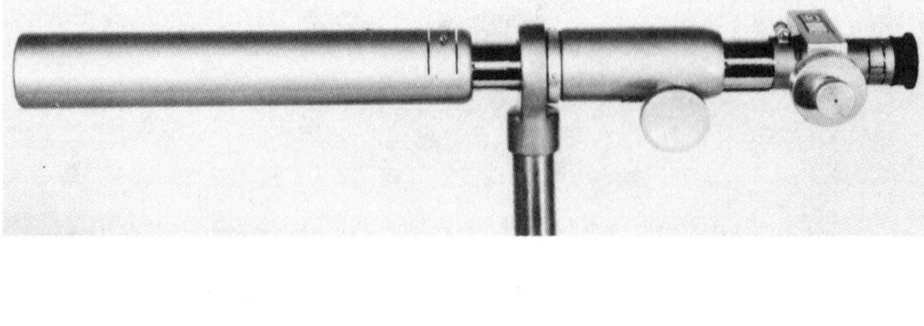

Figure 6-4. A long distance microscope with relay lenses which permit viewing a sample with a magnification of 20X or 50X at a distance of 10" or 16". Courtesy of Gaertner Scientific Corp.

A

B

C

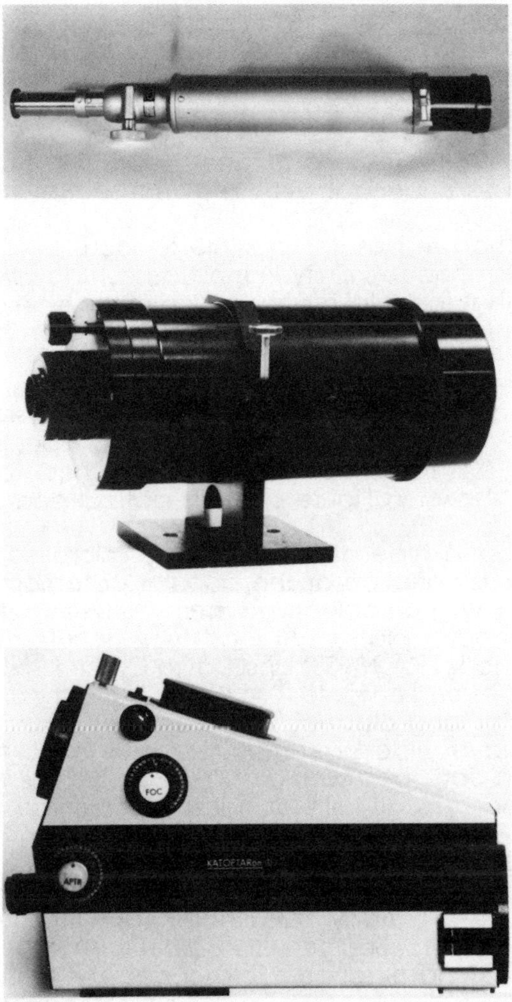

Figure 6-5. Three types of telescopes which have sufficiently short near focus that they can serve as excellent long distance microscopes: (A) a dioptric (with a lens objective) telescope, model M524, Courtesy of Gaertner Scientific Corp., (B) a catadioptric (with lens and mirror objective) telescope, courtesy of Davro Optical Systems, Inc., and (C) a catoptric (with mirror objective) telescope, courtesy of META-Gerätetechnik, Cologne, West Germany.

The instruments used for microscopy at a greater distance are essentially adaptations of telescope or telephoto optics. They may use refractive (dioptric) elements, reflective (catoptric) elements, or a combination of reflective and refractive (catadioptric) elements as objective lenses; examples of these are shown in Figure 6-5. These instruments unlike the fixed tube length microscope described above, are focused by changing the spacing between the objective lens and the eyepiece. This focusing method makes it is possible to focus on objects from a near focus distance to infinity in most cases. However, it must be emphasized that the total magnification of this type of microscope decreases as the point of focus moves away from the near focus position.

Few refractive telescopes focus close enough to be considered a long distance microscope, however, Gaertner does offer several special instruments of this type, one of which, Model M524, will focus from 80 cm (32 inches) to infinity. This instrument is shown in Figure 6-5a; its optical characteristics are given in Table 6-1.

The second type of long distance microscopes, shown in Figure 6-5b is a variation of the popular catadioptric telescope system, that is, with an objective system consisting of both a mirror and lens (corrector plate). At least two manufacturers, Questar (Box C, New Hope, PA 18938) and Davro Optical Systems, Inc., (5th and Cannon Ave. Lansdale, PA 19446) have models specifically constructed for long distance microscopical applications. These instruments have wide apertures that give excellent resolution. They are both equipped with a built-in Barlow lens which may be inserted in the optical path at will to double the magnification without removing the eye from the eyepiece. The characteristics of these instruments are given in Table 6-1.

The model LDM-1 manufactured by META-Geratetechnik of Cologne, West Germany (available from Infinity Optics, 706 Mohawk Dr., Suite 15, Boulder, CO 80303) is an excellent example of a reflective, long distance microscope. This instrument, shown schematically in Figure 6-5c, will accommodate any of the standard monocular or binocular bodies manufactured by Hertel & Reuss (also available from Infinity Optics). The LDM-1 also includes depth of field control and provisions for normal incident illumination. Characteristics of this instrument are given in Table 6-1.

These instruments require a stable mounting or stand for optimum viewing. Even with a stable mount it is sometimes difficult to locate the area of interest since the field of view is relatively small. Hyzer (1982) has described a simple method for locating the area of interest, wherein the light from a small flashlight, aimed into the eyepiece of the microscope, illuminates the area of the sample that is in the field of view.

The set-up of the long distance long distance microscope is essentially the same as the equivalent telescope. Care must be taken to assure that the objective lens to sample distance is within the focal range of the instrument. As mentioned above, the closer the sample is to the near focus of the instrument, the greater the magnification for a given eyepiece.

If ambient light is sufficiently bright, it will generally provide excellent illumination, otherwise the sample can be illuminated with a spot light. Often this latter illumination must be remotely located along with the microscope. If the sample is incandescent, the microscope should be equipped with a neutral density filter (see Chapter VII, Section 6) to protect the eye; for very bright sources a telescope solar filter may be required.

IV. REMOTE MICROSCOPES

Another method for the observation of a sample that is physically remote from the observer makes use of the remote microscope. The remote microscope may be thought of as a standard instrument in which the eyepieces have been separated

TABLE 6-1
GENERAL SPECIFICATIONS FOR SELECTED LONG DISTANCE MICROSCOPES

SPECIFICATION	M522B[1]	M-1[2]	77[3]	LDM-1[4]
Objective type	Lens	--Catadioptric--		Mirror
Aperture (mm)	40	89	77	42
Near focus (cm)	80	56	48	150
Far focus (cm)	inf.	196	177	inf.
Magnification[5]	3.2X	2.2X	2.4X	1.5X
Resolution[6]	-	128	-	270

1. Gaertner Scientific Corp., 1201 Wrightwood Ave., Chicago, IL 60614.
2. Questar Model M 1, Questar, Box C, New Hope, PA 18938.
3. Model 77 Micro Telescope, Davro Optical Systems, Inc., 5th and Cannon Ave., Lansdale, PA 19446
4. Katoptaron LDM-1 and LDM-1 with PB-1, Available from Infinity Optics, 706 Mohawk Dr., Suite 15, Boulder, CO 80303.
5. Magnification of the objective system at the near focus. Total magnification is that value time the magnification of the eyepiece. Magnification decreases as the far focus is approached.
6. Resolution in line pairs per mm at the near focus.

from the rest of the instrument and the light from the objective lens conducted to the eyepieces via one or more relay lenses and mirrors/prisms or fiber optics.

The most important of the remote microscopes is the borescope. These microscopes are used for the inspection of the interiors of all types of industrial products. This instrument, shown in Figure 6-6, makes use of relay lenses mounted in a rigid metal tube, or flexible fiber optics to place the objective lens and suitable illumination at some distance from the eyepiece. Instruments are available for examinations as remote as 5.64 m (222 in) and for openings as small as 1.75 mm (0.069 in). The focus and viewing direction are adjustable in most borescopes. Borescopes with coherent fiber optics for imaging may be used in the inspection of curved tubes with a minimum radius of curvature as small as 12 mm (0.472 in). It must be emphasized that there are a wide variety of borescope and not all of the above features are available on every instrument.

A second application of the remote microscopes is in the examination of radioactive or other toxic materials which would pose a health hazard to the microscopist.

Alpha or beta emitting radioactive materials are generally biologically hazardous but emit particles that are easily contained by a relatively thin layer of material, therefore the remote portion of the microscope is only the stage area which is enclosed in a "glove box" that precludes the escape of any of the sample material. The eyepieces, and camera are on the exterior of the glove box. Adjustments in the stage area are achieved by manual operation using the gloves. Such instruments are generally adapted from existing microscopes by building a custom, tightly fitting enclosure around the sample area and including sufficient, strategically located glove ports to permit operation of the various controls. Samples are typically inserted and removed through a double-door antechamber for appropriate decontamination procedures. A typical example of such a modified instrument is shown in Figure 6-7.

A remote microscope installation for gamma or neutron emitting materials requires considerable separation and isolation of the sample material from the microscopist. Such a system places a remote objective lens and stage assembly in a "hot cell" with leaded concrete walls and leaded glass windows to provide shielding from the radiation. The eyepieces, remote controls to operate the microscope, and photomicrographic facilities are located on the exterior of the cell; one such instrument is shown in Figure 6-8. Operation of these instruments is monitored either through a leaded glass window or by video monitor. Sample handling is accomplished with remote manipulators.

The microscopy of radioactive materials requires the objective lens to be very near the sample where it is subject to intense radiation. To avoid the rapid reduction of image quality

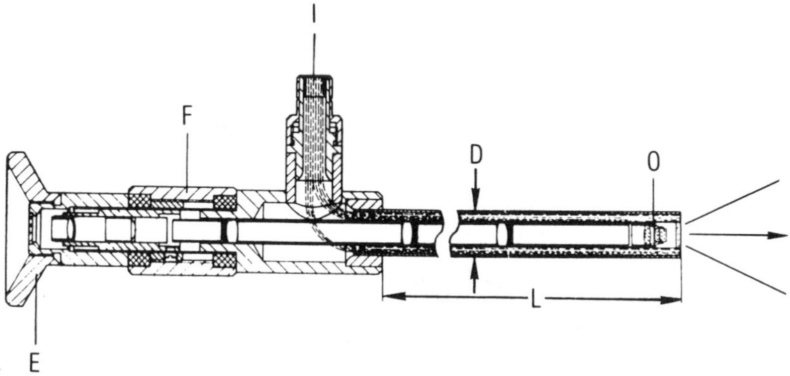

Figure 6-6. Cross section of a fiber optic illuminated, relay lens borescope. These instruments may range in active diameter, D, from 1.7 mm (0.067") to 12 mm (0.472") and in active length, L, from 100mm (4.0") to 2150mm (85"). The area of interest is viewed through the eyepiece, E, and focusing is accomplished with the ring, F. Illumination is provided via fiber optics through the port, I. The viewing direction is forward in this illustration, however, right angle, forward oblique or rear oblique viewing is possible with special heads. Some instruments are equipped with a rotatable prism in front of the objective lens, O, which allows a range of viewing angle. Courtesy of Titan Tool Supply Co., Inc.

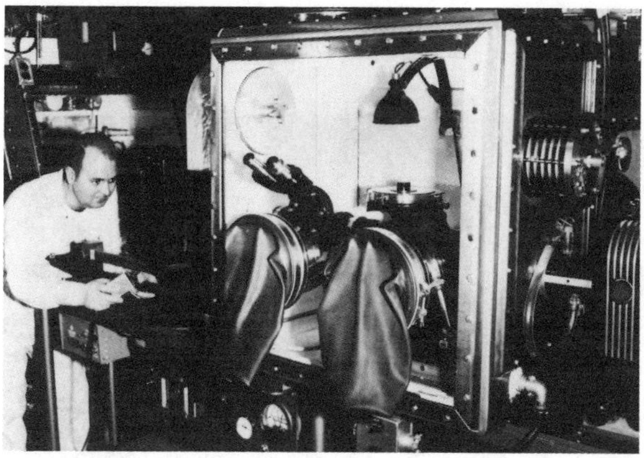

Figure 6-7. Remote microscope for the examination of samples that emit α and β radiation. The stage area is isolated by metal and glass and the samples are manipulated with the rubber gloves shown in the front of the instrument. The interior atmosphere is isolated by filters and vents to an external exhaust system. Similar arrangements can be used for samples that are chemical or biological hazards. Courtesy of Oak Ridge National Laboratory.

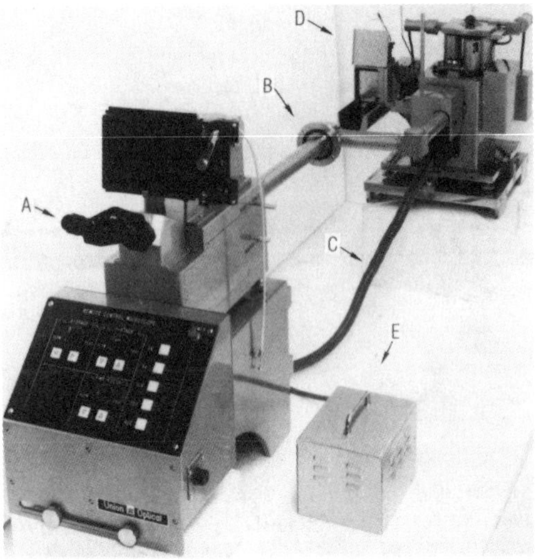

Figure 6-8. A hot cell microscope for the examination of samples that emit highly penetrating γ radiation. This mock-up arrangement shows the arrangement of the essential parts. (A) is the control console including the visual and photographic optics which are located outside the hot cell. (B) is the relay lens system with an "O" ring interface seal; this tube conducts the image information between the hot cell optics and the control console. (C) is the electrical interconnect and (D) is the optical system that resides within the hot cell. Courtesy of Unitron, Inc.

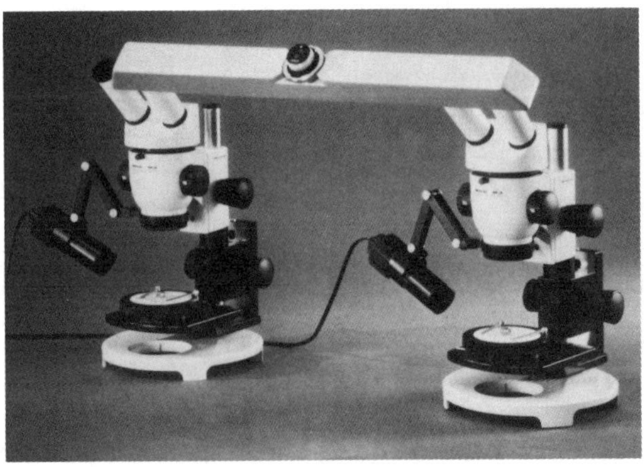

Figure 6-9. An example of a comparison bridge which permits the connection of two similar microscopes for comparison microscopy. Courtesy of Wild Leitz USA, Inc.

due to "browning" or darkening of the glass in the lenses, specially designed radiation resistant objective lenses must be used.

Remote microscopes suitable for hot cell work are available from Leitz and Unitron. The function of these instruments is straightforward, however the details of operation are unique to each instrument, as well as to the specific installation. Therefore, the instructions for their operation are beyond the scope of this book.

V. COMPARISON MICROSCOPES

A. General

The comparison microscope is an arrangement consisting of two identical microscopes joined with a optical bridge that allows the simultaneous observation of two objects in a single image. The images of each microscope may be relegated to one-half of the field of view or they may be superimposed. In the latter case, color filters in the respective optical paths may be used to provide identity for the two images. Transmitted, oblique or normal incident lluminators are available on various instruments. While brightfield or oblique illumination is most common, other forms of illumination such as polarized light can be obtained.

The most well known application of comparison microscopes has been in the field of forensics, particularly in the comparison of bullets. Thus, a given bullet can be associated with a given firearm by comparison with a bullet known to have been fired from that weapon. Typical magnifications for this purpose range from 10X to 300X.

Leitz and Zeiss supply comparison eyepiece bridges which may be used to connect two identical microscopes, however the attachment interfaces usually limit the microscopes that may be so joined to specific models of their own manufacture. A typical comparison bridge arrangement is shown in Figure 6-9. American Optical has also supplied a comparison bridge instrument in the past that was identified as their Duo-Star industrial comparison microscope. Cambridge, the successor to American Optical, now offers a similar instrument. Wild has produced comparison bridges for their compound instruments and currently offers a comparison bridge for their stereo instruments.

Leitz, Cambridge and Unitron also offer complete forensic, comparison microscopes. These instruments have matching revolving nosepieces, integrally mounted on the comparison bridge as shown in Figure 6-10. Located under the objective nosepieces are matching stage assemblies, equipped with mechanical stages and fitted for bullet/cartridge manipulators. The Leitz manipulator is unique in that it permits examining the sample with its axis of rotation at any angle between vertical and

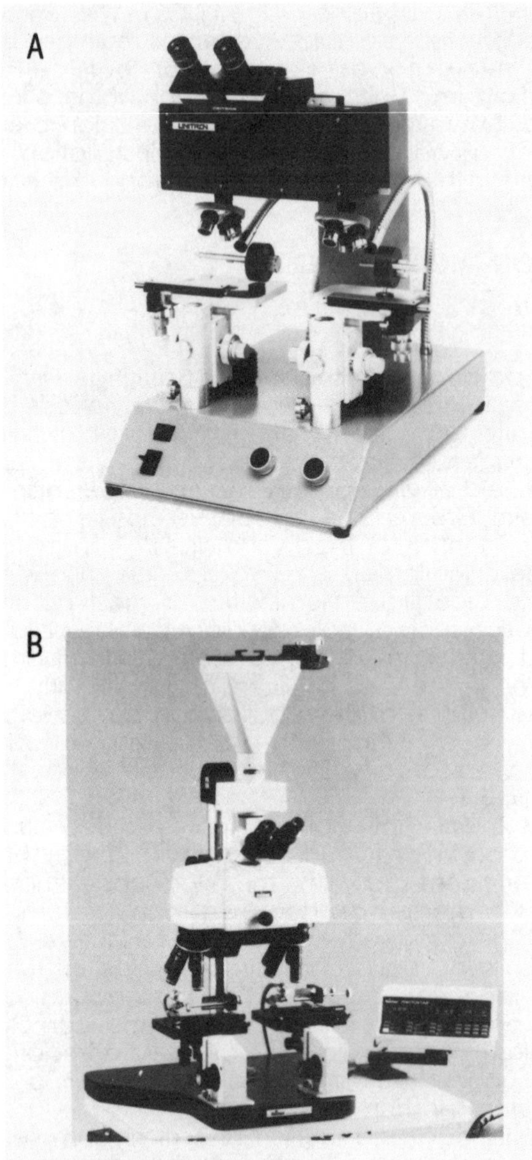

Figure 6-10. Two examples of comparison microscopes. Courtesy of Unitron Inc. and Cambridge Instruments Inc., Optical Systems Division.

horizontal. Dual lamp or fiber optic, oblique illuminators complete these instruments.

Needham (1958) has suggested, in addition to criminological uses, a number of applications for the comparison microscope; among these are the comparison of: 1) hairs, fibers, papers and textiles, 2) powdered drugs, 3) starches, 4) crystals, 5) cells, organisms and small insects, 6) tissue sections, and 7) optical components.

B. Procedure For Comparison Microscopy

Due to the variety of comparison microscopes available it is not possible to cover the unique aspects of each of them. However, a generalized procedure is given below as a guide. This procedure should be satisfactory for either a comparison microscope or two identical microscopes connected with a comparison bridge.

SET-UP OF A COMPARISON MICROSCOPE

1. Determine which arm of the comparison bridge has a magnification adjustment. Designate the microscope affixed to that arm as "B". The other microscope is "A".
2. Turn on the illumination to microscope A.
3. Remove the half stop for microscope A from the optical path.
4. Place a stage micrometer on the A stage and focus it carefully.
5. Calibrate the eyepiece reticle for the specific objective lens of the A microscope, as described in Chapter 5, Section II A.
6. Turn off the A illumination and turn on the B illumination.
7. Remove the half stop for microscope A from the optical path.
8. Place the stage micrometer on the B stage and, using the same magnification objective lens as used on the A microscope, carefully focus the stage micrometer.
9. Adjust the magnification of the comparison bridge B so that the eyepiece reticle calibration is identical to that for microscope A.
10. Again turn on the illumination for microscope A, insert both half stops and and match the observed illumination for the two microscopes.
11. Mount comparison samples on the stages of the two microscopes using special stage accessories as necessary and focus each sample.
12. Remove half stops and by translation and rotation of one or both samples, place the images in the center of view and accurately superimpose them.

13. Compare the samples either superimposed or with the half stops in place.
14. Repeat steps 2 through 10 each time the magnification of the two microscopes is changed.

C. Mask and Wafer Alignment Microscopes

A relatively recent application of comparison microscopy involves the alignment of a semiconductor wafer and the corresponding masks used in the processing of that wafer. Such alignment may be made on one side or both sides of the wafer simultaneously, depending on the sophistication of the alignment instrument. Also it is necessary to make comparisons of two masks to confirm their identical nature. Since the dimensions and location of the various opaque area on these masks are crucial to the ultimate functioning of the integrated circuit these alignments and comparisons must be precise; this can be achieved by direct optical comparison. These instruments are generally equipped with incident illumination, however transmitted illumination may be used with masks.

Unitron also offers a small alignment microscope for attachment to handling instruments involved in the accurate alignment of masks or electronic components. This instrument has an adjustable separation spacing between the objective lenses of 20 mm to 100 mm. Zeiss also supplies the MJM mask alignment microscope that is intended for incorporation into special purpose machines. The separation for the objective lenses of the MJM range from 20 mm to 120 mm.

Requirements for mask alignment and the associated instrumentation change rapidly in this fast paced and competitive field, therefore only three representatives of the more complex instruments will be mentioned. Leitz has two mask comparison instruments, one a largely manually operated instrument, and the Mask Comparator MVG Automatic. The latter instrument uses a computer for both control and data analysis. Masks, reticles, glass wafers, silicon wafers or any combination of these can be compared. Samples ranging from 2.5 to 7 inches can be accepted. Since some of the above samples are opaque and the others transparent, these instruments are equipped with both transmitted and incident illumination.

Karl Suss (represented by Karl Suss America Inc. P.O. Box 157, Waterbury Center, VT 05677) is just one example of a manufacturer of a wide range of mask aligning instruments dedicated to the solid state electronics fabrication industry. The set-up and operation of mask comparators will not be discussed due to their uniqueness and complexity.

VI. DOUBLE VIEW MICROSCOPES

The dual aligning microscope, Figure 6-11, is a unique combination of two incident light microscopes that permits separate or simultaneous viewing of both sides of a relatively thin, flat object at exactly the same x and y coordinates within a tolerance of 0.0025 μm (0.0001 inch). Unitron currently supplies one of these instruments. Another version of this instrument was available from Reichert (originally offered by American Optical). Total magnifications available using matched objective lenses are 50 to 100X for the Unitron instrument and 40 to 200X for the Reichert system.

Typical applications for these instruments include the examination of an object for top-to-bottom continuity of cracks or other defects, missing leads on integrated circuits in printed wiring boards, non-through bores and hole displacements in printed wiring boards. These instruments prove useful in the examination of integrated circuit wafers for special applications.

VII. HOLE INSPECTION MICROSCOPES

The hole inspection microscope is very important in the examination and evaluation of the holes in printed wiring boards used in electronic assemblies. The interior of these holes that accept the leads of the electronic components are often plated with copper to provide electrical interconnection from one side of the board to the other. Defects or cracks in the plating of these holes can cause delayed failures in the electronics.

The hole microscope can take the form of a simple magnifier or a compound microscope but the resemblance ends there. This instrument both in operation and image is quite different from the traditional microscope. Its image may best be described as that of a "fisheye lens" which images a right cylindrical hole as the interior of a truncated cone. Generally the depth of field is so great that a large portion of a hole is in focus. By comparison a standard microscope would show only a narrow circle of the hole in focus. Even a stereo microscope typically produces only a poor image of a hole because its aperture is markedly reduced by the upper edge of the hole.

Another unique aspect of the hole microscope image is that as the printed wiring board is moved across the microscope stage in a given direction, the conical image of the hole appears to tilt in that direction thereby giving a different prospective to the image.

Two, hand-held hole microscopes are available. The Zoom-A-Bore (Available from Titan Tool Supply Co. Inc., 68 Comet Ave., Buffalo, NY 14216) is focusable from 1 inch to 20 feet and may be used on holes from 0.2 to 5 inches in diameter. This device,

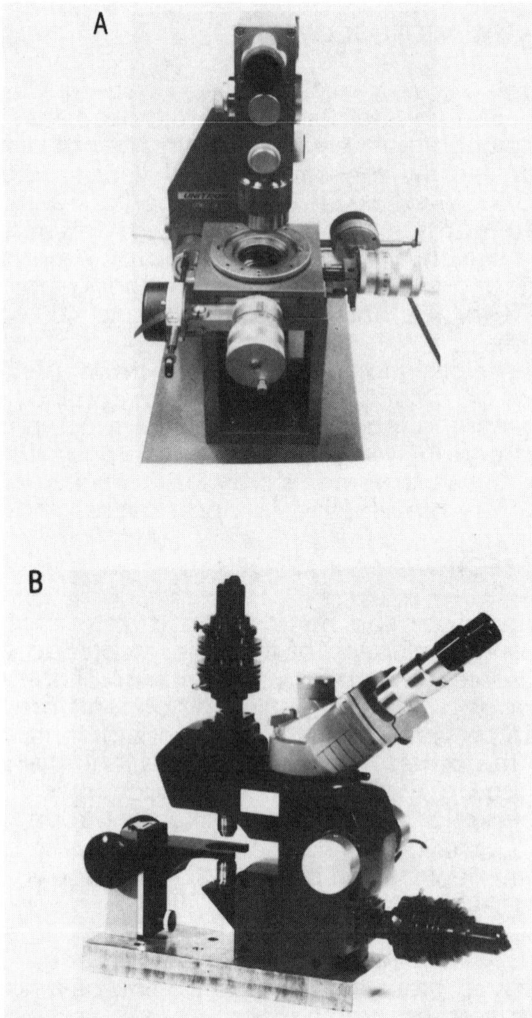

Figure 6-11. Two Double view microscopes. In these instruments both sides of a sample can be viewed either simultaneously or independently with an alignment accuracy of 2.5 μ meter (0.0001"). The lower instrument, formerly manufactured by American Optical Co., is no longer available but is shown in the interest of completeness. Courtesy of Unitron, Inc. and Cambridge Instruments Inc., Optical Systems Division.

shown in Figure 6-12, has no facility for illumination but a flashlight or a light table at the opposite end of the hole may be used to augment ambient light. The Hole Inspector (available from Zi-Tech Division: Aikenwood Corp., 2151 Park Blvd. Box 26, Palo Alto, CA 94302) has a fixed focus that is quite suitable for printed wiring board inspection. This device differs from the rest in that not one image is seen but rather a cluster of nine, each of which shows the hole in a slightly different perspective; this is shown in Figure 6-13.

A larger hole inspection microscope in the form of a shop microscope is available from Unitron (it was formerly marketed as an EPOI instrument). This instrument, shown in Figure 6-14, is designed to inspect holes from 0.002 to 0.375 inch in diameter and up to 6 inches in depth. The microscope body is supported on a post mounted on the illuminated base. The rack and pinion knob does not affect focus as with the compound microscope but rather serves as a zoom magnification adjustment. The maximum usable magnification is about 35X. Focusing of the instrument is accomplished by rotating the focusing eyepiece adjustment. Set-up of this instrument is des-cribed in the procedure below, using the steps preceded with an asterisk.

A very acceptable hole inspection microscope may be easily produced from a standard compound microscope if it is equipped with a focusable Bertrand lens (see Chapter 2, Section III C) or a focusing eyepiece telescope (See Chapter 7, Section VIII). This procedure is given below.

SET-UP OF A HOLE MICROSCOPE[1] OR A COMPOUND MICROSCOPE WITH BERTRAND LENS[2] AS A HOLE MICROSCOPE

1. If the microscope is a polarizing instrument, remove the analyzer from the light path.
2. *Turn on the transmitted light source.
3. Place a 40 to 50X objective lens and the Bertrand lens into the light path.
4. *Using the coarse focus knob adjust the microscope until the objective lens is about 1/4 inch from the sample. Focus the image with the Bertrand lens focus knob.Adjust the magnification of the image with the coarse focus knob.
5. Adjust the condenser focus to optimize the lighting.
6. *External illumination may be used if the illuminator direction is very close to the optical axis of the microscope.

[1]The steps preceded with an asterisk (*) may be used to set-up the Unitron Hole Inspection Microscope. In this case, substitute "eyepiece" for "Bertrand lens" in the procedure.

[2]If the microscope is not equipped with a Bertrand lens a focusing eyepiece telescope may be substituted for the standard eyepiece and the Bertrand lens.

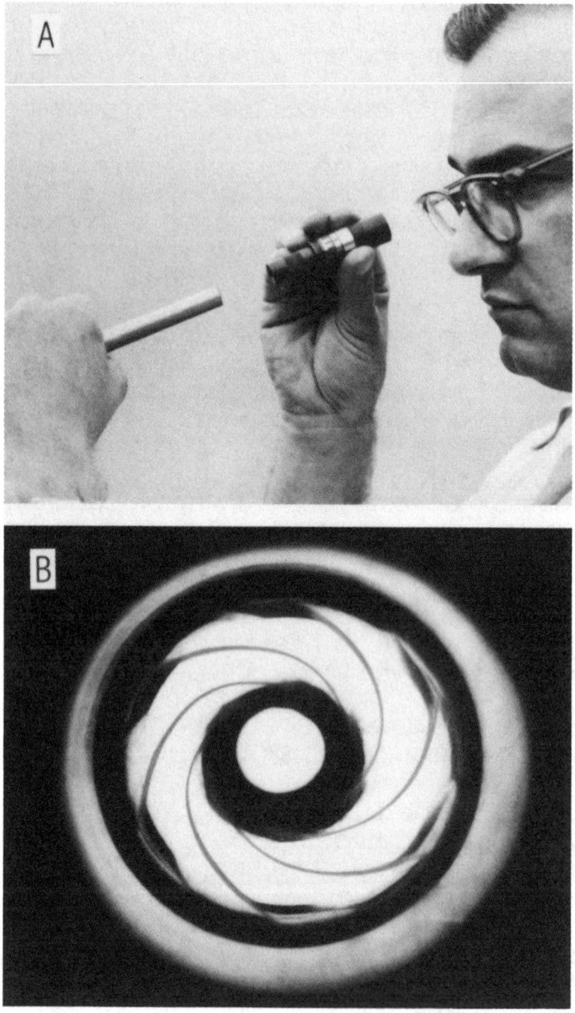

Figure 6-12. The Zoom-A-Bore hole inspection microscope can be used for inspection of holes from 5 mm (3/16") to 125 mm (5"), ranging in depth from 25mm (1") to 6 m (20'). This microscope relies on external illumination. (A) shows the instrument in use and (B) show the inside of a rifle barrel as seen through the Zoom-A-Bore. Courtesy of Titan Tool Supply Co., Inc.

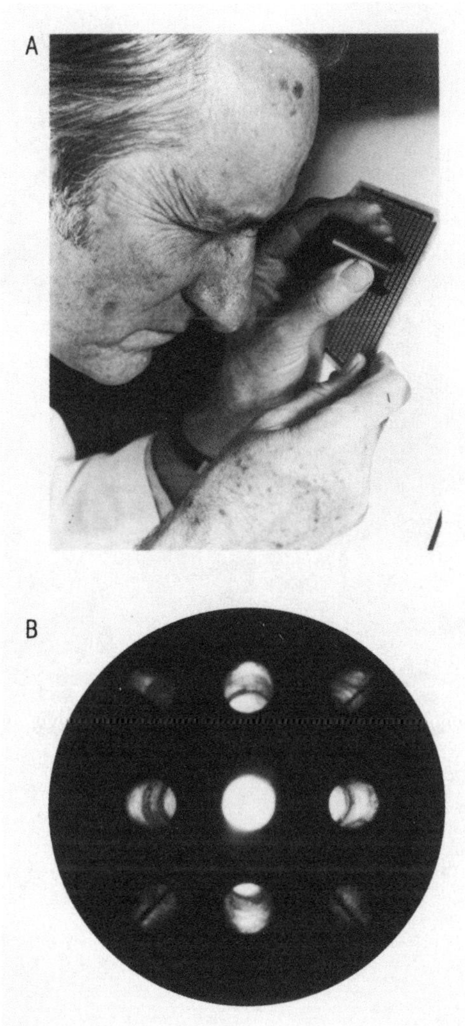

Figure 6-13. The ZI-TECH hole inspector is a unique hole inspection microscope. This instrument has a long truncated prism through which the hole is viewed. The resulting image provide a direct view of the hole surrounded by eight tilted views of the hole. Courtesy of ZI-TECH Instrument Corp.

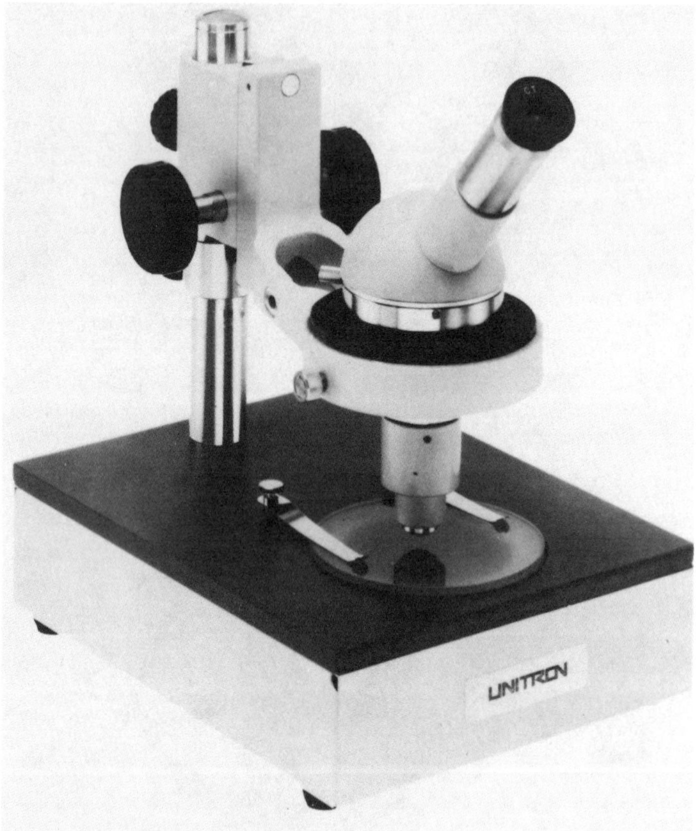

Figure 6-14. The Unitron hole inspection microscope is bench type instrument that is ideally suited to the examination of through holes in printed wiring boards. Focusing is achieved by adjusting the eyepiece, while the knob usually used for focusing adjusts the magnification of the image. Courtesy of Unitron, Inc.

VIII. PROJECTION MICROSCOPES

The projection microscope is a primary teaching aid in the biomedical field. A typical instrument, shown in Figure 6-15, is a transmitted, brightfield microscope with a high intensity lamp and an exit prism for redirecting the image to a projection screen. With a quartz halogen or xenon source the microscope can project a 2 m diameter image that does not require a totally darkened room for an acceptably bright image.

Since the illumination for a projection mcroscope is a high intensity incandescent source, a beam stop that may be shut when the sample is not being viewed and an efficient infrared filter for the illuminating beam are necessary to minimize thermal damage to the sample. To minimize stray light in the room the illuminator and condenser systems are enclosed.

Objective lenses are available from about 6X to 40X; oil immersion lenses are not generally offered. Total magnifications may range as high as 3000X at the projected image.

The projection lenses are several specially designed, low power eyepieces that will give reasonable projection distances of 5 to 9 meters. The projection prism, mounted on or beyond the projection eyepiece lens, redirects the image forming beam toward the screen. The prism may be rotated horizontally and tilted vertically. The prism may be removed for visual or photographic applications, at reduced illumination levels.

IX. SHOP MICROSCOPES

Unlike many of the sophisticated instruments described above and elsewhere in this book, there are a number of important, "workhorse" type microscopes that receive continuous, daily usage in machine shops and other heavy industrial areas. Among the more important of these instruments are the toolmakers's microscope and the profile projector.

The toolmaker's microscope is a very versatile instrument; usually having transmitted, incident and oblique illumination capabilities, a translating and revolving stage with micrometer read-outs. Various types of stage fixtures are available to permit the measurement of a wide range of objects in a variety of orientations. The toolmaker's microscope is capable of accepting a wide range of objective lenses, however, the lower powered ones with large working distances are most commonly used. The eyepiece may be equipped with an almost unending selection of special reticles for the measurement of a wide range of objects. Measurement eyepieces of either the double image or filar micrometer type, and protractor eyepieces round out the selection of visual measuring systems. An example of a well outfitted toolmaker's microscope is shown in Figure 6-16.

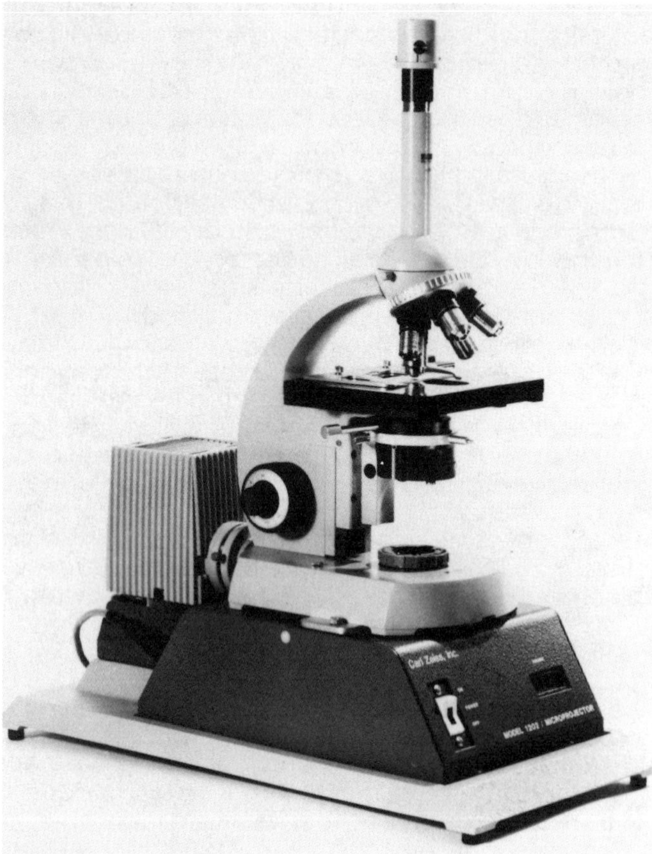

Figure 6-15. A projector microscope consisting of a standard compound microscope fitted with a high intensity illuminator and special eyepiece fitted with a prism to permit horizontal projection onto a screen for multiple viewing. Courtesy of Carl Zeiss, Inc., New York.

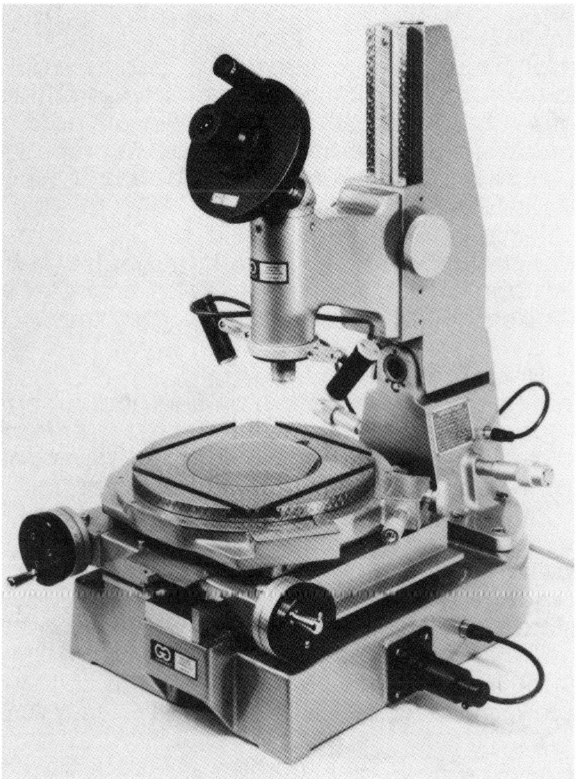

Figure 6-16. A toolmakers' microscope consisting of a microscope with a protractor eyepiece, a support column with adjustable tilt, and a mechanical stage that permits accurate and precise measurements in both lateral directions as well as angular measurements. It may be equipped with a number of stage accessories for holding various types of workpieces. Courtesy of Gaertner Scientific Corp.

Toolmaker's microscopes are available from a number of manufacturers including Gaertner Scientific Corp. (1201 Wrightwood Ave., Chicago, IL 60614), Leitz, Mitutoyo Mfg. Co. Ltd., Tokyo, Japan (Available from distributors in The U.S.),Nikon, and Unitron.

Another important microscope used in the manufacturing area is the profile projector. This instrument consists of: high intensity illuminators, both transmitted and incident; a sample holding stage; and a simple microscope that projects an image of the sample or its profile onto a built-in screen. The profile projector also has a wide variety of stage accessories to accommodate a large number of samples. The instrument can be purchased in several configurations, as shown in Figure 6-17.

Since the profile projector is a simple microscope, the magnifications are generally low, ranging from 5X to 100X. Some manufacturers also offer a zoom lens. Each projector lens require a specific condenser lens and, as a result, they are usually sold as a matched pair. Some of the manufacturers of profile projectors are Mitutoyo, Nikon and Unitron.

The centering microscope is a unique and important version of the microscope used for precision alignment of vertical milling machines, large jig borers, boring mills, lathes, electrical discharge machines, tape controlled machines and drill presses. This device, as shown in Figure 6-18, is a microscope that is precisely aligned to a hardened shank which mounts in the collet of the machine. In this way, positions of edges, centers, contours, irregular shapes and layout lines may be accurately obtained. Focusing of the centering microscope is accomplished by changing the distance between the microscope and the object to be centered. A number of these instruments with magnifications of 10X to 200X are available from Titan Tool Supply Co., Inc., 68 Comet Ave., P.O. Box 182, Buffalo, NY 14216.

A final word is in order regarding the care and cleaning of shop microscopes, in light of their environment. Routine maintenance is important for both the optical and the mechanical parts of the instrument. When not in use the instrument should be covered to prevent accumulation of dirt and oils on the instrument. When cleaning of the optics is necessary, all of the dust and dirt should be carefully removed first, using a jet of compressed "air" only from an aerosol duster. If the lens has oil on its surface, it should be cleaned with a cotton swab lightly soaked with xylene and then dried with a second swab. Then clean the lens thoroughly as described in Chapter 10, Section VI.

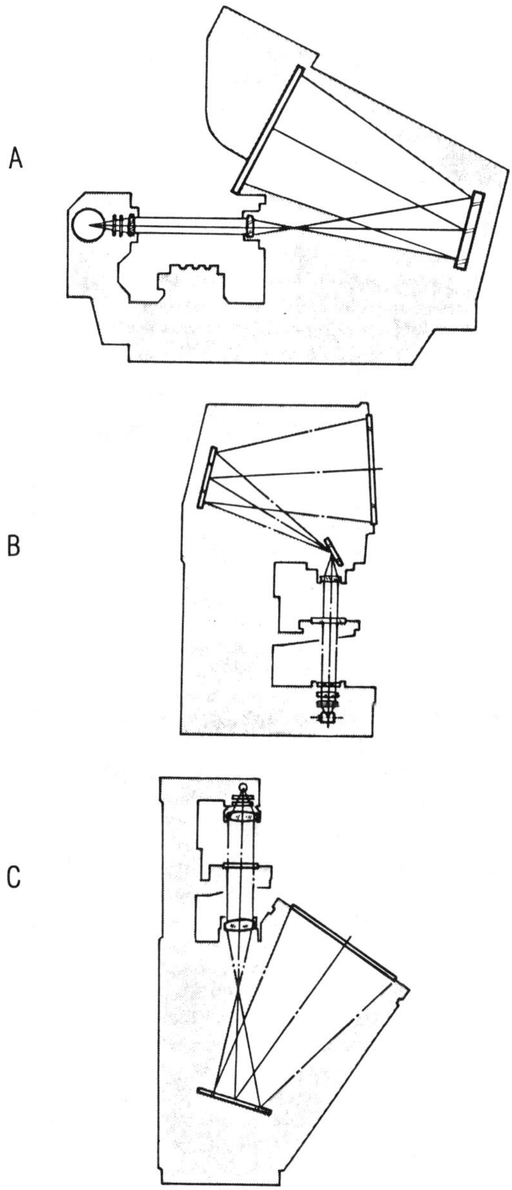

Figure 6-17. Profile projectors are basically a simple microscope with a large mechanical stage and a bright illuminator that projects the image of the workpiece onto a large viewing screen. Precise and accurate angular and lateral measurements may be made using the cross ruling and the angular scale on the projection screen. Courtesy of MTI Corp.

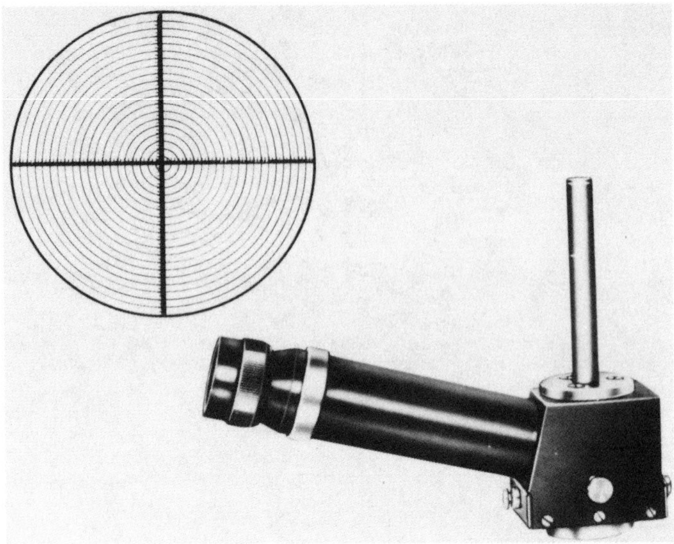

Figure 6-18. A centering microscope is an indispensable tool for the precision alignment of various machine tools such as milling machines, jig borers, boring mills, electrical discharge machines, and drill presses, either manual or computer driven. When the shaft of the microscope is mounted in the chuck of the instrument and the work piece viewed through the eyepiece, edges, centers, contours and layout lines may be accurately located. The inset shows a typical reticle for the centering microscope. Courtesy of Titan Tool Supply Co., Inc.

REFERENCES

Hyzer, W. G., Industrial Research & Development, 200, Feb. (1982).

MIL-M-83524, Microscope, Optical, Monocular, Hand-Held, Portable, Militarized, 200X Magnification, for Field Inspection of Optical Fibers, 18 October (1985).

Needham, G. H., "The Practical Use of The Microscope," p 87, C. C. Thomas, (1958).

7

Accessories

I. INTRODUCTION

The light microscope is a very powerful analytical instrument by itself, however, a wealth of accessories have been designed that multiply the utility of the instrument in a wide range of disciplines. Much of the usefulness of the microscope lies in the fact that the sample may be examined under ambient conditions of temperature and pressure but even these conditions may be altered with appropriate accessories.

The accessories discussed in this chapter are categorized with respect to their relationship to the microscope; specifically, the various accessories may be considered to replace or augment the stand, stage, illumination, objective lens, or eyepiece. Filters and similar accessories may be an adjunct to either the illumination or the stand and will be treated as a separate class.

Accessories may be available from the manufacturer of the microscope, from an after-market source or may be fabricated by the user. The descriptions of accessories from the first two sources will include their availability at press time. Those more important accessories that are no longer available will also be listed and so indicated because examples of these may be found hidden in storage areas or may be available as used items. Many fine accessories have been fabricated by the microscopists and a number of these are listed with their respective references.

II. ACCESSORIES FOR THE STAND

The accessories for the stand cover a broad range of items from vibration isolation supports to those which fit directly on some portion of the stand.

A. Vibration Isolation

In those microscopy labs afflicted with mechanical vibration whether from automotive/railroad traffic or building instabilities it is necessary to decouple the microscope from the vibration if high magnification photomicrographs are to be made. A simple way to evaluate such vibration is to place a petri dish filled with water on the bench and observe the surface of water in oblique illumination for ripples due to the vibration. Any significant disturbance of the water surface is cause for concern.

The isolation of the microscope can be achieved by using an vibration damping system between the vibration source and the instrument. Kinetic Systems, (20 Arboretum Road, P.O. Box K, Roslindale, MA 02131) offers several systems from complete, vibration-free tables to tabletop isolation platforms. Newport Corporation (18235 Mt. Baldy Circle, P.O. Box 8020, Fountain Valley, CA 92728) also offers several series of tabletop isolation platforms. Diagnostic Instruments, Inc. (6510 Burroughs, Sterling Hts. MI 48078) also offer a pneumatically isolated isolation platform. In severe vibration locations it may be necessary to test several different systems to select that one which will most effectively reduce the vibration.

B. Powered Focusing Accessories

Focusing aids are a group of stand accessories that can significantly reduce fatigue for the microscopist who must make observations for long periods of time. Wild supplies a motor-focused stage which affixes to the post of their M8 stereomicroscope. The stage elevation is controlled by a foot pedal arrangement.

McCrone (Accessories & Components Div., 850 Pasquinelli Dr., Westmont, IL 60559) offers a similar device, the Electra-Focus, that clamps to the microscope stand and attaches to the focus knob. A two-way foot switch controls a motor drive which will raise or lower the focal point of the instrument, thereby leaving the hands free for manipulation of the sample. This accessory does not interfere with the normal operation of the microscope. This device is best used for stereomicroscopes but could be used on compound microscopes at lower magnifications as well.

Velmex, Inc. (P.O. Box 78, Holcomb, NY 14469) supplies a complete motorized focusing stand to which a microscope body can be attached. The microscope can be moved up or down with manual or pedal switches to achieve focus at two different speeds, one for quick location of focus position and the other for critical focus. Dynamic braking takes place the instant that the switch is released. They also supply motorized X, XY or XYZ stages to further the utility of this stand. Again, such systems are most easily used with stereo or other low power incident illumination microscopes.

The most interesting of these focusing accessories is the Auto F Focus System, TIC-AF1 supplied by Technical Instruments Company (348 Sixth St., San Francisco, CA 94103). This system consists of a signal detector assembly, a focus servo assembly and an electronics control module. In operation the microscopist first focuses on the sample and then depresses the auto-focus, lock-on button. The autofocus system then will maintain correct focus through a range of $\pm$0.5 mm. This system may be retrofitted to most modern microscopes.

C. Multiple Viewing Accessories

For two or more persons to carry out a meaningful dialog about the microstructure of a sample it is necessary either to study photomicrographs/drawings or, better, to examine the actual sample image simultaneously. Photomicrographic accessories are of sufficient importance and diversity that they have been made the subject of Chapter 8. The various other means for group viewing of the sample will be described in this section.

The camera lucida or the drawing attachment, Figure 7-1, is a simple means of producing drawings of a sample. The camera lucida may be configured either as a stand accessory or as an eyepiece accessory. In operation, the image of the sample is viewed in the normal fashion with the image of the drawing surface superimposed on that image by means of a projection lens and mirrors. The image of the sample may then be traced on the drawing surface. The image size can be changed at the drawing surface with the objective lens and on certain camera lucidas with the projection lens that has zoom capability. The relative intensities of the sample and drawing surface images can be changed by adjusting the microscope illuminator or the room lighting. Camera lucida attachments are available from most major manufacturers.

A word is in order regarding the relative virtues of the photomicrograph and the camera lucida drawing. It is true that the photomicrograph records the image faithfully and in color, if desired, however the drawings are to be preferred if a specific detail is to be singled out for emphasis.

Microprojection is also a method for multiple viewing that is described in Chapter 6, Section VIII. Projection images can also be viewed on a screen that is an integral part of the microscope. For example, the ground glass focusing screen used for photomicrography can be used for multiple viewing of the

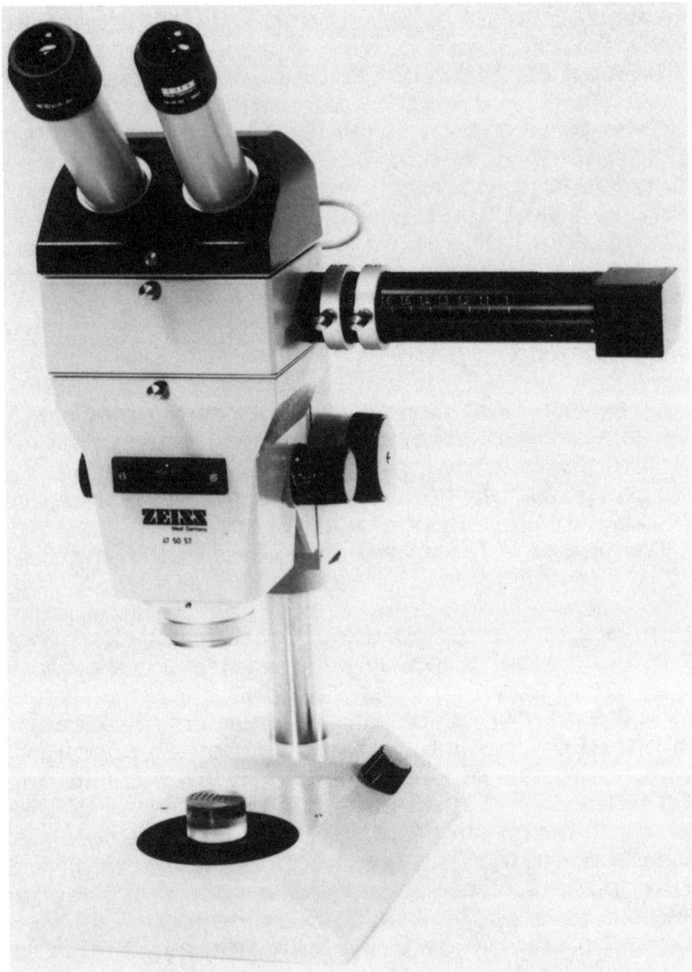

Figure 7-1. The camera lucida or drawing attachment permits simultaneous viewing of the sample and a drawing surface located beneath the attachment at the side of the microscope. The structure of the sample may then be traced on the drawing surface with pen or pencil. Courtesy of Carl Zeiss, Inc., New York.

sample. This screen is generally of limited value for this purpose because of the poor image resolution and brightness. The modern viewing screen, shown in Figure 7-2, is a significant improvement over the older ground glass because brighter sources of illumination, such as the xenon burner, are available and the ground glass screen is augmented with a Fresnel lens that acts as a field lens. Nevertheless, it is highly desirable to darken the room somewhat for viewing the screen. Most major manufacturers can provide viewing screens for their research instruments.

For the highest quality, multiple view images, both in resolution and brightness, a discussion bridge or a multiple binocular head, as shown in Figure 7-3, is recommended. The image is split by prisms between two or more binocular viewing heads. Since the discussion bridge produces only virtual images of the sample, a moveable pointer that is in focus with the sample image is often included to permit the mutual identification of an area of interest. Discussion bridges are available from most major manufacturers but they are commonly designed to fit a specific stand. The only additional requirements for the use of the discussion bridge are a somewhat higher level of illumination and a sturdy bench, free from vibration.

The most recent method for multiple viewing uses a black and white or color video camera that is attached to the camera port of the microscope and that is connected to one or more video monitors of any desired size. This arrangement, shown in Figure 7-4, is referred to as closed circuit television or simply, CCTV. The lighter cameras may be mounted directly on the vertical tube of the microscope, however the larger, heavier cameras should be affixed to a camera stand. Some camera units are equipped with a pointer for locating areas of interest in the image.

The resolution of CCTV is somewhat limited and is comparable to that with the ground glass screen. Since the image is composed of a number of horizontal lines, ranging from the 525 line, NTSC standard of U.S. commercial TV to 1000 line high resolution systems, the vertical resolution is limited. The horizontal resolution is, on the other hand, somewhat better.

The CCTV system magnifies the image presented to the sensitive surface in the camera by the ratio of the diameter of that sensitive surface to the diagonal diameter of the monitor tube; this ratio times the magnification of the microscope at that image plane gives the total magnification of the system. More accurate values for the magnification can be determined by measuring the image of a stage micrometer on the monitor. This procedure is analogous to that described in Chapter 5, Section II A for an eyepiece reticle.

CCTV projected microscopy has at least two unique capabilities to consider. Any image presented to the camera may be electronically enhanced to provide increased contrast as described in Chapter 4, Section X. Secondly, the resulting image

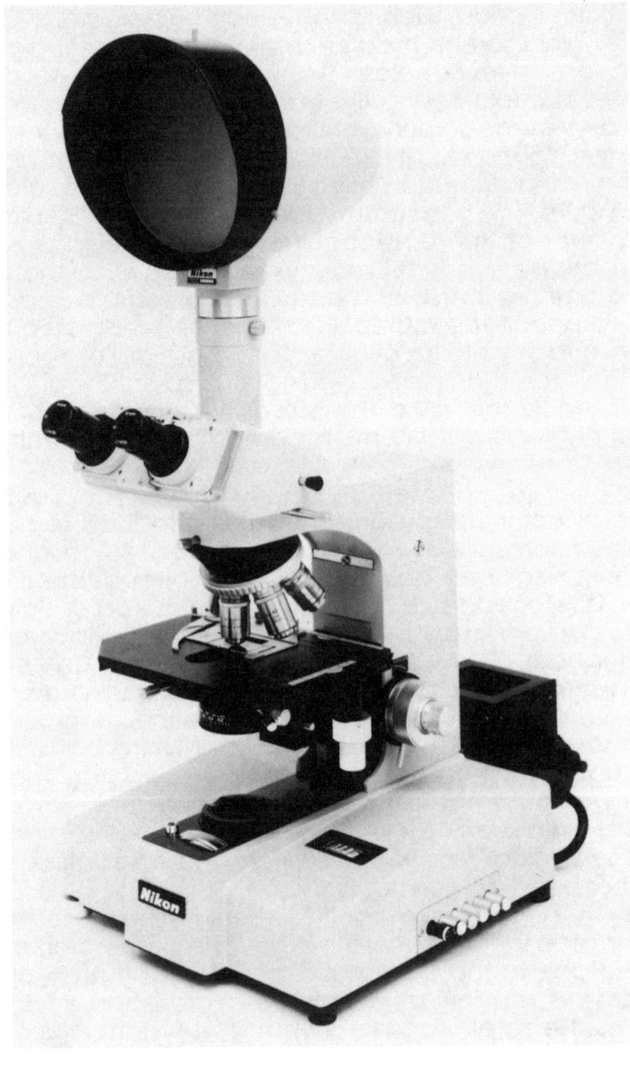

Figure 7-2. Microscope with viewing screen permits easy viewing by several people. Courtesy of Nikon, Inc.

A

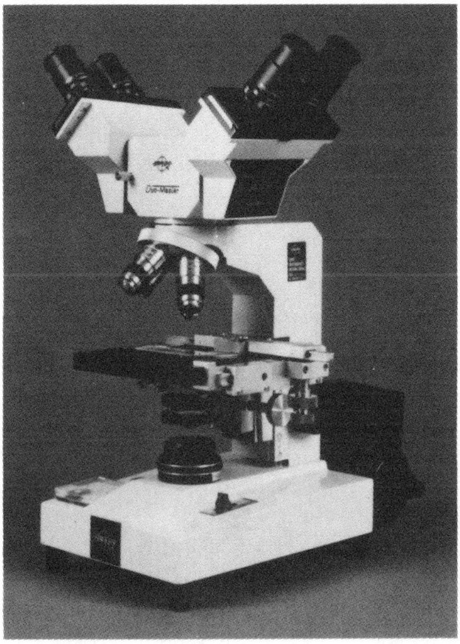

B

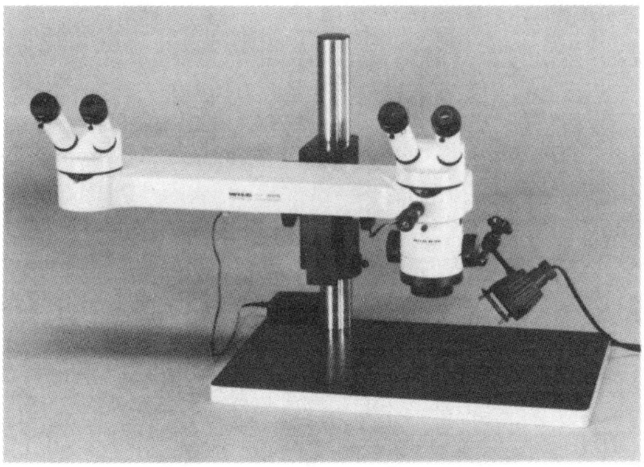

Figure 7-3. Simultaneous observation of a sample by two persons can be accomplished with a double head system, (A), or discussion bridge, (B). Courtesy of Swift and Wild Leitz USA, Inc.

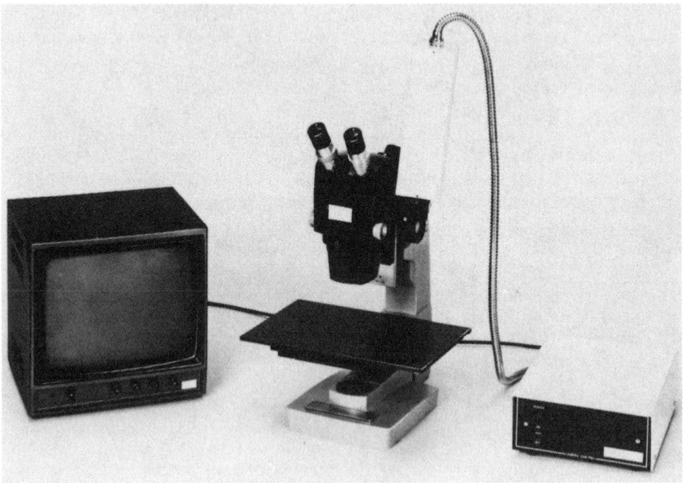

Figure 7-4. The most recent means for multiple viewing of a microscopical image uses closed circuit television, (CCTV) and one or more television monitors. Courtesy of Cambridge Instruments Inc., Optical Systems Division.

may be observed anywhere the video information can be transmitted and on as many monitors as desired.

D. Miscellaneous

The Bertrand Lens is an extremely useful accessory for centration of phase contrast objective lenses and the easy adjustment of the aperture diaphragm. Modulation Optics, Inc. (100 Forest Drive at East Hills, Greenvale, NY 11548) offers a Bertrand lens attachment for a number of microscope models. Once installed the Bertrand lens may be inserted into the optical path at will and is focusable. The more common uses of the Bertrand lens are described in Chapter 2, Section III C and Chapter 4, Section IV B and C. A rather unusual application for the Bertrand lens is given in Chapter 6, Section VII.

The base supplied with most microscopes is adequate for most purposes. However, it is very common that special applications of a stereo microscope require a unique base or support. In those cases where the microscope manufacturer cannot supply the necessary base, an appropriate one can be purchased from Diagnostic Instruments, Inc. An example of such a base is shown in Figure 7-5; this base has a special upper support with linear ball bearing motion that will support a microscope and large format camera while permitting vertical movement for focusing.

III. ILLUMINATION ACCESSORIES

Most modern compound microscopes and some stereo microscope are equipped with built-in illumination specifically suited to the instrument. This arrangement is convenient, minimizes the possibility of incorrectly adjusted illumination and thereby reduces the time required to conduct a study. For the remaining stereo microscopes, the older compound microscopes, and the macrographic instruments an extensive assortment of external illuminators is available from a number of sources.

The ideal external illuminator for compound microscopes, especially when used for photomicrography, is one that has a focusable condenser lens and a field diaphragm so that correct Koehler illumination can be obtained. The set-up and use of these illuminators are described in Chapter 2, Section IV B. Excellent examples of such illuminators are available, for example: (1) the American Optical Microlamp, models 370-A and 735-A and the General Purpose Illuminator from Bausch and Lomb.

Illuminators for critical normal incident photomacrography require, as a minimum, a focusable condenser lens; this is discussed in Chapter 9, Section IV C. The illuminators described above for the compound microscope are ideal, however there are a number of less sophisticated focusing illuminators that will serve admirably.

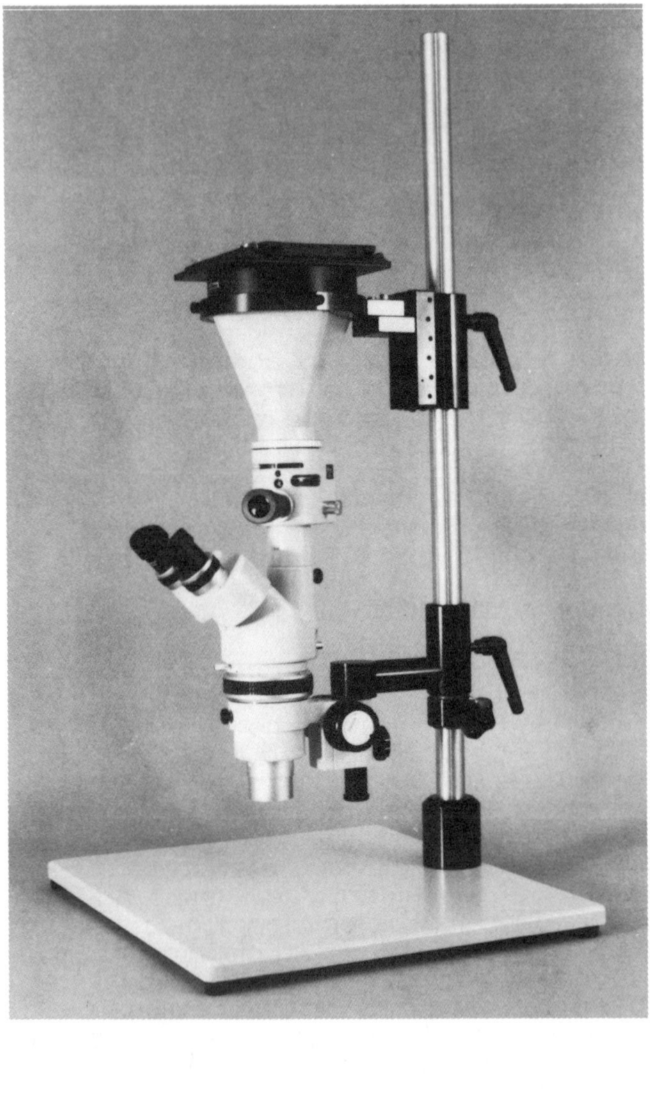

Figure 7-5. A special microscope stand that provides support for the camera. A linear bearing allows vertical motion of the camera as the microscope is focused up and down. Courtesy of Diagnostic Instruments, Inc.

Requirements are least stringent for oblique incident illumination in photomacrography and stereo microscopy; in these cases the illumination that the achieves the desired results is the best by definition. A wide range of illuminators from a number of manufacturers is available for this purpose.

Fiber optic illuminators are excellent illuminators for small areas. These usually consist of an variable intensity illuminator housing to which is attached one or two fiber optic cables. The fiber optic cables are usually armored with a flexible metal sheath that will remain in position, once located. Some of these illuminators are also equipped with an adjustable lens at the end of each cable for focusing of the illumination to match the sample size. A fiber optic cable terminated in a ring illuminator that surrounds the objective lens of a stereo microscope is very useful for illuminating surface detail. Fiber optic illuminators are available from a number of sources including American Volpi Corp. (26 Aurelius Ave., Auburn, NY 13021) and Dolan-Jenner Industries, Inc. (Blueberry Hill Industrial park, Box 1020 Woburn, MA 01801).

Flash lamps are brilliant sources of daylight illumination for photomicrography or photomacrography and are particularly useful for moving objects such as microorganisms because of their very short flash duration. However, it is that short period of illumination that makes it difficult to predict how well the finished photo is illuminated.

One illumination adaptor must be mentioned in closing; namely the Leitz Variolum. This device fits between one of the Leitz illuminators and one of their compound microscopes and provides advanced control of the illumination. The most important features that this device provides are: homogenization of the illumination, even with poorly centered lamps; continuous adjustment of light intensity without changing the color temperature of the illumination; adjustment of color temperature by progressive insertion of conversion filters; and continuous variation of the light color.

IV. STAGE ACCESSORIES

It should not be surprising to the reader that the largest group of accessories for the light microscope is associated with the stage. The sample is usually observed under ambient conditions however, it was early in the history of the microscopethat researchers realized that it could be augmented with accessories to permit measurements at high and low temperatures as well as in various other environments. Furthermore, stage accessories for rapid and easy manipulation as well as for quantitative and crystallographic measurement of the sample will be described. This section will provide a selected compilation of both commercially available accessories and user developed devices for many of these applications. Every attempt has been made to assure that the commercial accessories described are available at press time, however, discontinued devices of special interest and wide distribution have been included on the chance that they are available from a storage area or that they might be purchased in a used condition.

A. Sample Translation Stages

The sample translation stages include a wide range of accessories that permit x and y translation, in the plane of the stage, and sometimes z translation, along the optical axis of the microscope.

Most microscope manufacturers provide versions of the mechanical stage for their instruments, as shown in Figure 7-6; these provide simple x and y translation of the sample by means of manual knobs. The newer mechanical stages often combine both translation knobs on a single shaft that is ergonomically located near the bench surface for easy operation. Most mechanical stages are also equipped with metric scales and a vernier that permit the location of a desired feature on the sample to about 0.1 mm in both orthogonal directions. Those mechanical stages for revolving stages must be carefully designed for minimum overhang on the edge of the stage and a low profile above the stage to avoid collision with objective lenses on the revolving nosepiece. Edmund Scientific (101 E. Gloucester Pike, Barrington, NJ 08007) offers two excellent aftermarket mechanical stages that can be added to most existing microscopes without much effort.

Standard Stage Co. (790 Lucerne Drive, Sunnyvale, CA 94086) supplies several aftermarket mechanical stages which will accept very large samples up to 6" in diameter. Very precise X-Y and revolving mechanical stages are available from a number of sources such as Lansing Research Corporation (P.O. Box 730, Ithaca, NY 14850) Melles Griot (1770 Kettering St., Irvine, CA 92714) and Rolyn Optics Co. (706 Arrowgrand Circle, Covina, CA 91722). These aftermarket stages generally require some machining to properly affix them to the microscope.

Phipps (1966) has described the construction of a pneumatically driven stage for the scanning of a sample and Stevenson (1951) has published information on a stage which provides X,Y and Z (focus) motions as an aid for photomicrography.

The construction of two novel, magnetically geared stages have been described by Katz (1966). In this application two small magnets are rotated in such a way that the magnetic stainless steel stage is smoothly translated without the need for a rack and pinion. The first stage has the potential for the orthogonal translation of quite large samples; for example, Katz used such a stage for the examination of nuclear tracks on glass photographic plates. His second application of this technology is a magnetically geared revolving stage for attachment to an existing mechanical stage.

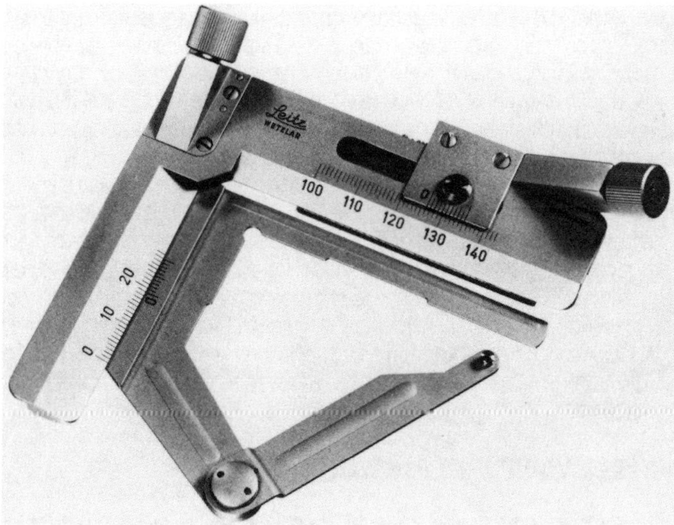

Figure 7-6. The mechanical stage allow precise, orthogonal translation of the sample on the stage of the microscope. The two scales with their respective verniers permit relocation of a detail on a sample with an accuracy of 0.1 mm. Courtesy of Wild Leitz USA, Inc.

The logical extension of the manually operated mechanical stage is to be found in the motor driven and computer controlled versions. Leitz has provided a mechanism for attachment to their mechanical stage which translated the stage with a synchronous motor. Zeiss makes available a number of computer controlled stages for various models of their instruments. These latter devices can provide both scanning and positioning capabilities with a precision of ±0.1 mm.

Notable, after-market computerized stages are available from Melles Griot, ITP Inc. (495 Mercury Drive, Sunnyvale, CA 94086), and New England Affiliated Technologies, Inc. (620 Essex St., Lawrence, MA 01841) for custom attachment to existing instruments. However, due to the size and weight of these devices the microscope should generally be one of the more rigid research models. A wide variety of stages are possible, with all translations or rotations accomplished by computer controlled stepper motors. A typical computer controlled stage is shown in Figure 7-7.

The Piezo-Flex stage (Available from Wye Creek Designs, 5319 Wye Creek Drive, Frederick, MD 21701) is one of the more unique after-market mechanical stages for the light microscope. This stage provides X-Y translation of the sample with a precision of ±0.001 μm and without backlash using electrically driven piezo ceramic inserts. Total displacements in the two orthogonal directions however, are limited to 150 μm. Integral flexures are used which provide smooth motion without mechanical backlash or "bearing noise".

B. Stages For Quantitative Analysis

One of the most important applications of the microscope in the area of materials analysis is measurement of sample dimensions. Concurrent with the development of the microscope, continuing attention has been given to the design, construction, and manufacture of accessories which simplify and speed up this measuring process. Undoubtedly, the culmination of this effort is in the computerized accessory systems which can translate the sample on the stage, collect the image information and automatically compute and print out the desired information; this systems is discussed in Chapter 5, Section IV D. However, a number of the devices for this purpose, both commercially and user fabricated, will be discussed here. Most of the commercial instruments are no longer manufactured, but where they exist, they constitute a valuable resource for the occasional or limited quantitative analysis.

As described in Chapter 5, Section IV, a number of quantitative measurements may be made on a sample by lineal analysis. Lineal analysis may be accomplished by translating the sample and ascribing a measured length to each phase or structure and from that calculating a volume fraction for each.

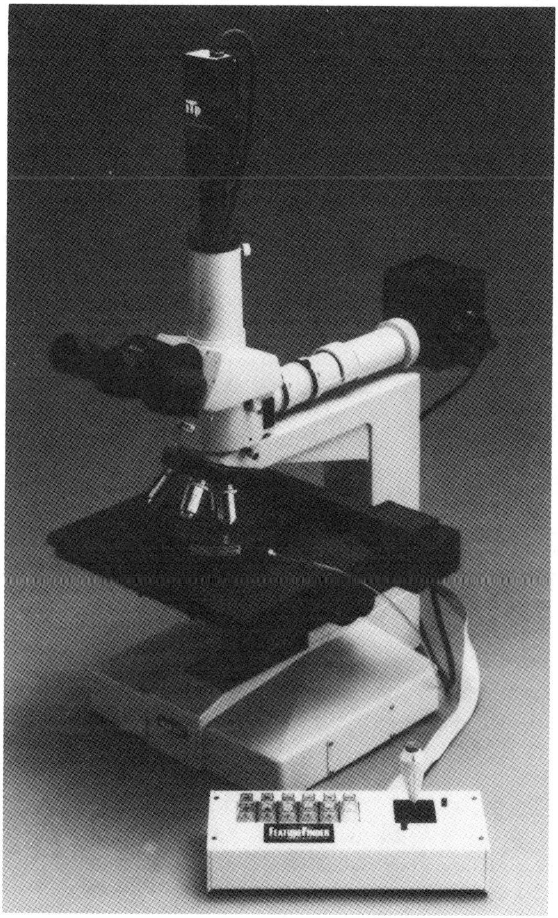

Figure 7-7. An example of a computer controlled 6"X 6" stage. The stage position is controlled with an IBM or compatible computer with a precision of 1 m. Stage coordinates are displayed on the computer screen. Stage speed is 2" per min. Courtesy of Mr. Bob Kinder, ITP Inc.

One of the early instruments for automating such measurements was developed by Wentworth (1923); a commercial version of this instrument is marketed by Eberbach Corp.(P.O. Box 1024, Ann Arbor, MI 48106). This device consists of a mechanical stage which is translated by any one of 5 coaxially arranged micrometer knobs, each of which can be assigned to a specific phase in the sample. In operation, the micrometer assigned to phase X is rotated while phase X is being traversed, using a cross wire eyepiece as a reference point, the other phases being traversed with their respective micrometers. The sum of all the micrometer values constitutes the total traverse and the individual micrometer values divided by the total traverse gives the length fraction traversed in each phase. The volume fractions are proportional to these length fractions. Leitz also supplied a similar device, their Integrating Stage which allowed the analysis of up to six phases. Hurlbut (1937) has described the construction of a motorized mechanical stage with provision for simultaneously translating the sample and engaging a mechanical counter. The drive had seven such counters permitting the lineal analysis of as many phases.

Point counting is a second means of performing quantitative measurements of microscopic structures; the utility of this method is described in Chapter 5, Section IV. This procedure requires the translation of the sample in uniform steps and assigning each one of these locations to a specific phase. The number of locations observed for a given phase divided by the total number of locations measured is proportional to the volume fraction of that phase. Chayes (1949, 1955) and Rosenfeld (1954) have described point counting stages based on modifications of a standard mechanical stage. Ford (1954) has described a point counting system incorporating an electrical system that permits the recording of the specific phase at each point.

Swift has offered an electrically-triggered traversing stage which has a switch for each phase as does the Ford device. However, in this case the microscopist notes the phase under the cross wire of the eyepiece and presses the switch for the corresponding phase, this records the presence of that phase on its counter and translates the stage to the next measurement point.

C. Sample Orientation Stages

Sample orientation stages are devices for the positioning of samples, for stereo photomicrography or for conducting crystallographic determinations but these may be used for various other purposes, as well. The simplest of these stages is the tilting stage. The McCrone tilting stage was designed to tilt a sample between two successive photomicrographs thereby producing a pair of pictures to give a stereo image, as described in Chapter 9, Section IV E.

McNeil (1956) has designed a rather complex tilting stage which will allow a number of surfaces on a sample to be oriented normal to the microscope axis for observation and photomicrography. The sample may be translated in three orthogonal axes on the tilting member thereby further increasing the utility of this stage.

The simplest of the accessories for crystallographic examinations is the single axis (or spindle) stage. Edwards (1961) has developed a device for rotating a transparent crystalline "whisker" while examining it with the polarizing microscope. In operation the whisker is affixed to the end of a manually rotatable shaft. The accessory provides for immersing the sample in refractive index fluid during examination. In this way, a small crystal can be rotated around the axis of the apparatus to orient one of its principle axes for refractive index measurement. A similar device, the Hartshorne Rotation Apparatus is available from McCrone.

Reichert supplies a very sophisticated version of the single axis stage, the micro-refractometer-spindle stage, shown in Figure 7-8. The device actually consists of two spindles, one for the transparent, crystalline sample and one for a standard material. By viewing the two materials in an appropriate refractive index medium and rotating their respective spindles the optical properties of the sample may be determined. This device may also be used for the orientation of samples to be examined in an x-ray diffraction camera.

The ultimate in sample orientation accessories is the universal stage, see Figure 7-9. These devices are used in conjunction with a petrographic microscope in the determination of the optical characteristics of non-opaque, anisotropic materials. The universal stage is used in the examination of mineral, rock and ceramic materials either as crushed grains in a refractive index medium or as ground thin sections mounted in Canada balsam. Glass hemispheres are affixed to both sides of the sample to minimize total internal reflection at large tilt angles. Special objective lenses are required as the upper glass hemisphere acts as a remote front lens.

The universal stages are supplied in three levels of complexity, the 3 axis, the 4 axis and the 5 axis devices. The description of the axes of rotation for the several stages, given in Table 7-1 indicates the great flexibility available for the orientation of a sample. The adjustment and use of the universal stage is quite involved; the most comprehensive text on the subject is by Emmons (1943). The "Manual For The Use Of The Universal Stage In Optical Crystallography" (1966) published by the U.S. Naval Propellant Plant has a wealth of practical information regarding this unique accessory.

Figure 7-8. The Micro-refractometer spindle stage is used for the rapid and accurate determination of the refractive index as well as other optical crystallographic parameters for individual grains. The stage also permits the orientation of single crystals on the light microscope for further evaluation by means of X-ray diffraction. Courtesy of Cambridge Instruments Inc., Scientific Instruments Division.

Figure 7-9. A five axis universal stage is used to determine the optical constants of minerals or other crystalline materials. It may also be used to perform a quantitative determination of preferred orientations of the various constituents in a rock thin section. Courtesy of Wild Leitz USA, Ina.

TABLE 7-1
DESCRIPTION OF THE AXES OF THE VARIOUS UNIVERSAL STAGES

AXIS NAME	ABBR.*	AXES PER STAGE 3	4	5	DESCRIPTION
Inner Vertical	IV	X	X	X	Rotates 360° around a vertical axis.
Inner East-West	IEW			X	Rotates on an east-west axis, tilts north-south.
North-South	NS	X	X		Rotates on a north-south axis, tilts east-west.
Outer East-West	OEW	X	X	X	Rotates on an east-west axis, tilts north-south.
Outer Vertical	OV	X	X	X	Rotates 360° around a vertical axis.
Microscope Stage	M	X	X	X	Rotates 360° around a vertical axis.

*Abbreviated Terms used by Emmons (loc. cit.)

A 3 axis stage has been produced by Vickers. Four axis universal stages are available from Leitz, Nikon, and Zeiss. Leitz also produces a 5 axis instrument. Leitz produces two unique accessories for their universal stages, the Waldman hollow glass sphere which allows the crystallographic examination of large crystals and the Emmons circulation cell, a temperature controlled lower hemisphere for the universal stage which permits double variation, refractive index measurements, see Emmons, loc. cit.

D. Micromanipulators and Microinjectors

The accessories described above provide for movement of the sample relative to the optical axis of the microscope. It is also important to consider the means for the precise manipulation of other devices relative to the sample under observation, i.e., the micromanipulator. Such devices are used in the microelectronics industry to position small electrical test probes on the surface of integrated circuits so that their electrical properties may be measured. These are typically used at magnifications from 100 to 500X. Micromanipulators are also very important and have a wide range of applications in the biological field, for such purposes as the injection of DNA into cell nuclei, embryo splitting, and cell enucleation. These procedures are often carried out at magnifications that require oil immersion objective lenses.

Micromanipulators designed for biological applications are available from E. Leitz, Inc.; Carl Zeiss, Inc.; Emerson (available from Arthur H. Thomas Co., P.O. Box 779, Philadelphia, PA 19105); Eppendorf Inc. (45635 Northport Loop East, Fremont, CA 94538); Ted Pella, Inc. (P.O. Box 510, Tustin, CA 92681); and Ernest F. Fullam Inc. (900 Albany Shaker Road, Latham, NY 12110). The Eppendorf microinjector when affixed to their micromanipulator, Figure 7-10a, permits the injection of femtoliter (10^{-12} ml) quantities of substances directly into individual living cells under pressure via a very fine glass capillary.

Micromanipulators for the electronics industry are usually a part of a complete system consisting of a base that supports a revolving and translating stage, an integral, incident light microscope and a circular bench surrounding the stage on which the micromanipulators are positioned. A typical instrument is shown in Figure 7-10B. For simple studies, individual micromanipulators could be adapted to a standard incident light microscope if supported by a suitable stand.

Some suppliers of the microelectronics type micromanipulators are Micromanipulator Sales and Service Inc. (Box 2245, Escondido, CA 92025); Alessi Industries, Inc. (35 Parker, Irvine, CA 92718); Wentworth Laboratories, Inc. (500 Federal Rd., Brookfield, CT 06804); Signatone Corp.(3687 Enoch St., Santa Clara, CA 95051); and Ernest F. Fullam, Inc.

E. Hot Stages

The hot stage accessories can range in maximum temperature capability from slightly above ambient to as high as 1800°C. The lower temperature devices are suitable for study of living cultures and for the determination of organic compound melting points. The higher temperature devices, see Figure 7-11, are used for dynamic examination of high temperature physical properties and kinetics of high temperature phase formation for metals, ceramics, and minerals.

The sources of heat for the stage are several: (1) self heating of the sample, (2) conductive heating from the sample support, (3) convective or radiative heating from a surrounding heater element, and (4) electron beam heating.

Ideally the atmosphere in the hot stage should be optimized for the particular sample and the requirements of the study, however, it is more often based on minimizing heater-atmosphere reactions or reducing heat transfer to the stage shell, i.e., with a vacuum.

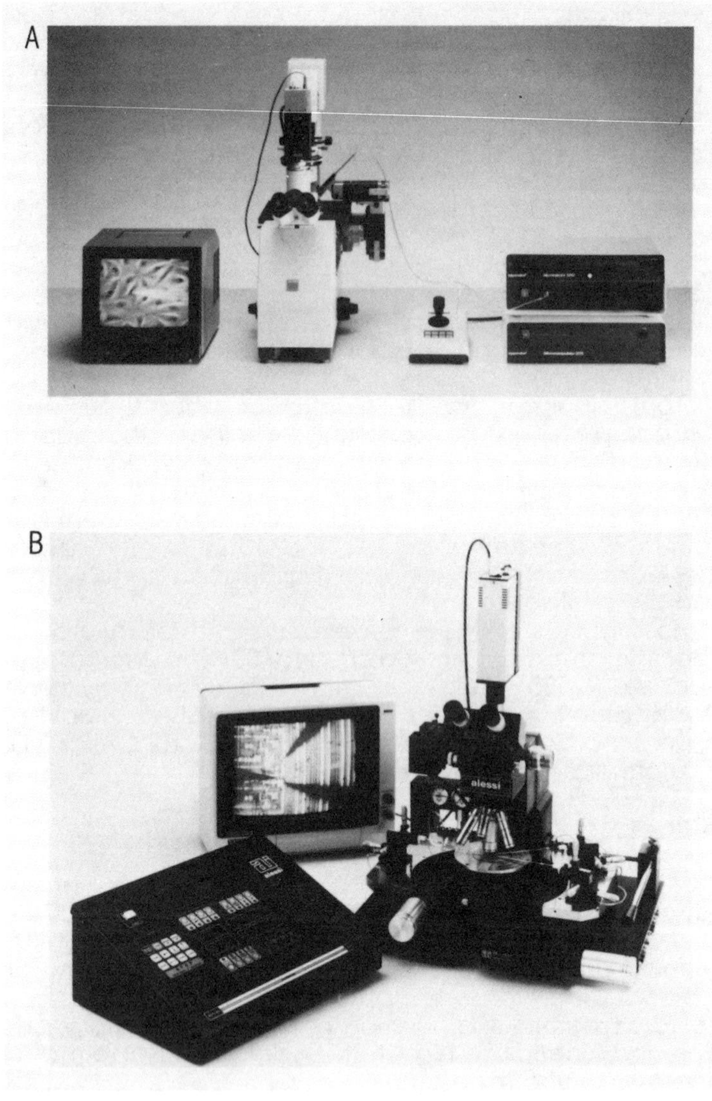

Figure 7-10. Examples of microprobe systems for precise placement of probes of various types on the sample. (A) is a microprobe system equipped to position a microinjector for biomedical research, Courtesy of Eppendorf Inc. and Carl Zeiss, Inc., New York. (B) is a microprobe system that places one or more metal probes on a microelectronic device thereby permitting the measurement of specific electrical parameters. Courtesy of Alessi Inc.

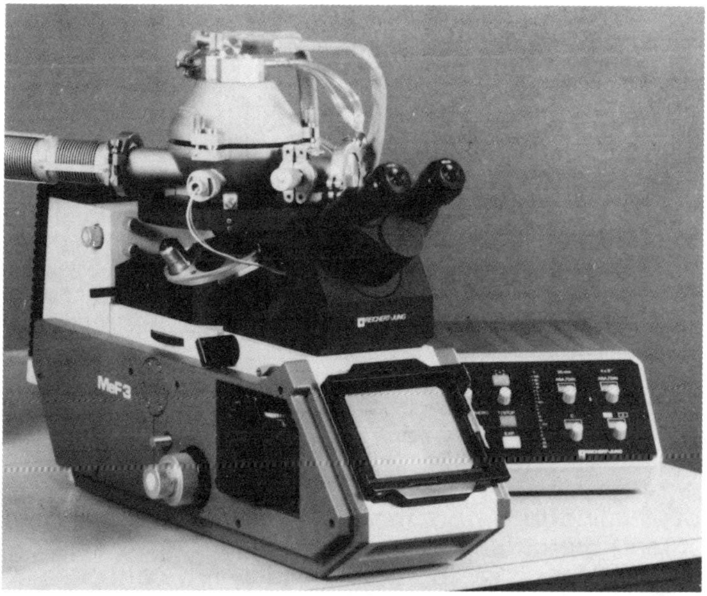

Figure 7-11. A commercially available, vacuum, hot stage for the evaluation of phase changes, grain growth and reaction kinetics. The large electrical leads for heating enter the top of the furnace and the flexible vacuum tube enters from the left. The small wires exiting the bottom of the furnace are the thermocouple leads. The knob at the bottom right of the furnace allows the operator to change cover glasses in the furnace when they become fogged with evaporated coatings. Courtesy of Cambridge Instruments Inc., Scientific Instruments Division.

Heating stages have been designed for transmitted light illumination, incident light illumination and in some cases for both types of illumination. Samples for transmitted light illumination at the higher temperatures must be individual granules or self supporting since any cement or adhesive binder would char or decompose and render the sample useless.

Since the objective lens is the most critically designed part of the microscope and it must operate as near as practical to the hot stage to yield the highest resolution, it is imperative that it be protected from any damaging effects of the heat from the stage. this may be achieved by physically isolating the objective lens from the hot sample and by the introduction of an optical quality window in the stage shell. On the contrary, long working distances limit the resolving power of the objective lens and manufacturers of hot stages have often designed special lenses with optimized N.A. for the required working distance. The thickness and composition of the furnace window is also taken into consideration of the objective lens to minimize spherical aberration. These special objective lenses have N.A. ranging from 0.45 to 0.60 giving useful magnifications from 450 to 600X. An unusual hot stage objective lens was designed by Olson, Brixner and Smith (1962) that uses a 4 inch diameter, ellipsoidal, objective mirror. This mirror has a relatively large N.A. of 0.47 because it is mounted within the hot stage.

Since there is such a great diversity in hot stages, they will be arranged in Table 7-2 by temperature range with the reference author or manufacturer listed. Other important considerations such as type of heating, the ambient atmosphere for the sample, the type of illumination, and the objective lens working distance requirements are also listed.

Because heating stages vary widely it is not possible to give a generalized procedure for their operation. Some of the more important considerations in hot stage microscopy are given by Brenden, Newkirk, and Bates (1960), Lozinskii (1961),and Smallman and Ashbee (1966). McCrone (1957) Discusses the application of the hot stage to the identification of compounds by their melting point. The main consideration with regard to the microscope is adequate protection of the instrument against excessive heat. For example, at the higher temperatures it may be necessary to limit the magnification used thereby taking advantage of the longer working distance of the lower power objective lenses. Some researchers move the objective lens away from the the hot stage window with the coarse or fine focus knob except during actual observations. Those microscope manufacturers who offer hot stage accessories usually make available the long working distance objective lenses. Appendix I lists most of the available long working distance objective lenses. Another approach to minimizing damage to the microscope is to use a long working distance microscope, See Chapter 6, Section III.

The contrast of the sample image will degrade progressively as temperatures increase above ~500°C because some of the thermal radiation emitted from the sample is in the visible region, that is to say, the sample becomes self luminous. The effect of this self luminous radiation can be minimized by using a green filter between the sample and the eye, see Section IV B.

This chapter would not be complete without mentioning two Unitron instruments, although not stage accessories, they constitute perhaps the ultimate in high temperature optical microscopy. The Unitron Model HM-4 Research Installation is a complete microscope console equipped to provide sample temperatures as high as 1900°C in vacuum. Likewise, the Model HM-100 Ultra-High Temperature console is capable of temperatures of up to 2500°C using electron beam heating. Image contrast is maintained through the use of a narrow bandpass filter between the sample and the eye. Nikon also produces a unique high temperature installation, their model QM High Temperature Microhardness Tester. This system permits the measurement of microhardness, see Section G, at temperatures from ambient to 1600°C.

F. Cold Stage

The cold stage is suitable for a number of applications in both the biomedical and the materials fields. For example, the study of effects of low temperature on microorganisms and the study of the effectivity of cryogenic storage on donor organs are suitable studies in biology and medicine. In the materials sciences area the cold stage is useful for study of low temperature phase transformations in metals and the evaluation of polymers below their glass transition temperatures.

Table 7-3 lists a number of commercially available and user constructed cold stages. although these accessories generally pose little problem to the instrument, one must always be concerned with the objective lens which is near any stage attachment. While damage to the objective lens is not the severe problem that exists with the hot stage, icing of the the cold stage window and the outer objective lens is a hindrance to clear observations. To minimize icing, the common solutions are isolation those surfaces from humid air, for example with sections of a toy balloon, or flooding these areas with very dry gas such as the vapor from a liquid nitrogen dewar. Some researchers move the objective lens away from the the cold stage window with the coarse or fine focus knob except during actual observations.

Cooling is commonly accomplished using liquefied gasses such liquid carbon dioxide, $T_{sublimation}$ = -78.5°C, liquid nitrogen (LN2), b.p.= -195.8°C, liquid hydrogen, b.p.= -252.9°C, liquid helium, b.p. = -268.9°C. Liquid nitrogen is generally the choice for studies within its temperature range because of its less difficult storage requirements, lower cost, and relative availability.

Cryogenic fluids used for the cooling stage must be properly controlled since if they are spilled or splattered they could possibly produce thermal shock to glass elements of the microscope and "burns" to the operator. Liquid hydrogen should be used only with extreme care because there is always the great danger of the build-up of explosive concentrations of the evaporated gas.

An interesting method of cooling a sample in a cold stage makes use of thermoelectric or Peltier coolers. These devices may be considered as a solid state refrigerator or heat pump; the pumping action is achieved by passing a d.c. electric current through the cooler. The heat that is removed from the sample is then usually removed from the opposite side of the cooler by a recirculating water supply. It is worthwhile to emphasize that if the polarity of the power is reversed the sample is then heated. Thus, this stage can be used as a hot/cold stage for temperatures limited to about ± 30 to 50° from ambient.

For the lower temperature cold stages, the illumination can introduce an appreciable quantity of heat, therefore it is important to include an infrared or heat absorbing filter, see Section IV C, in the illumination system.

G. Stages for Microdetermination of Material Properties

Technological demands both in the biomedical and materials fields continue to require the measurement of properties on progressively smaller samples. Regardless of whether the sample is the entire object such as a semiconductor chip or just a small select area of a larger object, the measurement of these properties can probably be best measured on an appropriately configured light microscope. The various examinations can be categorized as physical, chemical or mechanical properties measurements. This section will cover a number of stage accessories suitable for making these measurements.

The optical microscope outfitted with special accessories has become an important research and production tool in the realm of physical properties testing. Microhardness is one of the physical properties of much interest to the materials scientist. The microhardness measurements involve the microscope stage to some degree, however the unique aspects of the equipment rest in the accessories to the objective lens and therefore will be discussed in that section.

TABLE 7-2
HOT STAGES

Manufacturer* or Reference	Temperature Range	Working* Dist (mm)	Atmosphere*	Illumination*	Remarks*
Lowndes (1986)	RT - 40	Normal	A	Tr	(1)
Incu-Stage	27 - 43	Normal	A	Tr	(1)
Zeiss -	35 - 43	Normal	A	Tr	
MD - 5001/5002	5 - 55	Normal	A	Tr	(1,2)
Zeiss -	30 - 60	Normal	A	Tr	
Linkam - CO 60	RT - 60	Normal	A	Tr&In	(1)
Leitz - 80	-20 - 80	Normal	A	Tr	(1)
Sensortek	-20 - 100	Normal	A	Tr	(1,2)
Kaukler (1984)	-30 - 120	11	X	Tr	(1)
Hunt(1966)	-100 - 200	N/A	X	Tr	(3)
Mettler - FP-52	RT - 300	7	A	Tr	(1)
Thomas - 40	30 - 350	8	A	Tr	(4)
Leitz - 350	-20 - 350	6(a)	A	Tr	(4)
Sturmer, (1957).	RT - 400	Normal	A	Tr&In	
Greeson (1966)	RT - 400	7	V	In	
Hawkins (1970)	RT - 450	14	V	In	(5)
Unitron - MHS	RT - 500	5.8	V,I	Tr&In	(4)
Tynan (1975)	-243 - 527	N/A	V	In	(4)
Linkam - TH 600	-196 - 600	6(b)	A,I	Tr&In	(1,6)
Ogilvie (1955)	RT - 1000	N/A	V	In	(7)
Bridge (1957)	RT - 1200	N/A	A	Tr	(4)
Leitz - 1350	RT - 1350	4	I	Tr& In	(4)
Scott (1961)	RT - 1400	12.8	V	Tr&In	(4)
Linkam - TH 1500	RT - 1500	8(c)	A,I	Tr&In	(1,6)
Derge (1970)	RT - 1600	(d)	A,I	Tr&In	(4)
Leitz - 1750	RT - 1750	4	V,I	In	(4)
Unitron - HTL-4V	RT - 1800	5.8	V	In	(1)
Reichert-Vacutherm	RT - 1800	8	V	In	(4)
Saller (1955)	RT - 1800	N/A	V	In	(1)

*Incu Stage and Thomas 40 available from Arthur H. Thomas Co., P.O. 779, Philadelphia, PA 19105.

Carl Zeiss, Inc. One Zeiss Dr., Thornwood, NY 10594.

MD 5001/5002 and Linkam, available from Micro Devices, Inc., Suite A-103, Benjamin Fox Pavilion, Jenkintown, PA 19046.

E. Leitz, Inc. 24 Link Dr., Rockleigh, NJ 07647.

Mettler Instrument Corporation, P.O. Box 71, Hightstown, NJ08520.

Sensortek Inc., 154 Huron Ave., Clifton, NJ 07013.

Unitron Instruments Inc. 101 Crossways Park West, Woodbury, NY 11797. Reichert-Jung, available from Cambridge Instruments Inc.,P.O. Box 123, Buffalo, NY 14240.

Normal	- Any air, transmitted light, objective lens may be used.
(a)	- A long intercept condenser cap must be used.
(b)	- The condenser working distance is 16 mm.
(c)	- The condenser working distance is 11 mm.
(d)	- Designed for use with a stereo microscope.
A	- Air atmosphere
I	- Inert gas atmosphere.
In	- Incident light illumination.
Tr	- transmitted light illumination.
V	- Vacuum environment.
X	- Sample contained within a glass cell.
1	- Equipped with temperature controller and readout.
2	- Peltier heating/cooling - optional laminar flow gas curtain
3	- Stage provides a temperature gradient for sample.
4	- Equipped with temperature readout.
5	- Has quenching capability.
6	- Also available in models with X-Y sample manipulation.
7	- Electron beam heating yields heating rates of 2000°C/min

Carl Zeiss offers an example of a stage accessory for the measurement of a physical property; specifically this device permits the determination of the refractive index of liquids. This stage is supplied with two measuring bodies, each equipped with a lenticular concavity to contain the liquid of interest for different refractive index ranges. In operation, a scale in the accessory is viewed through the liquid of interest. A line on the eyepiece reticle is used as a fiducial point, for the scale. The measurements are carried out at a wavelength of 589 nm but conversion tables are included for 486, 546 and 656 nm. The range of R.I. measurable is from 1.300 to 2.000 with a precision of $\pm$0.001.

Microchemistry makes use of the microscope in the identification of very small quantities of inorganic compounds, minerals, organic compounds and drugs. In this section however, the emphasis is upon the microscope as a means of monitoring a chemical process. Locquin and Langeron (1978) have described a microelectrophoresis cell, attributed to Choucroun for mounting on the microscope stage This system is used to study the speed of electrophoretic motion of a cell or particle. A complete, commercially available microelectrophoresis system is available from Zeiss as their Cytopherometer. Wall and Roberts (1963) have described a microscopic cell for the observation of selective corrosion of metals.

The microscope with appropriate accessories also forms an excellent device for the determination of mechanical properties of micro samples. Kong et al.(1983) have described a tensile strainer stage for polymeric composites and Andrews (1957) has developed an extensometer for polymeric fibers. Foster (1965) has described a stage accessory and hydraulic power system that will apply a measured pressure from 0 to 30 lb to a sample on the stage. Such a system is suitable for studying deformation on the polished surface of a metallurgical sample.

H. Special Environmental Stages

The type of environment required for a special stage varies with the application. In the biological field Zeiss offers the Dvorak-Stoller controlled environment culture chamber. In the materials field, the variety of special environmental chambers for the microscope stage is somewhat greater. Sippel (1965) and Harmer et al.

TABLE 7-3
COLD STAGES

Manufacturer* or Reference	Temperature Range	Working* Dist (mm)	Atmosphere*	Illumination*	Remarks*
Hooss (1964)	-3 - 50	T.E.	A	In	(1)
Leitz - 80	-20 - 80	LCO2	A	Tr	(2)
Leitz 350	-20 - 350	LCO2	A	Tr&In	(2)
Zeiss	-30 - 70				
Kaukler (1984)	-30 - 120			Tr	
Markussen (1965)	-50 - 100	T.E.	A	Tr	(1)
Hildebrandt (1976)	-60 - RT	LN2	A,I,V	Tr	(2)
Hunt (1966)	-100 - 200			Tr	(2)
Monier (1950)	-130 - 35	LA	A	In	(2,3)
Rhodes (1950)	-140 - RT	LA	A	Tr	(2,3)
Rosenthal & Rall (1984)	-150 - 40	LN	A,I	Tr	(4)
Clothier (1967)	-185 - RT	LN2	A	Tr	(2,3)
Cech (1950)	-195 - 1200	LN2			(2)
Linkam - TH 600	-196 - 600	LN2	A,I	Tr&In	(2,5)
Tynan (1975)	-243 - 527	LHe	V	In	(3)
Kessler (1971)	-257 - 27	CT	V	In	(2,3)
Cosier (1983)	-263 - 77	LN2/LHe	V	Tr	(2,6)
Isaacs & Pops (1964)	-269 - 20	LHe	V	In	(2)
Edmonds (1977)	-270 - 72	LN2/LHe	I	In	(2)
Carroll (1967)	-272	LHe	I	In	

*E. Leitz, Inc. 24 Link Dr., Rockleigh, NJ 07647.

Carl Zeiss, Inc. One Zeiss Dr., Thornwood, NY 10594.

Linkam, available from Micro Devices, Inc., Suite A-103, Benjamin Fox Pavilion, Jenkintown, PA 19046.

CT - Cryotip, supplied by Air Products and Chemicals, Inc.
T.E. - Thermoelectric cooler
LA - Liquid air
LCO2 - Liquid carbon dioxide
LN - Liquid nitrogen
LHe - Liquid helium
A - Air
I - Inert gas
V - Vacuum
(1) - With recirculating water and power supplies and contact thermometer.
(2) - With temperature readout.
(3) - Temperature controlled by amount of coolant gas flowing; this flow is adjusted by an immersion heater in coolant.
(4) - Temperature control by a digitally controlled heater at sample.
(5) - Temperature digitally controlled by gas flow switching.
(6) - Uses a continuous- flow cryostat.

(1968) have described vacuum chambers to study electroluminescence of rocks and minerals but suitable for other materials as well.

Leitz manufacturers a dual purpose accessory, their Contrasting Device, which can be used for electroluminescence studies, as described above. In addition, this stage is used to bombard a polished metallurgical sample with reactive gas ions to produce a thin reaction layer on the surface of the sample grains. This reaction is allowed to progress until the thickness of the layers is sufficient to produce interference colors when observed in normal incident illumination. The thickness is uniform on each grain but different in thickness, for each grain orientation, Since different thicknesses produce different colors, the process produces a color contrasting of the sample structure.

Leitz also manufacturers a stage accessory, their Metal Coating Chamber, which permits the direct observation of the metal coating of a sample in preparation for study in a scanning electron microscope. The usual coating metal for this application is gold.

I. Miscellaneous

One of the most important accessories for the microscope stage is the stage micrometer, used for calibrating eyepiece reticles, see Chapter 5, Section II A. The stage micrometer is available in either metric or English units and with or without cover glass. The typical stage micrometer is uncertified and may have errors as large as 1%, however, they may be certified by a competent metrology laboratory. Precision stage micrometers may also be purchased that are certified accurate to ±0.05 μm for each 0.01 μm division and to ±0.1 μm for the entire scale when used at a specified temperature of 20°C.

The remaining stage accessory to be discussed is the field finder slide, such as the Lovins Micro-Slide Field Finder (Available from Teledyne Gurley, Division of Teledyne Industries, Inc., 514 Fulton St., Troy, NY. 12181), shown in Figure 7-12. This device, designed for the transmitted light microscope, consists of a precision, rectangular coordinate system printed on a precisely dimensioned microscope slide. The coordinate system is graduated in intervals of 0.1 mm with each millimeter numbered or lettered, thus making it possible to locate an area of interest on a sample with great accuracy. The procedure for using a field finder is given in Table 7-4. Ted Pella also offers a similar unit.

V. OBJECTIVE LENS ACCESSORIES

Accessories designed for replacing the usual objective lens of the light microscope are the topic of this section. They are usually equipped with an RMS thread, thus they may be mounted directly on the base of body tube or on the revolving nosepiece. For dimensional details of this thread type, see U.S. Department of Commerce, National Bureau of Standards Screw-Thread Standards for Federal Services, Handbook H28 (1957), Part 3, (Reprinted April 1962 with corrections).

TABLE 7-4
LOCATING AN AREA OF INTEREST WITH A FIELD FINDER SLIDE

1. Center the area of interest at the cross wire of the eyepiece, using a mechanical stage. Record the magnification of the objective lens used.
2. Carefully remove the sample slide, taking care not to disturb the position of the stage.
3. Substitute the field finder slide for the sample slide.
4. Using a 5X or 10X objective lens, note the coordinates on the finder stage at the cross wire of the eyepiece. Record these coordinates along with the objective lens magnification for that particular area of interest.
5. To relocate that area of interest, reverse steps 1-4.

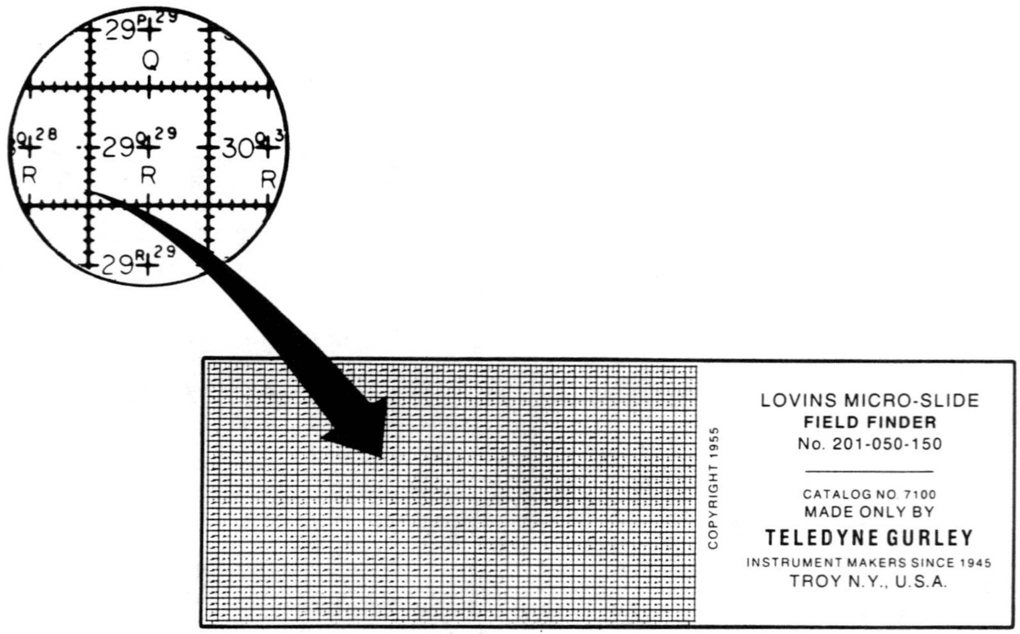

Figure 7-12. The Lovins field finder is an example of several microscopical slides available for relocation of specific areas on a microscopical sample. Courtesy of Teledyne Gurley.

A. Microhardness Testers

Hardness of a material is usually determined by forcing an indenting tool into the surface of a sample under prescribed conditions. The dimensions of the indentation can be related to the hardness of the material. When the applied load is $\leq$200 ponds (grams force) the procedure is referred to as microhardness testing and is usually accomplished with a microscope and appropriate attachments. The attachments are a means of applying a load to the sample that is usually attached to the stage and a diamond indenter that alternates in position with the objective lens.

To perform a microhardness test, a sample is mounted or clamped on the stage and a site for the hardness indentation chosen using the objective lens. The revolving nosepiece is then rotated to bring the diamond into position. The stage is moved toward the indenter at a predetermined rate until the sample contacts the diamond with the desired load. After a prescribed time the sample is moved away from the indenter and the objective lens returned to its former position. The indentation dimensions then may be accurately measured using a micrometer eyepiece. An excellent discussion of microhardness measurement is given by Gahm (1968).

Bergsman (1948), Grodzinski (1951) and Lloyd and Norris (1951) have described the construction of microhardness testers. Several manufacturers, including McCrone, Leitz, Reichert and Zeiss produce microhardness testers which may be considered accessories to an incident light or metallographic microscope; a representative of this group of microhardness testers is shown in Figure 7-13. Vickers had produced a microhardness tester for some time.

The McCrone instrument is designed for an inverted stand or metallographic microscope. In this instrument, a diamond indenter substitutes for one of the objective lenses in the revolving nosepiece and is made parfocal and concentric with the other objective lenses. The sample is mounted on a balance beam arrangement on the stage of the instrument; this arrangement allows a desired weight to be added to counterbalance the sample thus determining the effective indenter load. Once the area for the indentation has been located using one of the objective lenses, the indenter is rotated into position and contact with the sample is made using the fine focus knob. A lamp indicates when the indenter has fully entered the sample. The indenter is allowed to remain against the sample for a prescribed length of time then withdrawn using the fine focus mechanism. An objective lens is rotated into position to permit measurement of the indentation. Vickers, Knoop and Berkovich indenters as well as a 2 mm diameter steel ball for Brinell testing are available. Indentation loads from 50 to 2000 grams are available.

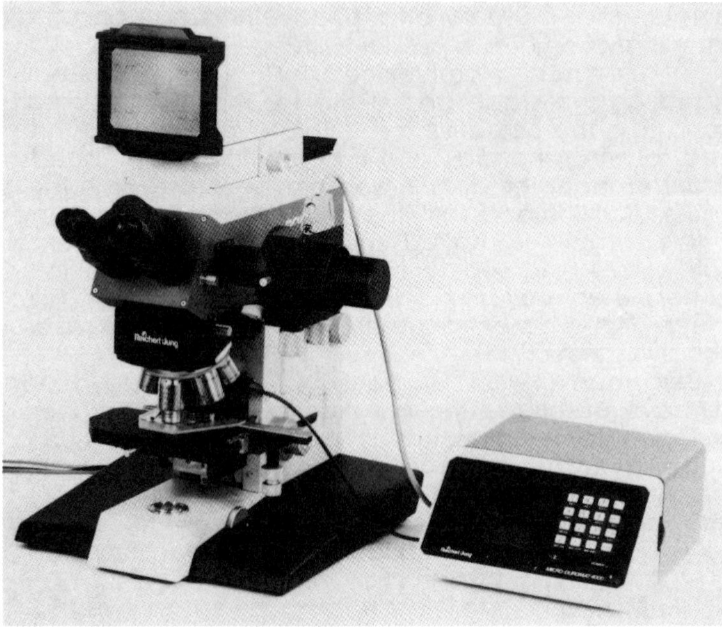

Figure 7-13. An example of a microhardness tester attachment for a metallurgical microscope. The indentor mechanism is the objective-lens-sized device in position above the sample. The large attachment to the right side of the eyepieces is the optical measurement system. This instrument has a load range from 0.05 to 200 p (grams force) and uses either a Vickers or Knoop indentor. Load, approach speed, load gradient and dwell time are programmable and reproducible. Courtesy of Cambridge Instruments Inc., Scientific Instruments Division.

The Reichert Micro hardness tester for the MeF1 and 2 was similar to the McCrone device in that it is designed for the inverted (Reichert) microscope and that the load is applied via the fine focus knob. However, in this device all of the mechanism is an accessory to the objective lens. The sample is held in position with a simple clamp. This instrument has a lever mounted on the side that moves either an integral objective lens or a Vickers diamond indenter into an identical position facing the sample. If the objective lens is in position, the location on the sample for the indentation is seen in the eyepiece; if the indenter is in position, the scale of an internal torsion balance may be observed. In operation this Reichert microhardness tester is located over the desired area of interest then the lever is changed and the indenter positioned over that spot. The indenter is applied to the sample with the fine focus knob while the load is determined by watching the torsion balance scale in the eyepiece. When the desired load is achieved, the fine focus is held in position for a prescribed time and then turned to release the load and retract the indenter before moving the lever to the viewing position for measurement of the indentation.

The Reichert microhardness tester for the MeF3 is similar to the Leitz and Zeiss units in that the indenter force is supplied pneumatically from a remote unit. These microhardness testers have varying degrees of automation; their various capabilities are listed in Table 7-5. A similar unit was also produced by Vickers.

The operation of the various microhardness testers is quite specific to the particular instrument, therefore the exact procedure must be obtained from the appropriate instruction manual. A few words of caution are in order for the proper operation of any microhardness tester. The general microscopical requirement for low external vibration levels is very important for consistent microhardness results. For the same reason the microscope and the external pressure unit must be treated very gently and much care must be maintained to avoid any shocks or vibration induced by the microscopist during the time that the indenter is in contact with the sample.

B. Slide Marker

Selected areas on a sample may be relocated using a field finder as described in Section II D above or a slide marker. The slide (or object) marker is equipped with an RMS thread and substitutes for an objective lens in the revolving nosepiece. When an area is to be marked, the slide marker is rotated into position over the sample and pressed onto the sample. This procedure produces a uniform ink ring on the sample. Nikon supplies an object marker which produces a 1.8 mm circle on the sample while Polysciences, Inc. (400 Valley Rd. Warrington, PA 18976) offers a ring slide marker that produces a 1 mm circle on the sample.

TABLE 7-5
MICROHARDNESS TESTERS

SOURCE	INDENTER*	LOAD RANGE*	DWELL TIME	MISCELLANEOUS COMMENTS
Leitz	K,V	2-400 P	manual	Indentation rate adjust. for load.
Lloyd & Norris (1951)	V	0.1-50 P	manual	Approach rate = 0.016 mm/s.
McCrone	B,Br,K,V	50-2000 P	manual	
Reichert	K,V	0.05-200 P	0-9999 s	Adjustable load and approach rate.
Vickers	V	5-200 P	manual	No longer available.
Zeiss	K,V	5-200 P		

*B = Berkovich; Br = Brinell; K = Knoop; V = Vickers
P = grams force.

VI. FILTERS

Filters are necessary to modify the illumination for several purposes in optical microscopy: these include reduction of illumination level, increasing color contrast, infrared and ultra violet light absorption, and for such special applications as color photomicrography, polarized light illumination and fluorescence microscopy. Each of these applications will be discussed and selected filters listed.

A. Neutral Density Filters

The most common problem with illumination is that of incorrect intensity; this can be corrected with most incandescent lamps by simply adjusting the transformer output. However, the xenon lamp is a good example of a very bright source that cannot be reduced in power to any significant extent without affecting its stability. Therefore, if a xenon lamp is being used for visual examinations it is often necessary for eye comfort to introduce a filter in the illumination path. A neutral-density filter is ideally used for this purpose because it does not significantly change the color of the image. It is appropriate to mention once again that even though the aperture diaphragm can be closed down to reduce intensity, it should never be used for this purpose.

Glass absorption type neutral density filters are available, however, filters made by depositing a thin layer of a metal alloy on a fused silica substrate have the most uniform response with respect to wavelength. Various densities are available and 4 filters with optical densities (percent transmissions) of 0.1 (80%Tr), 0.5 (32%Tr), 1.0 (10%Tr) and 2.0(1%Tr), are suitable for most microscopical applications. These filters may be used individually or in combination to give a wide range of optical densities. This type of filter is quite stable and resistant to damage from heat and therefore may safely be placed in the illumination path prior to the condenser system.

B. Color Filters

Color filters are available in several forms. The most common forms are the absorption filters. These filters rely on carefully controlled additives to the substrate material to achieve the desired filtration. Generally the most stable and durable types are the glass absorption filters. Selected glass filters, such as green and daylight blue, are available from each of the major microscope manufacturers; these have the advantage that they fit the filter holders for their microscopes. Sources for other types of glass filters are Kopp Glass Inc.(2108 Palmer St., Pittsburgh, PA 15218), Edmund Scientific, Melles Griot, and Rolyn Optics Co.

Another common absorption filter is fabricated with a gelatin substrate; Eastman Kodak Co. produces a wide range of these for a number of applications. The gelatin filters are easily damaged by heat and therefore, should be protected by an infrared filter (see below). Also, the gelatin filters are soft and must be cleaned only with a lens brush. Do not use liquid cleaners on these filters. For less critical applications that do not require great uniformity of color, Kodak makes available a number of acetate based absorption filters.

For very sharply controlled transmission characteristics, the more expensive, interference filters are available for high-pass, low-pass or band pass applications. These achieve filtration through the use of multilayer, thin films, sometime augmented with broadband absorption layers. The better interference filters are sandwiched between glass or silica substrates with an epoxied aluminum edge ring that physically protects the thin film layers and also impedes the ingress of moisture that is detrimental to the layers. Interference filters may be cleaned with minimal amounts of lens cleaner and lens tissue.

The interference filters must be oriented normal to the illumination path to obtain the desired wavelength transmission. If the interference filter has one side with a metallic luster, that side should be oriented toward the illumination thereby minimizing the thermal load to the filter.

Color filters are used in four distinct applications in optical microscopy and photomicrography: modifying the illuminating light for a specific type of objective lens, providing increased contrast for colored samples, matching of the illumination color temperature to that of a given type color film, and fine tuning the tint of a given color or stain in a color photomicrograph.

The first application of color filters is to improve sharpness and contrast in microscopy and particularly B&W photomicrography when using achromatic objective lenses. It will be recalled from Chapter 2, Section III C 4 that an achromatic objective lens is fully corrected for only one color, usually yellow-green. Therefore, for the sharpest image with these objective lenses, the appropriate yellow-green filter should be used in the illumination. The best source for the correct filter to match any given objective lens is its manufacturer. The Kopp 4-65 or the Eastman Kodak 57A filters are generally suitable for this application.

This means of improving contrast is appropriate for colorless subjects such as metallurgical samples. On the contrary, the deep coloration of the filter will mask the subtle colors of a stained biological or medical sample and should not be used. Likewise, it is obviously inappropriate to use the filter with color film.

The second application of colored filters is in the enhancement of contrast in the B&W photomicrography of stained samples. Staining has reached a very high level of sophistication in the fields of biology and medicine and has found some application in geology, as well. In this procedure one or more specific stains are used to highlight specific structures in the sample. The color sensitivity of the eye and color films are generally sufficient that the color contrast of subtly stained structures is distinguishable, but it is very possible that two areas stained different colors may be imaged as the same shade of gray in a B&W photomicrograph. If, however, appropriate color filters are introduced in the illumination path it is possible to make one of stained areas appear lighter or darker than the other in the photomicrograph.

The general rule that guides the selection of the appropriate filter is that a filter of the same color as the stain produces minimum contrast with an otherwise light background and a filter which has the complimentary color will produce the maximum contrast with the background. Figure 7-14 lists both glass and gelatin filters suitable for either reducing or increasing the contrast for a given color stain. It must be remembered that the introduction of any color filter will reduce the intensity of illumination, therefore some increase in exposure time may be expected.

The third application of color filters is in the modification of the color temperature of the illumination to match that of the selected color film. Color films are available that are balanced for three

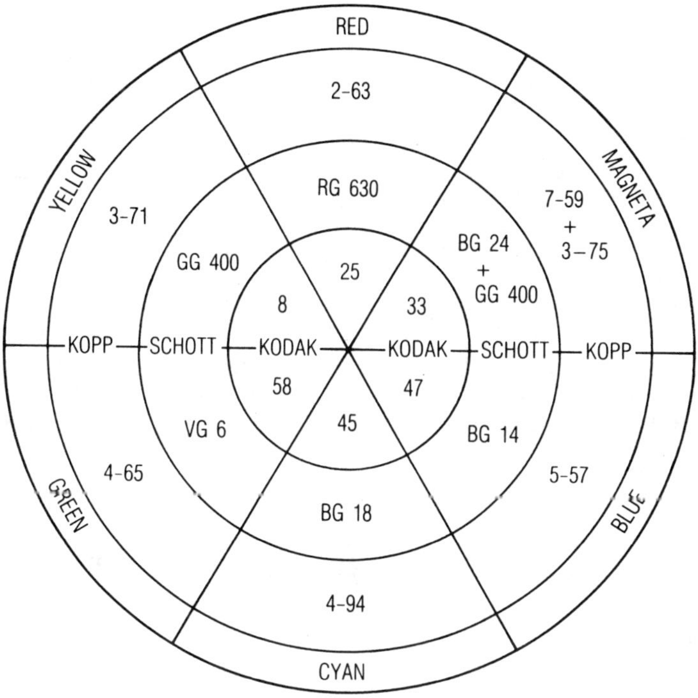

Figure 7-14. A Chart relating Kodak, Schott and Kopp color filters to their transmission colors. If a filter is used that matches the color of the structure of interest, the contrast in a black and white photomicrograph is reduced. If a filter is chosen that is on the opposite side of the chart from the color of the structure of interest, the contrast is enhanced for a black and white photomicrograph.

different color temperatures of illuminating light: 3200°K, 3400°K and daylight (5500°K).

The effective color temperature of the microscope illumination is primarily dependent on the type of illumination, See Chapter 2, Section III A. A typical incandescent lamp source has a color temperature ranging from 2800 to 3200°K at their rated operating voltage. A xenon arc lamp has a 5500°K color temperature.

The effective color temperature will be reduced by an aged or yellowed sample mounting medium. Also a diffuser plate, if present in the illuminator, may reduce the color temperature by as much as 200°K.

Any common illumination source can be modified to match any of the above film types by using the appropriate light balancing (or color temperature conversion) filter; an excellent selection of these is available from either Eastman Kodak Co. or Melles Griot. The Kodak 82 series filters effectively increase the color temperature of the illumination while the 81 series filters reduce the color temperature of the illumination. The latter series is not much used since most microscope illumination sources are generally too low in color temperature, however they are very useful for lowering the temperature of electronic flash lamps. Table 7-6 lists the nominal filters for the various combinations of illumination and film types.

TABLE 7-6
SELECTION OF COLOR BALANCING FILTERS*

ILLUMINATOR		FILM TYPE		
SOURCE	COLOR TEMP.	3200°K	3400°K	5500°K
Incandescent lamp,	2900°K	82B	82D	80A+82B
Incandescent lamp,	3000°K	82A	82C	80A+82A
Incandescent lamp,	3100°K	82	82B	80A+82
Incandescent lamp,	3200°K	--	82A	80A
Zirconium arc lamp,	3000°K	82A	82C	80A+82A
Halogen lamp,	3200°K	--	82A	80A
Xenon arc lamp,	5500°K	85B	85	--

*Filters available from Eastman Kodak Co.

The fourth application of color filters is in the fine tuning of color photomicrographs to closely match that visually observed in the microscope. This application is particularly important for published papers and for teaching purposes. These filters may also be used to enhance a particularly pale stain for emphasis. For example, a didymium glass filter is recommended to provide satisfactory color rendition when either eosin or fuchsin stains have been used on a biological sample. Another application for these filters is to correct for the slight greenish tinge of the heat absorbing filters, if one is used in the illumination.

It is important to carefully correct the color temperature and to be sure that the illumination system has a stable color temperature before attempting the correction of tints. The sample is photographed using each of the various densities of the various color filters until the desired effect is achieved. Suitable filters for this purpose should include the three primary colors; red, green, and blue and the intermediate colors red-green (yellow), red-blue, (magenta) and blue-green (cyan). A selection of all six colors in densities of 0.1, 0.2 and 0.5 are recommended by Kodak to cover most applications. These filters are available from the Eastman Kodak Co. as a series of six gelatin based Compensating Filters or as a series of acetate based filters designated the Color Print Viewing Filter Kit.

C. Heat Absorbing Filters

A major component of the illumination from an incandescent lamp is in the infrared portion of the spectrum. This heat can affect the gelatin color filters, polarizers of both the filter and prism types, the sample and even the mounting medium. To reduce the amount of infrared energy in the illumination, a heat absorbing glass filter should be used. These filters are available from most of the above mentioned sources.

If a heat absorbing glass filter is used in the illumination for color photomicrography, it will be necessary to correct the illumination for its slight inherent coloration in the visible spectrum with compensating filters.

For critical applications, infrared radiation can be more effectively removed from the illumination with a dichroic mirror; the arrangement for both heat reflecting and heat transmitting mirrors is shown in Figure 7-15. These mirrors are available from both Melles Griot and Rolyn Optics Co.

D. Ultraviolet Absorbing Filters

Certain illuminating sources such as the flash lamp, the xenon lamp, and the somewhat older carbon arc produce appreciable quantities of light in the ultraviolet portion of the spectrum. This light, if appreciably transmitted by the microscope optics, may affect the balance of certain types of color film. Suitable filters to remove ultraviolet light from the illumination are the Wratten UV absorbing filters, 2B or 2E, available from Kodak or the Schott colorless, sharp cut-off glass filters GG 395 or GG 385, available from either Melles Griot or Rolyn Optics Co.

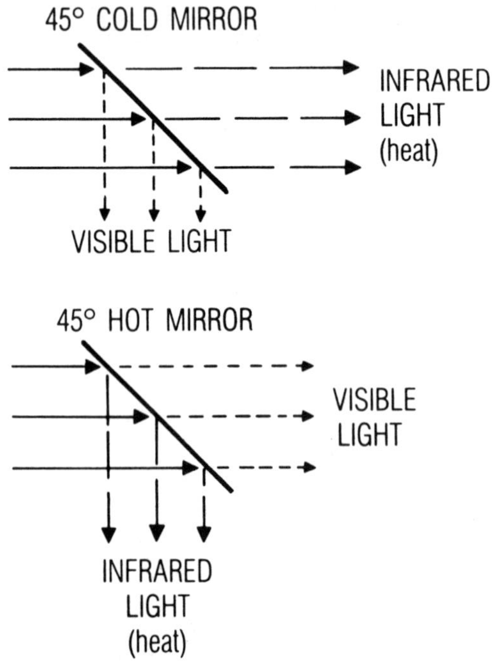

Figure 7-15. Proper placement of a cold or hot dichroic mirror in the illumination path of the microscope can significantly reduce the heating of the sample by the infrared radiation from the lamp.

E. Filters For Fluorescence Microscopy

Successful fluorescence microscopy, as described in Chapter 4, Section IX, depends upon the proper selection and quality of the filters used. The ideal excitation filter is one that passes only the wavelength region that produces the fluorescence in the sample and the ideal barrier filter is one that passes only the resulting fluorescent radiation.

The use of thin-film interference type filters has markedly improved fluorescence microscopy since they have much sharper cutoff characteristics than absorption filters, they can be tuned, during their manufacture, to closely match the maximum excitation and fluorescent wavelengths of a given fluorescing compound and they may have higher transmissions at their design wavelength than a comparable absorption filter. The only negative aspect of the interference filter systems is that since they are so finely tuned, a separate filter system may be required for each fluorochrome.

In the Leitz, Nikon, Olympus, Reichert and Zeiss incident light, fluorescence instruments, the excitation filter, a dichroic mirror and the barrier filter designed for a given fluochrome are combined in a single module, as shown in Figure 7-16. The dichroic mirror or beam splitter is also a thin-film device constructed so that both the reflection of the excitation radiation and the transmission of the fluorescent radiation are optimized simultaneously, thereby improving the image contrast ratio significantly. The Leitz Ploem, incident light, fluorescence module is such a system that is further augmented with a second barrier filter on an insertable slider.

Fluorescence microscopy is a dynamic field both from the standpoint of instrument design and the development of new fluorochromes. The development of a new fluorochrome merely triggers the development of a new filter module specifically tuned for that fluorochrome. Therefore, it is not productive to make an exhaustive listing of the available modules. Listed in Table 7-7 is a brief tabulation of simple excitation and barrier filter combinations that may be installed on either a transmitted or incident light microscope for an occasional observation that may not require a high performance system.

F. Polarizing Filters

The description and use of polarizing filters are given in Chapter 4, Section VI. Polarizing filters as accessories can be used to convert a transmitted or incident light, brightfield microscope into a moderately effective polarizing microscope; the major problem encountered is that the strain inherent in tyical objective lenses can severely reduce the effectiveness of the filters. However, for the occasional observation, this approach may be adequate.

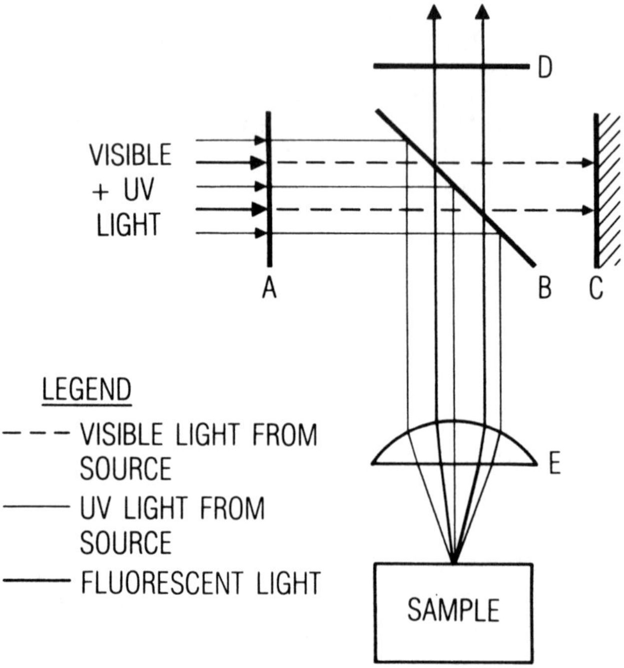

Figure 7-16. Illustration showing the improvement of image contrast in incident light, fluorescence microscopy that is achieved by addition of a dichroic mirror. These mirrors are designed to reflect the excitation light and transmit the fluorescent light. In this case the UV light from the illuminator and excitation filter, A, is reflected by the dichroic mirror, B, to the objective lens, E, and the sample, while any residual, visible light is preferentially transmitted by the mirror and absorbed by the black surface, C. The visible, fluorescent light generated in the sample is preferentially transmitted by B to the barrier filter, D, and then to the eye. The UV light reflected from the sample is reflected back to the illuminator.

More expensive polarizing filters are available but the simple filter polarizers available from Edmund Scientific are adequate for most purposes. Suitably sized pieces of the filter for the polarizer and analyzer may be cut from the bulk material with scissors. The polarizer should be mounted between the illuminator and the condenser system. For high intensity illuminators an infrared absorption filter should be used between the illuminator and the polarizing filter. The simplest location for the analyzer is on the exterior of the eyepiece.

The optimum extinction positions for crossed polarizing filters are achieved by inserting the polarizer in any random orientation in the microscope illumination between the illuminator and the condenser. The analyzer is then installed, usually as a cap over the eyepiece, and rotated about the axis of the microscope until the field of view is darkest: this condition is often referred to as "crossed polarizers". The polarizer is then rotated a few degrees about the axis of the microscope and the analyzer again crossed. This process is repeated until a position that gives the darkest field of view is obtained. These positions of polarizer and analyzer gives the best polarization effect for observation of samples. Mark these orientations of the polarizer and analyzer in some manner for future use.

G. Polarizing Microscope Accessories

Although not filters in the usual sense, the various polarizing microscope accessories do modify the light passing through the microscope and, for this reason, are included in this chapter. These devices are usually sized to fit the accessory slot of a specific instrument, therefore for any given microscope, the manufacturer is the best source for an accessory device.

The accessory slot is usually found in the body tube of the microscope just above the objective lens although some instruments include the slot in the revolving nosepiece. Some of the more elegant instruments have a rotatable cover for the slot when it is not in use.

The use of the accessories generally requires considerable training in optical crystallography and petrographic microscopy, therefore it is beyond the scope of this book to describe their detailed use. The reader is referred to such texts as Ehlers (1987) for detailed information on the interpretation of the microscopical images as modified by these devices. The various accessories are listed along with their more common applications in Table 7-8.

TABLE 7-7
ABSORPTION TYPE EXCITATION AND BARRIER FILTERS*
FOR SIMPLE FLUORESCENCE STUDIES

EXCITATION RANGE	EXCITATION FILTER[1]	BARRIER FILTER[2]
DEEP ULTRA-VIOLET[3]	(UG5ß)	3-74 or (GG395ß)
NEAR ULTRA-VIOLET	7-39 or (UG1) or (18A)	3-71 or (GG475ß)
BLUE-VIOLET	1-64 or (BG12)	3-70 or (GG495ß)
BLUE	5-61 or (BG14) or 47B	3-68 or (OG590ß) or (12) or (15)

*Filters are available from Kopp Glass Inc. except those in brackets which are available from Melles Griot or Rolyn Optics Co. and those in parentheses which are available from Eastman Kodak Co.

1. The excitation filter should be placed between the illuminator and the condenser. A heat absorbing filter may be necessary for the gelatin filters.
2. The barrier filter should be placed between the objective lens and the eye. The simplest location is on the exterior of the eyepiece.
3. Use a mercury vapor lamp with a quartz glass envelope as oblique incident illumination. Protect eyes from direct lamp illumination.

H. Diffusing Screens

While it is not a filter in the usual sense, the diffusing screen is placed in the optical path and does modify the light passing through it. The diffusing screen is used in the condenser system of some polarizing microscopes to eliminate the image of the lamp filament in the conoscopic image, see Chapter 5, Section V and to permit viewing of the sample image in the camera, see Chapter 8, Section III or a projection screen, see Section II C. Diffusing screens are of two types, ground glass and flashed opal glass. The ground glass is the more commonly used diffusing screen however, it presents an image with low brightness and resolution. The flashed opal screen has higher resolution but even poorer brightness.

VII. EYEPIECE ACCESSORIES

The eyepiece is the final element of the microscope before the eye, therefore it is not surprising that a major share of the eyepiece accessories provide measurement references for the entire optical system. The various measurement accessories as well as other specialized accessories will be discussed in this section.

A. Screw Micrometer Eyepiece

The screw micrometer eyepiece or filar eyepiece micrometer, shown in Figure 7-17A, has a composite reticle consisting of a fixed, graduated scale and a moveable index line that is coupled to a micrometer drum. One full revolution of the drum corresponds to one full division on the fixed scale. Since the drum is graduated into 100 divisions, each division on the drum corresponds to one-hundredth of a fixed scale division. Screw micrometer eyepieces are available from each of the major microscope manufacturers.

TABLE 7-8
POLARIZING MICROSCOPE ACCESSORIES

ACCESSORY	APPLICATION
1st order red compensator (gypsum plate) or (sensitive tint)	Determine the optical sign of a crystal or cleavage fragment. Determine direction of preferred orientation. Give color contrast to photomicrograph of non-isotropic material.
1st order red compensator in subparallel position	Determine optical directions in weakly birefringent materials.
Quarter-wave compensator (quarter undulation or mica plate)	Determine the optical sign of a crystal or cleavage fragment.
Quarter-wave compensator in subparallel position	Determine exact measurement of retardation using the Senarmont method.
Quartz wedge (1-4 order)	Determine the optical sign of a crystal or cleavage fragment.Qualitative measure ment of the path difference in sections of birefringent materials.
Babinet compensator - two quartz wedges in opposite orientations. Cemented together in opposite directions	Determine the optical sign of a crystal or cleavage fragment. "0" order fringe is at the middle of the compensator plate

A

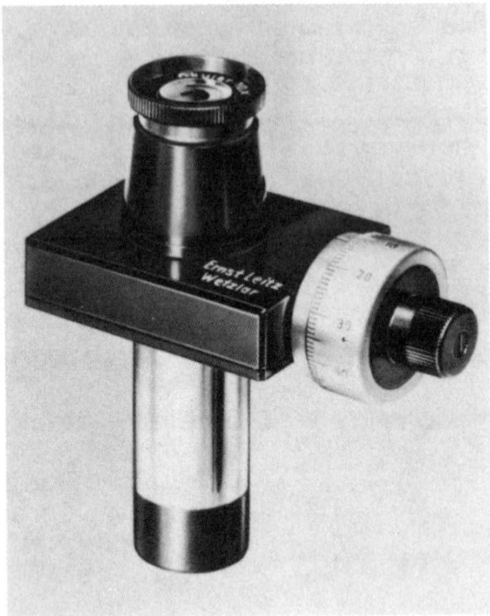

Figure 7-17. The screw micrometer eyepiece, (A), is the common device for visually making lateral dimensional measurements with the light microscope. Coincidence of the moveable eyepiece reticle line with the points of interest on the sample is established visually. The micrometer values corresponding to those points of interest on the sample are obtained by visually noting the coarse position of the moveable line in the reticle image and combining that value with that obtained visually from the micrometer drum. The digital micrometer eyepiece, (B), also requires the observer's eye to indicate coincidence of the moveable eyepiece reticle with the points of interest on the sample. However, the numerical values for each point of interest is given by the digital system. The difference between two readings is determined by the electronics and this value times the calibration factor for the optical system's magnification can be displayed directly as length measurements in either metric or English units.

B

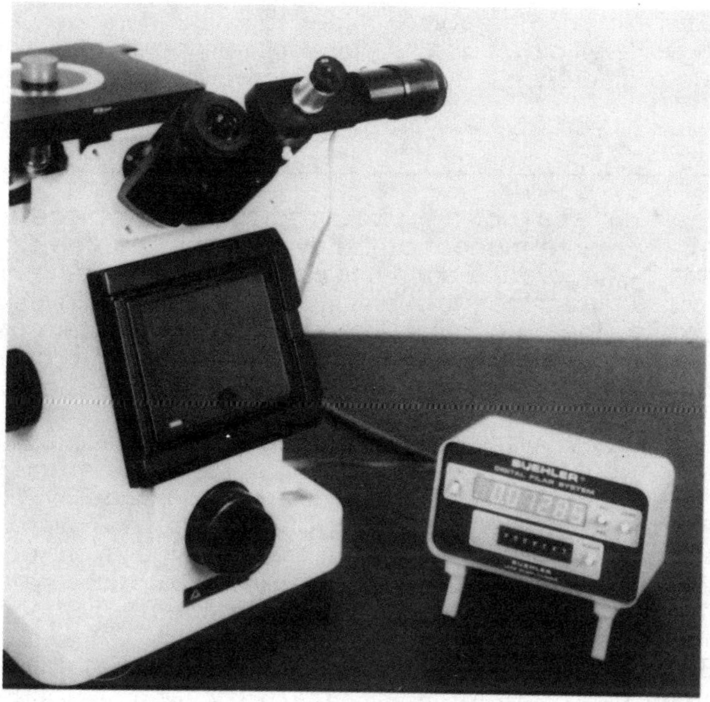

Figure 7-17: (continued)

Brace-Koehler rotatory compensator. Three ranges available are: λ/10, λ/20, and λ/30.	Measure path difference in sections of birefringent materials, quantitatively.
Berek tilting compensator. three ranges, I - IV orders I - X orders, and I - XXX orders.	Measure path difference in sections of birefringent materials, quantitatively. Also determine optical sign of a crystal or a cleavage fragment.
Ehringhaus tilting compensator. Two ranges: I - VII orders and "large path difference"	Measure path difference in sec tions of birefringent materials, quantitatively. Also, determine the optical sign of a crystal or a cleavage fragment.

The fixed scale of the screw micrometer eyepiece is calibrated in the same manner as the eyepiece reticle, see Chapter 5, Section II A, and each drum unit corresponds to one-hundredth of that value. The concerns, mentioned in Chapter 5 with regard to the setting of the mechanical tube length, zoom system (if any) and the cover glass correction ring must be strictly followed for greatest accuracy.

The Nikon screw micrometer eyepiece has a built-in zoom system which allows the eyepiece fixed scale values to be adjusted so that they correspond to integral values of the stage micrometer, thus providing a direct reading capability. Direct reading can also be achieved for other screw micrometer eyepieces by using a zoom system built into the microscope or changing the draw tube, if present. If either of these systems is used, it is imperative that the correct setting be used when screw micrometer measurements are made.

Before any measurement is undertaken with the screw micrometer eyepiece, the reticle should be correctly focused as described in Chapter 5, Section II A. The measurement of an object is accomplished by placing the moveable index line first on left side of an object and recording the position as the fixed scale value immediately to the left of the moveable index line followed by the reading from the drum. The index line is then shifted to the right side of the object and the position recorded, noting both the fixed scale and drum readings in the same manner as above. This procedure is illustrated in Figure 7-18. To minimize any affect of backlash in the screw thread, always approach all measurement points from the same direction.

A number of manufacturers, including Buehler (41 Waukegan Rd., Lake Bluff, IL 0044); Modulation Optics, Inc.(100 Forest Dr., Greenvale, NY 11548); and Wild Leitz, now offer digital readout

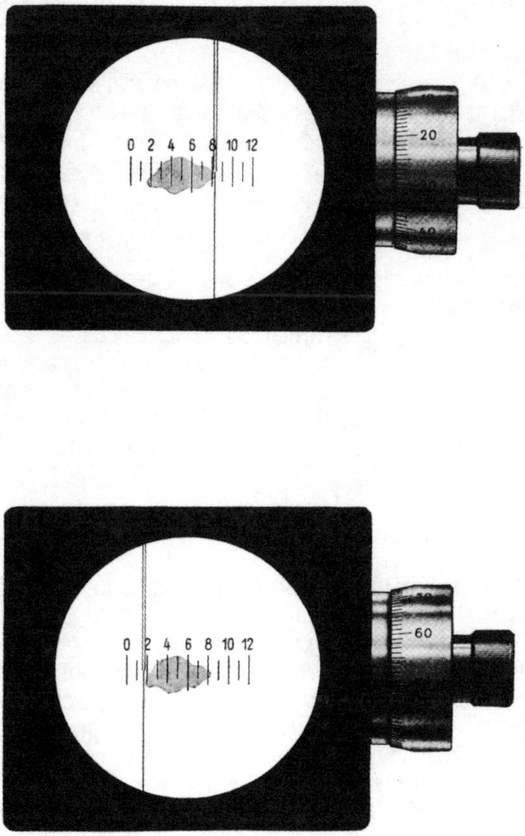

Figure 7-18. Illustration of the measurement of an object with a screw micrometer eyeplece. In (A) the moveable reticle line is moved from the right until it contacts the right side of the object image. A reading is taken at this point, namely 8, from the reticle scale and 0.27 from the drum = 8.27. The reticle is then moved by rotating the micrometer drum until it coincides with the left side of the object image. At that point, the position of the moveable reticle line is 1.67. The difference between the two reading is 8.27 - 1.67 = 6.60. This value times the calibration factor for the particular lens combination gives the object size. The drum must be turned in the same direction for each measurement to minimize errors resulting from any hysteresis in the screw drive mechanism.

versions of the screw micrometer eyepiece. In these accessories, the rotation of an ungraduated drum causes simultaneous translation of the index line and a proportional change on a digital display. These devices provide fast readings because they have a switch arrangement that produces an internal multiplication of the filar values by a calibration factor, thus giving a direct length reading on the display. One such digital micrometer eyepiece is shown in Figure 7-17B.

Leitz manufactured a multiple version of the screw micrometer eyepiece in the form of their integrating eyepiece. This accessory had six micrometer screws each of which caused the index line to be moved. The integrating eyepiece was used for quantitative analysis of multiphase materials where each micrometer spindle was assigned to a specific phase. In operation the index line was moved over a given phase by its respective micrometer screw. The composition of the material was then computed in the same way as for the integrating stage, see Chapter 5, Section IV B.

B. The Image Shearing Eyepiece

The image splitting or shearing eyepieces produced by Vickers Instruments Inc. and Watson and Sons Ltd. as shown in 7-19A and B are unique versions of the screw micrometer eyepiece that provide for increased precision and speed of measurement. The optical system of the image shearing eyepiece produces two identical images of the sample which can be split or sheared across each other by the rotation of the micrometer screw. The image shearing image is shown in Figure 7-19C, D, and E. The two images can be distinguished at will by inserting a built-in, dichroic filter into the optical train; in this way one image is tinted red and the other green. A digital readout/printout capability is also available. The Vickers image shearing system was also available as a module which was designed to fit a number of the Vickers bench microscopes. These eyepieces or systems are no longer manufactured.

The image shearing system is calibrated as described in Table 7-9. The concerns mentioned in Chapter 5, Section II A about setting the mechanical tube length, zoom system (if any) and the cover glass correction ring must be strictly followed for greatest accuracy. Two methods of measurement are commonly used with the image shearing system, these are described in Table 7-10.

C. Comparison Eyepiece

This accessory eyepiece variously known as "comparison", "turret", or "grain size" eyepieces make use of a number of fixed reticles in a revolving turret which may be individually compared with the sample image, a typical example is shown in Figure 7-20.

Unitron supplies a 10X turret eyepiece that is fitted with eight standard ASTM and JIS grain size charts for austinitic steels. Since the grain size charts are designed for comparison with the steel structures at a specific magnification, a graduated tube is also supplied with the eyepiece to adjust the total magnification of the microscope to 100X. Leitz and Zeiss have supplied a comparison eyepiece in the past.

A general procedure for using the comparison eyepiece is given in Table 7-11.

D. Eyepiece Telescope

The eyepiece telescope, also known as a telescopic eyepiece, auxiliary telescope, or centering telescope, is commonly used in phase contrast microscopy to center the phase annulus on the phase ring, see Chapter 4, Section IV. It can also serve as a desirable accessory for adjustment of the independent reference, dual beam interferometer, see Chapter 5, Section III D. This device is an adjustable, near-focus telescope which is normally used to examine the rear aperture of the objective lens. Thus, the application of this accessory is the same as for the Bertrand lens, see Chapter 5, Section V A.

To use the eyepiece telescope, it should be inserted in place of the eyepiece and adjusted, either with its draw-tube or its threaded focusing screw, to give a sharp image of the dark periphery of the objective lens rear aperture. The correct focus can be verified by very slightly moving the revolving nosepiece and noting that the image of the aperture moves in the same direction that the nosepiece is moved. The image of the objective rear aperture as viewed with the eyepiece telescope will appear about the same size as when viewed with the Bertrand lens.

E. Pointer Eyepiece

As the name implies this eyepiece is equipped with a pointer that is visible in the image. This pointer is used to highlight a given structure when a microscope is shared by two or more viewers. Usually, it is possible to externally position the pointer in the field of view. These are available from Reichert, Leitz and Zeiss. Needham (1958) has suggested that a simple fixed pointer eyepiece can be constructed by cementing an eyelash to the field limiting diaphragm of an eyepiece.

A

Figure 7-19. Examples of (A), the Watson image shearing eyepiece and (B), Vickers image shearing micromeasurement system. These devices permit rapid visual determination of feature sizes. The measurement principle is shown in (C) where the dark circle represents the two completely overlapping unsheared images of the object. In (D) the calibrated adjustment knob has been turned and the two images are displaced. In (E) the two images have been displaced until they just touch each other. At this point the scale value on the adjustment knob times the calibration factor for the particular objective lens give the object dimension in the shear direction. The Watson device is no longer available. Photos courtesy of Mr. Ivan Hills, Hacker Instruments Inc. and Mr. Michael Palmer, Vickers Instruments, Inc.

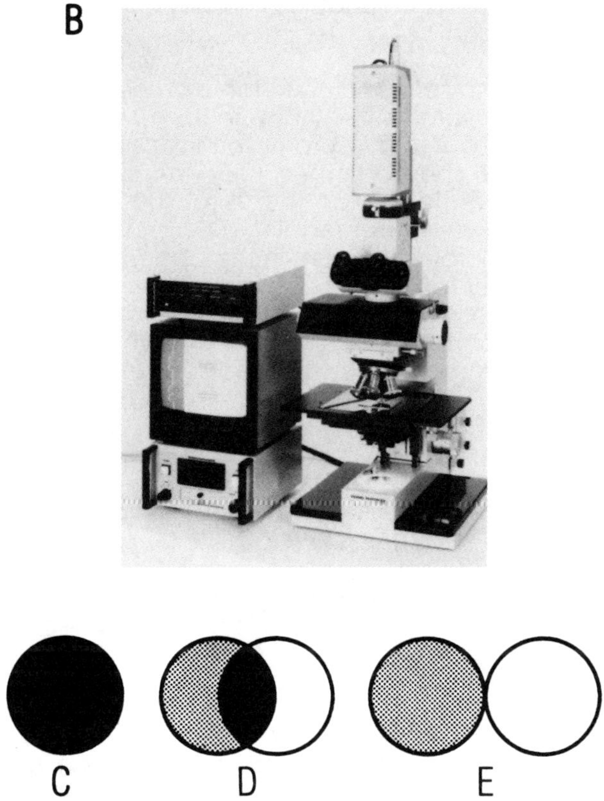

Figure 7-19: (continued)

TABLE 7-9
CALIBRATION OF THE IMAGE SHEARING EYEPIECE

1. Set up the microscope for brightfield observation and critically focus a stage micrometer image. The length of the micrometer should be exactly parallel to the shear direction. Tabulate the objective lens magnification, the mechanical tube length (if an adjustable draw tube is present), the zoom value (if present),and the cover glass correction setting (if present).
2. Set the shear to zero and insert the dichroic filter.
3. Rotate the micrometer drum so as to shear as many as possible of the stage micrometer lines. For example, If five lines can be sheared, then the number five line of one image will exactly overlap the "zero" line of the other image. Record the micrometer reading, D1.
4. Rotate the micrometer drum in the opposite direction until the same number of lines have been sheared. Record the micrometer reading, D2.
5. The calibration factor M = 2d/(D2-D1) where d is the number of stage micrometer lines sheared in one direction.

TABLE 7-10
IMAGE SHEARING MEASUREMENTS

Method A:
1. The two images are sheared in one direction until the structures just touch and the micrometer reading is recorded.
2. The two images are then sheared in the opposite direction until the structures just touch and the micrometer reading is recorded.
3. The structure's dimension is given by multiplying the difference in the reading two and reading one by the calibration factor from Table 7-9 and dividing by two.

Method B:
1. Set the system to zero shear, i.e., with images exactly superimposed and the micrometer reading recorded. NOTE: The value for zero shear is constant and once noted can be used without again referring to the micrometer scale.
2. The two images are sheared in either direction until the structures just touch and the micrometer reading recorded.
3. The structure's dimension is given by subtracting the two readings from steps one and two and multiplying by the calibration factor from Table 7-9.

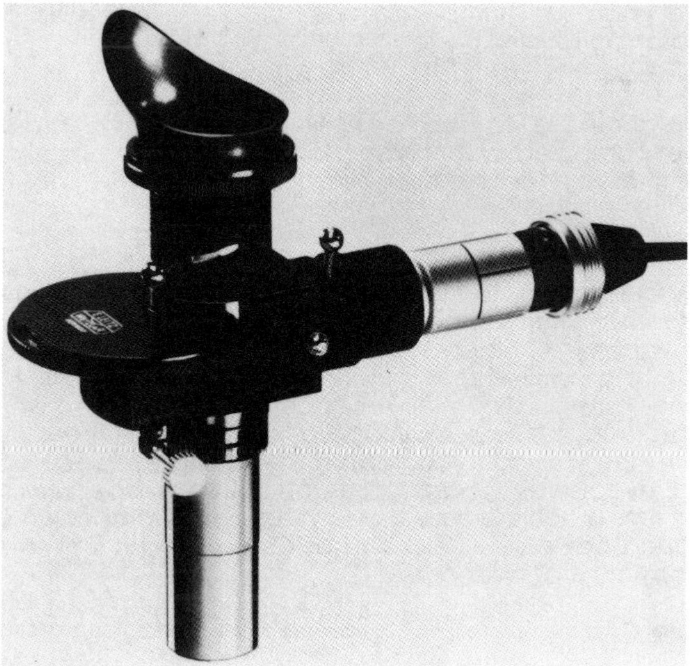

Figure 7-20. The comparison eyepiece has a turret which can bring a number of reticles into position for comparison with the sample image. This specific eyepiece is no longer available. Courtesy of Wild Leitz USA, Inc.

TABLE 7-11
SET-UP OF THE COMPARISON EYEPIECE

1. Carefully focus the reticle plate as described in Chapter 2, Section IV C.
2. A polished and etched steel sample is then carefully focused using the 10X objective lens, (100X total magnification).
3. Rotate the turret until a reticle image matches the grain size of the sample.
4. The grain size number for the sample is indicated on the matching reticle image.

An alternate to the pointer eyepiece is a cross wire eyepiece combined with a mechanical stage for locating the object of interest under the cross wire.

F. Setting Eyepiece

An alternative to the eyepiece telescope is offered by Watson for their interference objective lenses; this is referred to as a "setting eyepiece". This device consists of a 5X lens mounted by a swing arm to a collar that fits around the microscope body tube at the eyepiece level. When the rear aperture of the objective lens is to be observed the lens is swung up, directly over the eyepiece and the image viewed through both the setting eyepiece and the standard eyepiece. Sharp focus of the rear aperture of the objective lens is achieved by moving the collar up or down on the microscope body tube. Reichert has also offered a similar device as their auxiliary magnifier.

G. Pinhole Cap

The pinhole cap is a third alternative to either the Bertrand lens or the focusing telescope. It consists of a simple cover with a central hole ranging from 0.5 to 1.0 mm in diameter. This cap replaces the eyepiece and gives a very sharp image of the objective lens rear aperture but the image is small. The pinhole eyepiece is often used in optical crystallography or petrography when the interference figure is very diffuse in the Bertrand lens.

An efficient and inexpensive pinhole eyepiece can be constructed by drilling a hole in the center of one of the black plastic inserts used to protect the tube of the microscope during shipping. A 0.5 mm hole (#76 drill) to 1 mm hole (#60 drill) is ideal.

H. Protractor Eyepiece

Unitron supplies two protractor eyepieces that allow the measurement of angles in the plane of the sample. A knurled collar rotates a moveable reference line with respect to a fixed line and the resulting angle is read from a scale visible in the eyepiece. One of these eyepieces may be read to one degree while the other may be read to 5 minutes of arc with the aid of a vernier.

I. Wright Eyepiece

An accessory eyepiece was developed by Wright (1910) to permit the quantitative use of the quartz wedge, see Section VI G., in optical crystallographic measurements. This eyepiece is also used with such accessories as the quarter wave plate and the Nakamura half shadow plate to permit the analysis of elliptically polarized light from bireflective metallic and ore samples.

J. Soleil-Babinet Compensator

The Soleil-Babinet compensator is an eyepiece accessory that incorporates a continuously adjustable retardation plate. This accessory may be used to very accurately determine parameters in optical crystallography and ore microscope. It may also be used to measure the degree of elliptical polarization in the image and thus, can be used to determine the thicknesses of very thin surface films. The Soleil-Babinet compensator is available from Melles Griot.

REFERENCES

Andrews, E. H., J. Sci. Instrum., 34, 71, (1957).

Bergsman, E. B., Metal Progress, 54, 183, (1948).

Brenden, B. B., H. W. Newkirk, and J. L. Bates, J. Am. Cer. Soc., 43, 246, (1960).

Bridge, H., J. Sci. Instrum., 34, 136, (1957).

Carroll, K. J., D. H. Liebenberg, and W. C. Overton, Rev. Sci. Instrum., 38, #2, 260, (1967).

Cech, R. E., Rev. Sci. Instrum., 21, #8, 747, (1950).

Chayes, F., Am. Mineral., 34, 1, (1949).

Chayes, F., Am. Mineral., 40, 126, (1955).

Clothier, W. C., J. Sci. Instrum., 44, 535, (1967).

Cosier, J. and I. G. Wood, J. Physics E. Sci. Instrum., 16, 687, (1983).

Derge, G. and J. R. Shegog, ASTM Spec Tech Publ. 472, 132, June 1970.

Edmonds, D. T. and J. P. G. Mailer, J. Physics E. Sci Instrum., 10, 868, (1977).

Edwards, P. L., Rev. Sci. Inst., 32, 95, (1961).

Ehlers, E. G., "Optical Mineralogy: Theory and Techniques, Vol 1,Blackwell Pubns., (1987).

Emmons, R. C., "Universal Stage", Geol. Soc. Amer., Memoir 8, (1943).

Ford, I. H., J. Sci. Instruments, 31, 164, (1954).

Foster, A. D., J. Sci. Instrum., 42, 172, (1965).

Gahm, J., Optical Spectra, 2, #3, 28, (1968).

Greeson, R. L., W. L. Johns, and R. J. Jackson, Tech. Papers of the 19th Metallographic Group Meeting USAEC, ONRL-TM-1161, 146, (1966).

Grodzinski, P., J. Sci. Instrum., 28, 117, (1951).

Harmer, P. R., S. D. Ford, and R. A. Dugdale, J. Sci. Instrum., 1, Series 2, 59, (1968).

Hawkins, M. J. and J. Barford, J.Phys. E (Sci Instrum), 3, 11, 892, (1970).

Hildebrandt, W. H. and F. H. Cocks, Rev. Sci. Instrum., 47, #7, 866, (1976).

Hooss, Von E., Mikroskopie, 18, 269, (1964).

Hunt, J. D., K. A. Jackson, and H. Brown, Rev. Sci. Instrum., 37, 805, (1966).

Hurlbut Jr., C. S, Am. J. Sci., 237, 253, (1937).

Isaacs, L. L. and H. Pops, Trans. Metal. Soc. AIME, 230, 266,(1964).

Katz, R., Rev. Sci. Inst., 37, #10, (1966).

Kaukler, W. F., Rev. Sci. Instrum., 55, #10, 1643, (1984).

Kessler, G. and M. R. Rudman, Prakt. Metallogr., 8, #1, 40, (1971).

Kong, F. M., C. O. Pruneda, and R. J. Morgan, Compos. Technol. Rev., 5, #3, 95, (1983).

Lloyd, S. J. and D. J. Norris, J. Sci. Instrum., 28, 81, (1951).

Locquin, M. V. and Langeron, "Handbook of Microscopy," pp.90, Butterworths, London, (1978).

Lowndes, R. H. and M. B. Hallett, J. of Microscopy, 142, 371, (1986).

Lozinskii, M. G., High Temperature Metallography, Translated by L. Herdan, Pergamon Press, N.Y. (1961).

Markussen, J. and W. C. McCrone, The Microscope and Crystal Front, 14, #10, (1965).

McCrone, W. C., Fusion Methods in Chemical Microscopy, Interscience New York (1957).

McNeil, J. F., Metallurgia, 54, 674, (1956).

Monier, J. C. and R. J. Hocart, J. Sci. Instrum., 27, 302, (1950).

Needham, G. W., "The Practical Use of The Microscope," p 264, C. C. Thomas, (1958).

Ogilvie, G. J. and G. Brinson, J. Sci. Instrum., 32, 170, (1955).

Olson, D. M., B. B. Brixner, and M. C. Smith, Metal Progress, 81, #6, 75, (1962)

Phipps, L. W., J. Sci. Instruments, 43, 485, (1966).

Rhodes, R. G., J. Sci. Instrum., 27, 333, (1950).

Rosenfeld, M. A., Am. Mineral., 39, 834, (1954).

Rosenthal, M. and W. F. Rall, Med. & Bio. Eng. & Comp., 22, #5, 471, (1984).

Saller, H. A., R. F. Dickerson, and R. J. Carlson, Metal Progress, 67, 105, (1955).

Scott, W. D. and J. A. Pask, J. A. Cer. Soc., 44, 181, (1961).

Sippel, R. F., Rev. Sci. Instrum., 36, #11, 1556, (1965).

Smallman, R. E. and K. H. B. Ashbee, Modern Metallography, pp 1928, Pergamon Press, N.Y. (1966)

Stevenson, D. G., J. Sci. Instruments, 28, 275, (1951).

Sturmer, W., Photographie und Forschung, 7, #8, 233, (1957).

Tynan, E. E. and R. J. von Gutfeld, Rev. Sci. Instrum., 46, #5, 569, (1975).

Wall, R. and D. I. Roberts, Metallurgia, 68, 291, (1963).

Wentworth, C. K., Jour. of Geology, 31, 228, (1923).

Whitman, E. A., "Manual For The Use Of The Universal Stage In Optical Crystallography," NPP/RR 66-4, U.S. Naval Propellant Plant, Indian Head, MD, 7 July 1966.

8

Photomicrography

I. INTRODUCTION

A major goal of any scientific investigation is to communicate the findings to others; this is most effectively accomplished with the aid of pictures or drawings. To provide the necessary visual information, the camera lucida, the photographic camera and more recently the video camera have been combined with the microscope to provide images of suitably prepared samples; the camera lucida and the video camera have been discussed in Chapter 7, Section II C. This chapter deals with the instrumentation and the procedures for the preparation of a photograph of the sample image, the photomicrograph.

At the outset a review of the correct terminology is worthwhile. The photomicrograph is a large photographic image of a small object produced with the aid of a compound microscope at magnifications of 50X or larger. This term must not be used interchangeably with "microphotograph" which is a very small photographic image of a large object. Another term, discussed in more detail in Chapter 9 is "photomacrograph"; this is a large photographic image of a small object at magnifications less than 50X.

In principle, almost any microscope and any camera can be combined with varying degree of success. From these combinations photomicrographs can be generated that range from illegible to works of art. It must be understood that to prepare photomicrographs of consistent, publication quality, the essential instrumentation and materials are: (1) a stable work bench, (2) a suitably corrected microscope with properly aligned illumination, (3) a camera of the desired format, and (4) an appropriate photographic medium (film). An optimum photomicrograph requires a full understanding of the instrumentation, the photographic medium, and to some extent, the sample. Furthermore, the same kind of patience, care, keen insight and professionalism that produces a fine portrait or landscape is required if high quality photomicrographs suitable for publication are to be prepared. To be sure, these latter virtues cannot be acquired merely from a reading of this or any other text but should be strived for in the microscopist's own personal development.

Photomicrography has become a necessary element of industrial and scientific endeavors. However, the requirements for format, and film types may vary according to the specific needs. For example, 35 mm film may be desirable for the rapid, economical recording of sequential events, whereas the more expensive, instant format photomicrograph may be absolutely necessary where the timely adjustment of manufacturing processes depends on microscopically determined parameters.

II. MICROSCOPE CONSIDERATIONS

As stated above, almost any microscope can be used to produce a photomicrograph, in general the better the instrument, the better the potential for a quality photomicrograph. A camera may be attached to a microscope at any point where an eyepiece may be placed to obtain a visual image, this includes a monocular tube, one of the binocular eyepiece tubes, and a trinocular tube, see Figure 8-1. Special camera ports are available on some instruments. In those cases where photomicrography is the prime concern, instruments are available in which the camera is an integral part of the microscope body such as a photomicroscope, Figure 8-2 or a metallograph, Figure 8-3.

The principles and operation of the compound microscope, as described in Chapters 2 and 3, should be thoroughly understood before undertaking any serious photomicrography. Furthermore, the appropriate techniques described in Chapter 4, should be reviewed before photographically recording structures requiring contrast enhancement.

At this point it is appropriate to reemphasize that more care is required in the selection of the optics for photomicrography than for visual microscopy. This is due to the forgiving nature of the visual process that integrates images via the memory during observation of the sample. Thus, as we move the fine focus back and forth through optimum focus for each radial zone of the image, we tend to integrate these multiple images into a mental composite that constitutes an overall sharp image. With film you record only one image plane, except in the case of deep field microscopy, see Chapter 6, Section VIII. To make a particular image plane useful, photographically, its field should be as flat as possible. Thus, it is desirable to select plano objective lenses and use the correct and

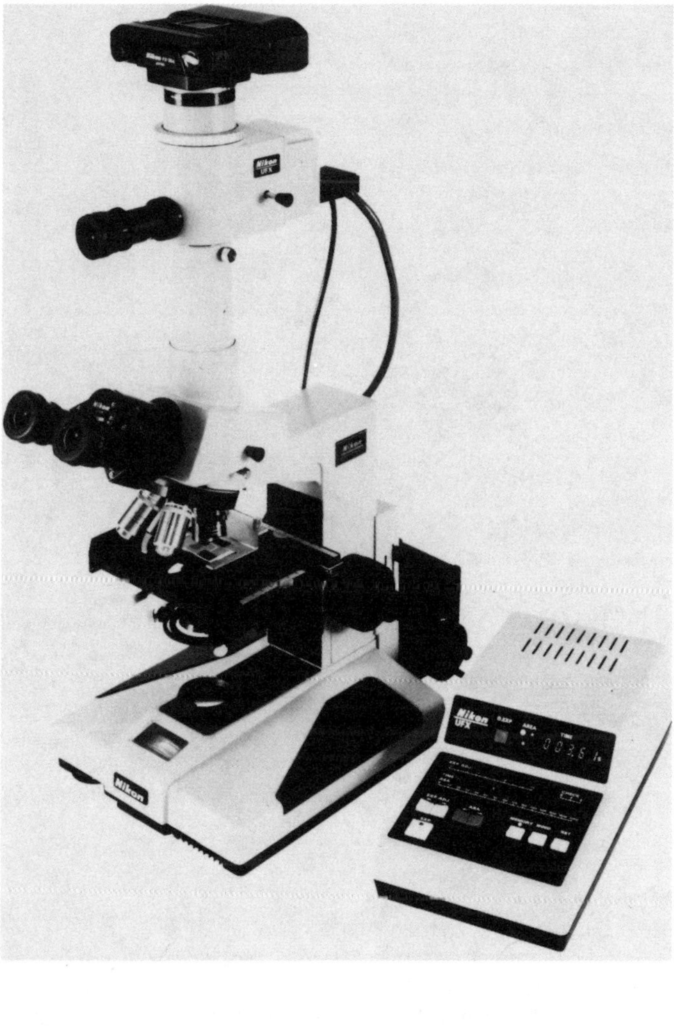

Figure 8-1. The trinocular body provides one of the more convenient means of attaching a camera to the microscope. In some units the image may be available for the camera at all times, however, the best arrangement uses a slider knob on the trinocular body to direct the image either to the binocular eyepieces or to the camera; in this way maximum intensity is available for either use. Courtesy of Nikon, Inc.

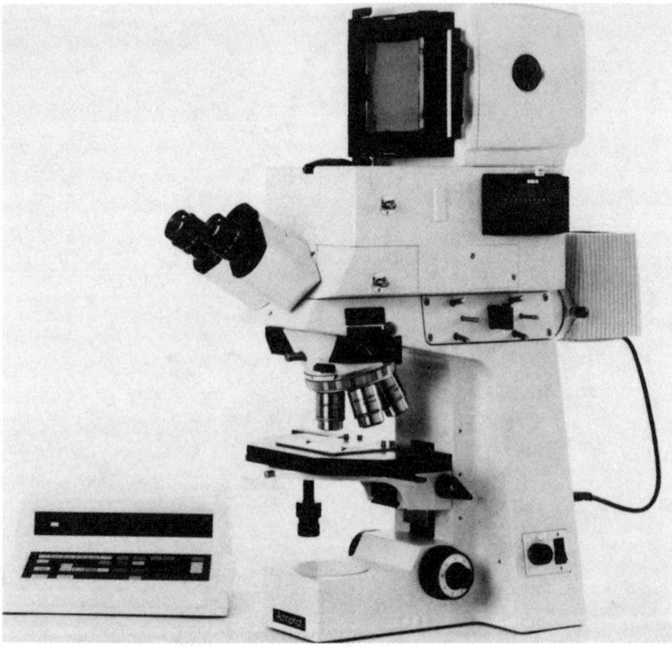

Figure 8-2. For greatest photomicrographic capability, the optimum instrument is the photomicroscope. This instrument incorporates within its body all of the necessary photographic optics for greater stability and convenience. Courtesy of Carl Zeiss, Inc., New York.

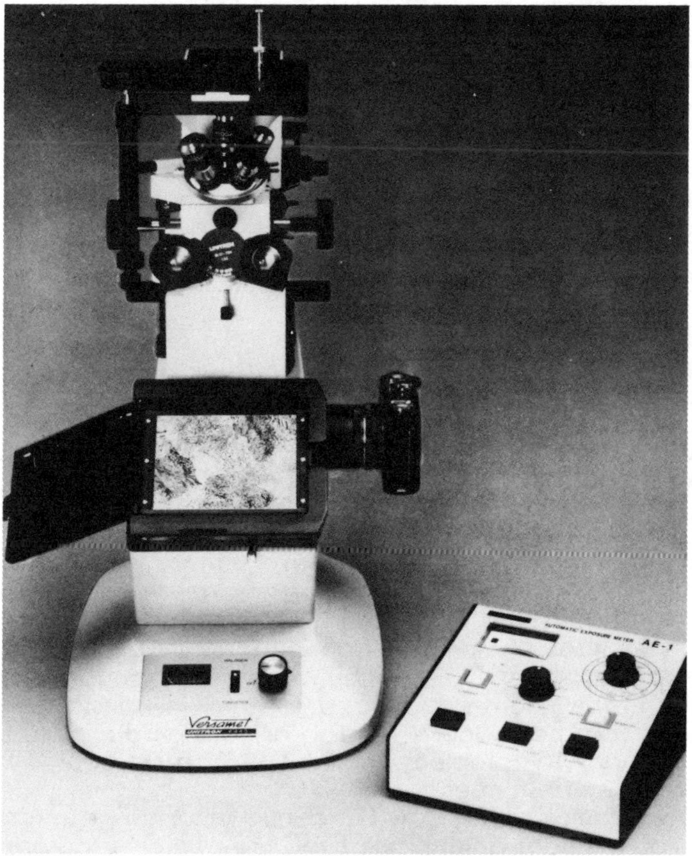

Figure 8-3. The metallograph is an incident light instrument that also has a built-in photographic capability. Courtesy of Unitron Instruments, Inc.

corresponding eyepieces. If plano objective lenses are not available, flatness of field can be improved, at the expense of resolution by reducing the N.A. of the condenser system; this is described in the set-up procedures later in this chapter. Some older microscopes are equipped with special photographic eyepieces that also improve the flatness of field.

After the appropriate optics have been selected, the main concern with regard to the microscope is image stability. Image stability is improved by reducing external and internal vibrations, improving the rigidity of the microscope/camera interface and finally, reducing drift in the focus.

External vibration can result from any source, ranging from railroad and street traffic to personnel movements near the instrument. The microscope table should be located on the ground floor, ideally concrete. If this is not possible, one of the vibration isolation devices described in Chapter 7, Section II should be used.

Movement of the photomicrographic system relative to the microscope rapidly degrades picture quality. This problem may be the result of external vibration acting on the large mass of a camera that is supported only by the relatively thin walled microscope tube. The best way to eliminate this motion is to use a camera support, such as shown in Figure 8-4. The microscope sits on the base of the support and the camera is affixed to the rigid support post. The camera is then optically connected to the microscope through a non-contacting, light-tight coupling.

Vibration internal to the microscope/camera combination is most often due to the camera and will be covered later in detail, however sources of vibration can infrequently be located in the microscope proper. Perhaps the most common source of vibration in the microscope stand is a built-in transformer for the illuminator that has a loose winding or core producing a 50 or 60 hertz vibration. This problem can be cured only by replacement of the defective transformer.

Focus drift is a very slow but continuous change in focus, that can seriously affect a very long exposure photomicrograph. Focus drift is most commonly observed on a microscope that is focused by changing the position of the body tube. The additional weight of the camera assembly may then cause the fine or coarse focus to creep. When a camera support is not possible or practical, a simple cure for this problem is to use shorter exposure times. In the extreme cases of focus drift, the focus mechanisms may require professional service.

III. THE CAMERA

An almost endless number of camera arrangements are possible for use in photomicrography. This section will cover the range from the simplest set-up using a snapshot camera to the

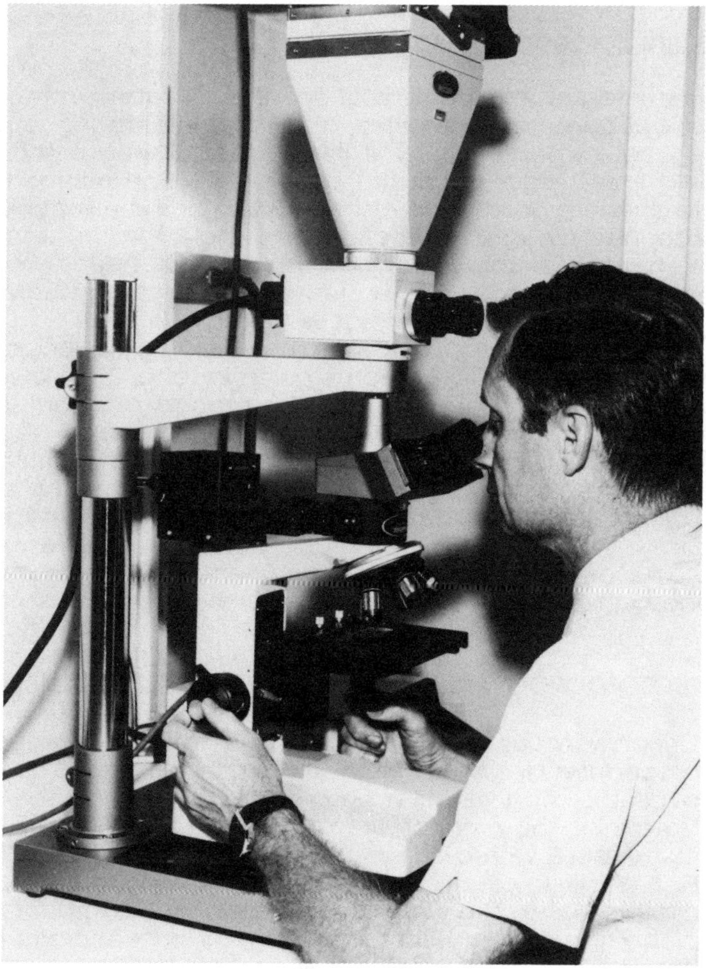

Figure 8-4. The camera on a trinocular head should be attached to a support to minimize stress on the vertical tube and to stabilize both the camera and the microscope.

most sophisticated photomicroscopes having a selection of camera capabilities. The major microscope manufacturers have a wide range of photomicrographic attachments and accessories for their instruments.

A. The Integral Lens Camera

The integral lens camera is any one of many consumer or professional cameras with either a fixed or interchangeable lens used for regular photography. If this camera is set at infinity focus and held very steadily over the eyepiece of the microscope so that the aperture of its lens is at the position of the exit pupil of the eyepiece, See Chapter 2, Section III E, a photomicrograph may be taken of the sample. Obviously, this is neither easy or convenient but does serve to illustrate the general simplicity of the process. The camera simply replaces the human eye.

A generalized procedure for using the integral lens camera for preparing a photomicrograph is given in Table 8-1. However, a few words of caution are in order. This process is recommended only for instruction or for very occasional use because of the difficulties of assuring that the image is in focus and determining exposure. Also blurring of the image due to motion of the camera is a problem. It should be noted that the microscope image does not usually fill the film frame; the image is circular and often with an out-of-focus periphery. These latter problem may be minimized by using the highest magnification eyepiece available.

TABLE 8-1
PHOTOMICROGRAPHY WITH AN INTEGRAL LENS CAMERA

1. Critically focus the microscope on the sample using the procedures described in Chapters 2, 3 and 4. Use the highest available magnification eyepiece.
2. Determine the distance from the top lens of the eyepiece to its exit pupil with a thin sheet of paper in a darkened room. First, gently lay the paper on the eyepiece and note a bright circle of light. Then slowly lift the paper away from the eyepiece and note that the circle first diminishes in diameter and beyond that point increases in size. The distance between the eyepiece and the paper when the circle is at it smallest dimension is the height of the exit pupil for that eyepiece.
3. Position the camera with a tripod or other support so that the camera lens is centered over the eyepiece of the microscope and spaced so that the camera lens surface is at the height of the exit pupil of the eyepiece. If the camera has a lens aperture adjustment, it should be set wide open.
4. Stray light may be excluded from the camera using black paper carefully positioned around the lens and the eyepiece.

5. Expose the film using the camera shutter taking care not to move or vibrate either the camera or the microscope. Correct exposure will be obtained directly if the camera is equipped with a spot exposure meter otherwise the exposure may be determined using the multiple exposure method described in Section V E.

Magnification for the integral lens camera image is given by the following expression where M is the total magnification at the film plane, fl is the focal length of the camera lens and M_t is the visual magnification as given below:

$$M = M_t \times fl/250 \qquad (40)$$

Polariod offers a very serviceable, albeit not inexpensive, integral lens camera attachment which is almost universal in application. The SX-70 Microscope Adaptor, Model 137 will mount on any microscope body tube with a diameter of 24 to 40mm by means of a self-centering clamp. Although the unit is fairly light in weight, it should be used with care on an inclined tube such as found on a binocular head. The adaptor unit as purchased is equipped with a Polariod SX-70 camera that can be removed at will for other general photographic purposes.

The SX-70 microscope adapter has a number of useful features; among these are: (1) it provides for through-the-lens focusing, (2) it provides for automatic exposure, and of course automatic film advance (3) it is easy to use, and (4) it produces color pictures. On the other hand, its image quality is not optimum and it must be used with high magnification eyepieces if the film frame is to be completely filled.

B. The Camera Without An Integral Lens

For highest quality photomicrography the cameras of choice rely on the eyepiece as the main image forming element in its system. These cameras range from relatively simple add-on units to large cameras stands and of course the built-in cameras of the photomicroscopes.

It will be recalled from Chapter 1, Section III that a real image is formed at a distance of 250mm (10in) above the eyepiece; if a sheet of film is placed at that plane a sharp image of the sample can be recorded. It is the information at this plane that is imaged by the non-integral-lens camera, either directly or modified by a relay lens.

The most commonly used camera formats are: 35 mm, 3-1/4 in. x 4-1/4 in., 4 in. x 5 in. and for certain applications 8 in. x 10 in. The most popular format for the biomedical field is 35 mm because it is relatively inexpensive, a wide range of film types are available, and frames can be advanced rapidly for sequential photomicrographs. The 4 in. x 5 in. format is generally accepted as the standard in the industrial and metallurgical fields because the larger format gives photomicrographs of sufficient size that enlargement is unnecessary and permits the use of the instant prints that are so desirable in the rush of the industrial world. The 3-1/4 in. X 4-1/4 in. fixed bellows format is also used industrially since it is less costly than 4 in. X 5 in. film.

The 35 mm cameras and their attachments are generally small and light in weight, with a fixed bellows length, therefore they can be mounted directly on the vertical tube of a trinocular head or on a camera port without any additional support. The entire 35 mm assembly usually consists of the microscope attachment, a right angle viewing and focusing tube, the shutter mechanism that may include exposure monitoring and control, a relay lens, fixed bellows, and the camera back. A fixed bellows, 35 mm camera is shown in Figure 8-5.

The viewing and focusing (v&f) tube is, as the name implies, the means of viewing the image and focusing the camera. A beam splitter or a beam diverter directs a portion or all of the image beam, respectively, to the reticle in the v&f tube. The reticle accurately defines the effective film plane of the camera and at the same time provides the outline of the camera frame so the photomicrograph can be properly composed.

The shutter is usually a manually or electronically operated iris type to minimize vibration. The more modern units can measure exposure times and use these to control the shutter speeds for automatic exposures. These automatic systems, to varying degrees, can monitor any desired spot or an entire field of view for determining the exposure. In the same manner, many of these automatic systems can compensate for either brightfield or dark-field images, adjust for reciprocity failure of the film, and may provide color balance information for manual adjustment of the illumination color temperature. Manual control of the shutter is also possible to produce identical exposures for the preparation of montages.

The purpose of the relay lens can be seen from Figure 8-6. If the 35 mm camera is placed so that its film plane is at the real image distance of 10 inches above the eyepiece, the film frame accepts only a small portion of the total image. To correct this problem, the relay lens is inserted in the light path above the eyepiece.

Normally, a simple 35 mm camera back with either manual or motorized film advance is all that is required. Often the camera backs are equipped with a dark slide which may be closed so that a camera with one type of film may be removed from the attachment without exposing the film and replacing it with another containing a different film type.

A single lens reflex (SLR) camera back would seem to be desirable for use with a simple microscope attachment since the

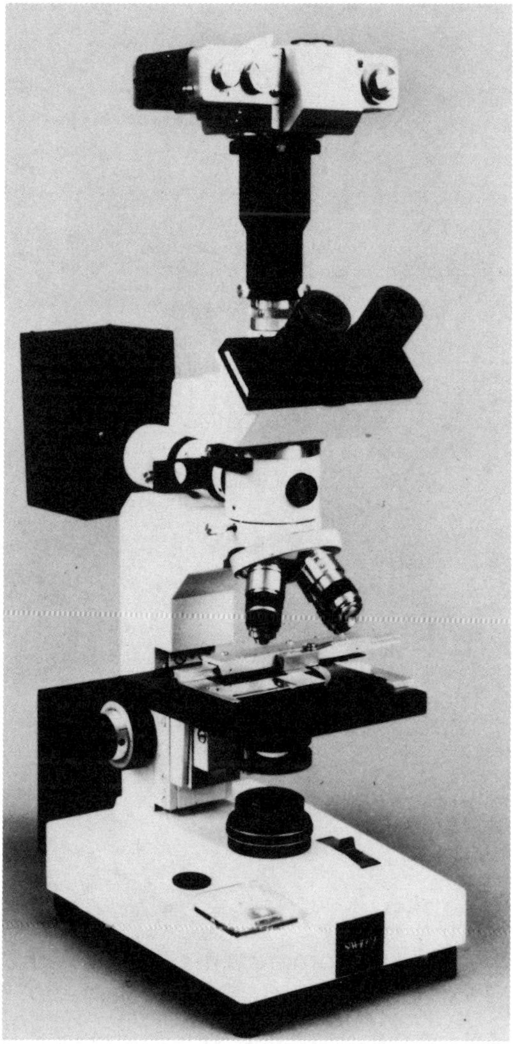

Figure 8-5. The 35 mm camera format is an economical approach to photomicrography both in terms of instrumentation and film. With automatic film advance it also permits taking rapid sequences. Courtesy of Swift instruments, Inc.

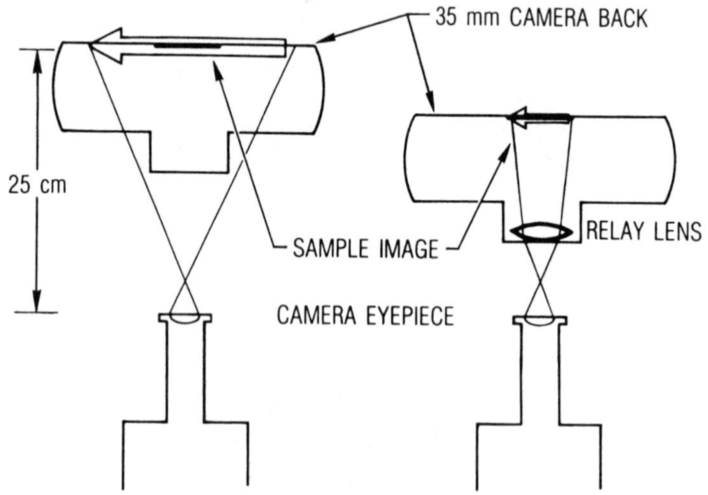

Figure 8-6. If a 35 mm camera is mounted at the point of the real image, 10 inches above the eyepiece, a photomicrograph can be recorded, as shown at the left, however, that image would represent only a very small portion of the visual image. The support tube for the 35 mm camera normally incorporates a relay lens, shown at the right, that both physically shortens the camera assembly and reduces the image size to match the 35 mm format.

right angle viewing tube and the shutter assembly would not be needed. However, the SLR viewer is generally of the ground glass type which does not permit critical focusing of fine structures. Furthermore, the vibration resulting from reflex mirror movement and the focal plane shutter are very difficult to minimize. For these reasons the SLR camera backs are not recommended for photomicrography. Nevertheless, such attachments are available from a number of the SLR camera manufacturers and from Edmund Scientific.

The larger format cameras are usually mounted on a separate support to reduce the weight resting on the microscope, minimize the effect of shutter vibration on the microscope and to provide additional stability for the entire assembly. There is no mechanical coupling between the camera attachment and the microscope; light leaks are eliminated by two nested cups, one surrounding the microscope eyepiece and the other affixed to the camera attachment. These cameras may be of either the fixed or flexible bellows type, but the flexible bellows gives more freedom in adjusting the magnification to a specified value.

The attachment for these larger cameras contains the shutter and possibly an exposure monitor and control. Focusing of a fixed bellows camera may be accomplished with a right angle v&f tube included in the attachment; the arrangement in this case is similar to that of the 35mm camera. The function of the various parts of the attachment are the same as for the 35mm camera above.

If the camera has a floxible bellows, as shown in Figure 8-7, the focusing must be accomplished on a ground glass screen either at the actual film plane or in a reflex housing, if present. The ground glass may be examined directly or in a clear, central area, if available, using an auxiliary focusing magnifier. Since the eyepiece to film plane distance can be varied with the bellows, the exposure time and thus the shutter speed is not a constant for a given intensity image. Therefore, the correct exposure must be measured at the actual film plane and the shutter speed set manually to match the determined exposure.

A very economical, but effective fixed bellows camera arrangement is available from Fischer Scientific. This camera has a simple attachment that can be used with most microscopes. Once the area of interest has been located on the sample, a focusing tube is placed on the attachment and the correct focus obtained for the camera, then it is replaced with the camera and the exposure made. The camera uses either 100 or 600 series Polaroid 3-1/4 in. X 4-1/4 in. film packs.

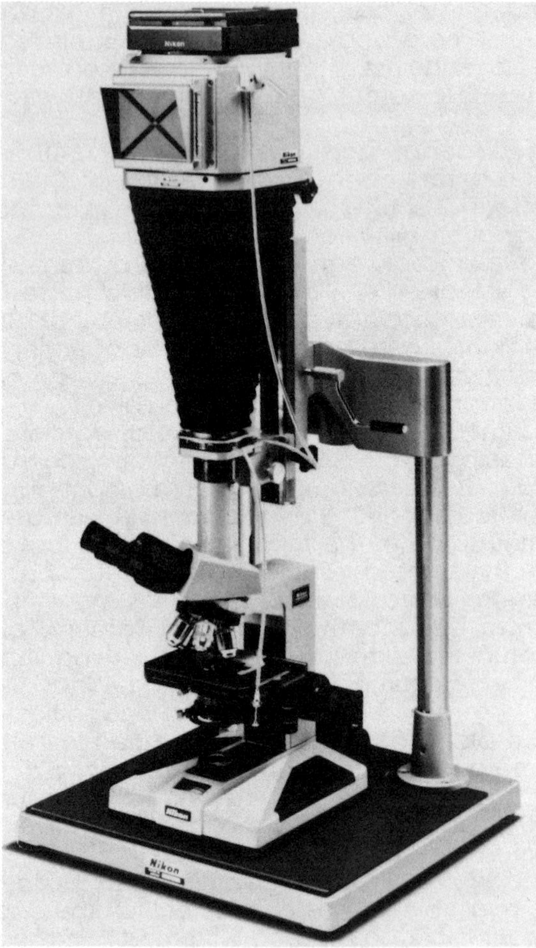

Figure 8-7. The large format camera has an image that usually does not require enlargement and that can provide immediate photomicrographs with instant process film. A flexible bellows, if a part of the system, can provide additional capability in composition and adjustment of magnification. Courtesy of Nikon, Inc.

IV. PHOTOGRAPHIC MEDIA

This section provides a brief review of the properties of photographic media of importance to the photomicrographer and a listing of the more important film types currently in use. No attempt will be made to discuss processing or printing procedures for the various media as this is beyond the scope of this text; in this regard, the reader is referred to Loveland (1971).

The term photographic media is used herein since modern photomicrography relies not only on the wet process photo materials but also on the instant process materials as well. To reduce the listing of these materials to a workable number, only specific wet process products of the Eastman Kodak Company and instant process products of the Polaroid Corporation will be discussed.

A. Wet Process Media

The wet process photographic materials consist of a light sensitive layer, the emulsion, coated on a polymeric sheet or substrate. The light sensitive material in the emulsion is a dispersion of fine silver halide grains. At one time glass plates were used as substrates to provide dimensional stability, however, newer polymeric materials such as ESTAR™ are quite stable for most purposes. These media are available as roll film in light tight canisters in the case of the 35mm film and as individual sheets in the case of the 4" x 5" film. The latter sheet film must be used in specially designed light tight film carriers.

The black and white (B&W), wet process media are commonly referred to as "negatives" since the processed negative is dark (or exposed) where the image is bright and clear (or unexposed) where there was no light in the image. When a print is made of the negative, another negative process is used; thus, a negative of a negative gives a positive or a direct representation of the image. For further information on the photographic process the reader is referred to Loveland (1971).

The properties of the B&W photographic medium of interest to the photomicrographer are film speed, resolution, graininess, contrast, reciprocity failure, and spectral sensitivity. Film speed is a measure of how sensitive the film is to light. In the United States, film speed is measured in ASA (more recently referred to as ISO) units. A film rated at 100 ASA is twice as fast in its reaction to light as a film with an ASA rating of 50. Speed is a very important consideration if the object of interest in the sample is moving, in this case the faster the film, the less likely that the image of the subject will be blurred due to movement. Conversely, if the subject is inanimate, speed is of little importance. Increased speed is of importance if the light level is very low as in the case of dark field and fluorescence microscopy. However, increased speed of a film is not without a penalty, the greater the speed, the larger the grain size of the silver halide and the poorer the resolution of the film. Typical slow films range from 10 to 35 ASA, medium speed films range from >35 to about 150 ASA and high speed films have ASA values >150.

Resolution of a film, as with the optical system, is the ability of a film to depict two closely spaced objects in the image as separate and distinct objects. Resolution is expressed as the number of line pairs/mm that are recognizable as distinct lines and spaces in a photomicrograph. Kodak groups resolution in several ranges in terms of resolving power(RP): low RP - <55 lines/mm, moderately low RP - 55-68 lines/mm, medium RP - 69-95 lines/mm, high RP - 96-135 lines/mm, very high RP - 136-225 lines/mm, and extremely high RP - >225 lines/mm.

Graininess is a measure of the size and distribution of the silver halide particles in the emulsion. Generally, the faster the speed of the film, the greater the graininess. While graininess is largely a property of a specific film type it can be affected by the development process. For example, a "fine grain" developer can minimize graininess but will generally reduce the final contrast and will require somewhat longer processing time. Graininess is categorized by Kodak as: coarse, moderately coarse, medium, fine, very fine, and extremely fine; only those in the latter two categories are commonly used in photomicrography.

As important as high resolution and fine grain are in a film type, it must be remembered that these are no substitute for a good microscope and proper microscopical technique. This not to say that photographic resolution cannot be used to advantage, but it must be used having full knowledge of the resolution of the microscope, the human eye, and the photographic emulsion. For example, when a sample with structural detail at 2000 line pairs/mm is examined at 40X, that detail at the film plane is 2000/40 = 50 line pairs/mm. Since the human eye typically resolves only about 10 line pairs/mm, a 5X enlargement of the negative will permit the visual resolution of the structure. It must be emphasized that the detail was already resolved by the microscope but the enlargement permitted the eye to observe it. The advantages of a high resolution film with fine grain are the greater probability of obtaining a good quality image and the ability of enlarging that image without degrading the overall image quality.

Contrast of a film type is another important consideration for the photomicrographer. For example, if the structure of a sample consists of only very bright areas and very black areas as the case in etched metal samples, a high contrast film will crisply display the structure on the photomicrograph. However, if the sample has a number of gray levels, a film with a low to moderate contrast is more suitable to fully register the various structures.

The response of a photographic medium to exposure and subsequent development is often described by a characteristic curve as shown in Figure 8-8. This curve expresses the darkening of

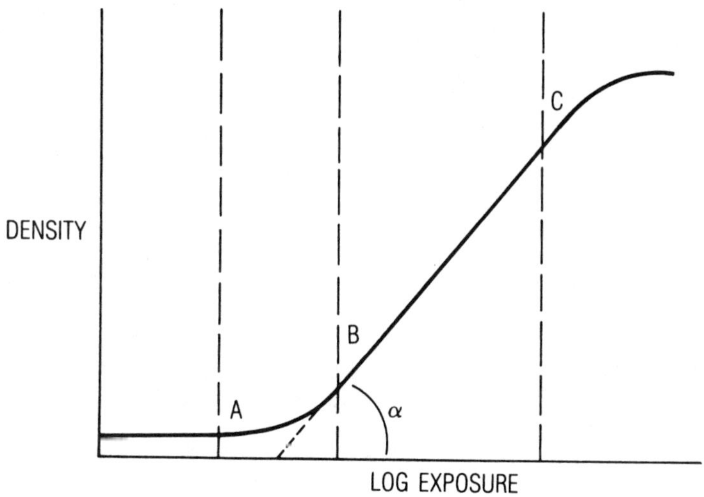

Figure 8-8. The gamma curve expresses the relationship between the exposure (time and intensity) and the density of the photographic negative that results. Area A represents the absorption of the film base and the background darkening inherent in the film. The area between A and B is the area of under exposure. The area between B and C is the linear region normally used for photography and beyond C is the area of overexposure. Alpha is the slope of the curve in the linear region.

the negative material, in terms of optical density, as a function of the log of exposure to light. Density is the log of the opacity of the developed film. A brief examination of this curve is worthwhile to understand the general effects of exposure. The effect of the exposure up to A on the curve is to produce essentially no darkening of the film. In this area a significant portion of the optical density is due to the absorption of the substrate material. In the area of the curve between A and B the slope of the curve is increasing with increasing exposure. The portion of the curve from B to C has a constant slope and the density of the exposed film increases linearly with the log of the exposure. The extent of this region varies with the film type and the development process used. Furthermore, in this area of the curve, if the exposures for all image areas are doubled, such as by doubling the exposure time, the densities of all the areas will be increased equally and there will be no change in the density separations between the various image areas. Above C is the region of overexposure where the slope decreases with increasing exposure. Eventually the slope is zero and further exposure does not produce additional darkening of the exposed film. The portion of the curve from A to C is used in the preparation of negatives.

The slope of the curve in the linear region is known as α and the tangent of α with the horizontal is γ. γ is a measure of the film contrast that can range from a low contrast value of 0.2 to a very high contrast of 4 and higher. γ can also be affected to some extent by the choice of developers in the processing, this is discussed more fully by Loveland (1971).

Reciprocity is a property of all photographic emulsions which is described by the Bunsen-Roscoe reciprocity law:

$$I\,t = K \tag{41}$$

where I is the intensity of illumination, t is time and K is a constant. This law states that the total amount of photochemical reaction in a process is the product of the intensity of the illumination and the time. While this law applies to a photochemical process it can only be related to darkening of the film through the characteristic curve. Since the characteristic curve is not linear over its entirety it is to be expected that the reciprocity relationship has a limited range.

The usefulness of the reciprocity relationship for the linear range is shown by illustration: if the illumination is doubled the exposure time is cut in half. Reciprocity failure or failure of the Bunsen-Roscoe reciprocity law occurs for very short or very long exposure times; the latter being the most important for photomicrography. Reciprocity exists for most B&W films anywhere between 1/5 and 1/1000 second exposures. For exposures greater than 1/5 second additional time will probably be required than would be indicated by the reciprocity relationship. Reciprocity failure is a common cause for underexposure of a negative when using filters because of the reduction in intensity of the illumination.

Black and white negative materials are divided into several classes of spectral sensitivity, of these only the orthochromatic and the panchromatic materials are used for visual photomicrography. The orthochromatic films are sensitive from the ultra violet through violet and blue to green. The panchromatic films are sensitive from the ultra violet through the entire visible spectrum. The panchromatic films are most popular with the biomedical microscopists because of the wide range of stain colors. Since metallurgical samples are generally colorless, orthochromatic films have found much favor in that field because the development of the film may be followed visually using a red safelight. Orthochromatic films also have some application with stained samples if they contain any yellow through red colors since these are imaged dark as contrasted with light violet through green. However, more flexibility in contrasting is possible with the panchromatic films when used with filters (see Chapter 7, Section VI) since this film is sensitive to all colors.

Color, wet process media are also of two general types, the reversal films and the negative films. The reversal films differ from the negative type films in that a transparency is produced upon development of the film which registers each color directly. It should be emphasized that there are no subjective adjustments required during the processing of a reversal film, therefore, if the color balance of the illumination was correct for the exposure, the transparency should register very close to the correct colors, even in the more subtle tones. Since there is no subjectivity required in the processing of this type film it is recommended for photomicrography. This type of film produces the projection transparancies.

The negative color film produces a negative in which each color in the image is recorded as the complimentary color. When a print is made the complimentary color of that on the negative results in the specific color in that area of the sample. The printing process has one step that requires a subjective correction of the illumination using color balancing filters. For normal snap-shot subjects, the printer has a good idea of how certain areas of the print appear and hence can make an intelligent choice for the balancing filters to give a natural looking print. However, in the case of the photomicrograph most printers would have no feeling for the proper correction. Therefore, unless a color reversal film, taken under identical conditions, accompanies the negative film there is little likelihood that the printer can achieve the proper print colors.

Prints can be prepared from the reversal film by preparing an internegative. It should be emphasized that printing from the internegative requires the subjective color balancing step as in the case of the color negative film but the original transparency can be used to judge the correct color balancing filters.

In addition to the properties discussed under B&W media, color films have an extra property to consider, namely, color correction. Each color film is balanced in its color response for a specific illumination color temperature. Therefore to obtain an image that is a faithful reproduction of the sample it is important to adjust the color temperature of the illumination to that for which the film was designed. Kodak balances their film for tungsten (incandescent) illumination at either 3400 K (type A) or 3200 K (type B) and at 5400 K for daylight. Furthermore, their professional films are further designated either type L which is balanced for tungsten light at 3200 K and long exposures ranging from 1/10 to 60 sec or type S which is balanced for daylight and short exposures of 1/10 sec or less. Balancing of the microscope illumination to match the film is described below.

Reciprocity failure for color films is more complicated than for the B&W films because the color film has three emulsion layers, each with its own reciprocity behavior. Most Kodak color reversal films are designed for short exposure times of 1/10 sec or less. If longer exposure times are necessary, color correction filters must be used; these however increase the exposure time and further complicate the situation.

B. Instant Process Media

The many improvements forthcoming since the introduction of the instant films by Edwin Land (Polaroid) in 1947 have established this process as a major contributor to the field of photomicrography. While the early roll film, camera backs were used to some extent for photomicrography, it is the more recent sheet and pack films that are proving to be so useful, especially where rapid response is needed in industry. Two new entries, the 35mm transparencies and the 8" X 10" color sheet film hold promise for further extending the capability of photomicrography. These products, without exception, require special Polaroid film backs or processors to produce the finished film or print.

The Polaroid B&W sheet and pack films are produced in a wide range of emulsions; of these several panchromatic types are very suitable for photomicrography. The unique process for the Polariod B&W film is worth reviewing, thereby giving the microscopist a understanding of the various manipulative steps required to obtain a finished print. We shall use, as an example, the 4 "X 5" sheet film. The sheet film package consists of an outer light-tight cover terminated at one end by a metal clip. Inside the package is the negative sheet, a spacer, the print sheet and a developer pod.

The Polaroid 4" X 5" sheet film camera back, shown in Figure 8-9A, is readied for use by placing the lever in the insert position. The sheet film package is inserted into the camera back as described on the package until a click is heard that indicates the metal clip has been properly seated. The film back should be firmly mounted on the bellows and the shutter closed. The back is

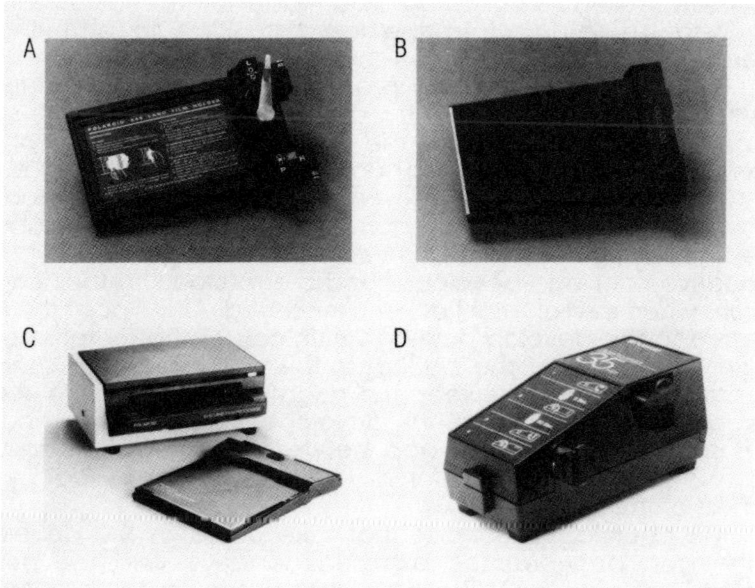

Figure 8-9. Polaroid Camera backs and processors. A is the 4 X 5 sheet film back, B is the 4 X 5 pack film back, C is the 8 X 10 sheet film processor and D is the 35 mm film processor. Photos C and D Courtesy of the Polariod Corporation.

prepared for exposure by gently pulling out the outer light-tight cover until it stops. The exposure is taken as described below, then the light-tight cover is returned to its original position. The lever is rotated to the process position to engage a pair of metal roller on the film package. The film package is then pulled out of the camera back using a firm uniform motion. When the metal clip reaches the rollers as indicated by a resistance to movement, the lever can be returned to the insert position to permit easy withdrawal of the package.

The physical, and photo processes taking place during the above steps are shown with the aid of Figure 8-10. When the sheet film package is inserted and the metal clip seated, the photographic negative is in place for an exposure. When the light-tight cover is withdrawn, the positive print and the developer pod are also withdrawn so that the negative material is now available to accept the image from the camera. After the negative is exposed, the positive print and developer pod are returned to their original positions when the light-tight cover is reinserted. At this point the lever is rotated to the develop position which causes the metal rollers to rest on either side of the package, their position established by spacers at the edge between the negative and the print. As the package is now withdrawn, the developer pod is broken and the developer is spread in the gap between the negative and the print. The layer of developer is very closely controlled by the thickness of the spacers.

The chemical processes that take place as the developer fills the gap between the exposed negative and the print is illustrated in Figure 8-11. The developer proceeds to develop the negative in the same manner as in the wet process; where the negative is fully exposed, all of the silver halide is reduced to black metallic silver, where the negative is partially exposed some of the silver halide is reduced and where the negative was unexposed none of the silver halide is reduced. But that is where the similarity to the wet process ends. Wherever there are unexposed silver halide grains present, these are dissolved in the developer. The resulting silver ions diffuse across the developer layer and react with heavy metal sulfides on the surface of the print to form silver sulfide which is immediately reduced to black silver on the print surface. Since the layer of developer is very thin and carefully controlled in thickness, this diffusion process happens uniformly across the entire film. The resulting print is dark where the negative receives no light, has various shades of gray where the negative receiver small amounts of light and is white where the negative was fully exposed. The print is thus a positive of the image as required.

After an appropriate development time, ranging from 15 sec to 1 min, the light-tight package is torn apart at the edge so that the negative and positive may be separated. All of the earlier positives required a coating of a quick drying fixer from a coater pack supplied with the film. However, some of the more recent film types

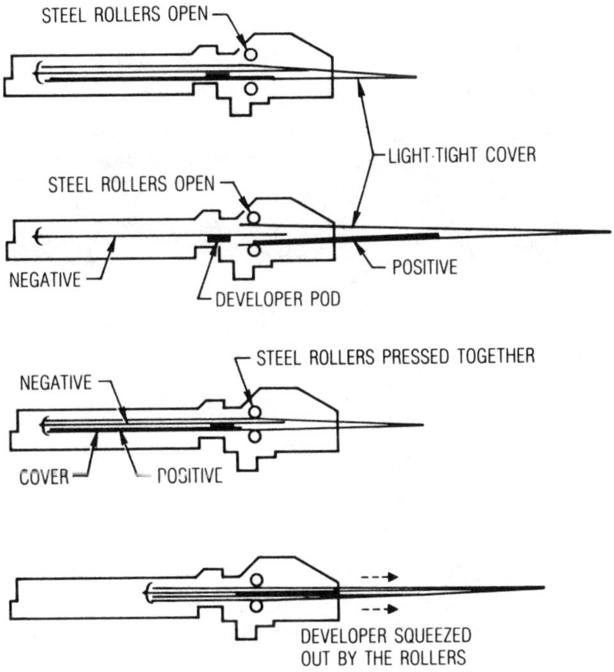

Figure 8-10. The Polaroid 4 X 5 sheet film process. The sheet film is loaded into the camera with the lever in the "load" position; in that position the steel rollers are open. The metal strip at the inner end of the film is captured and thus holds the interior portion of the film pack. To make an exposure, the light-tight cover is withdrawn until it stops. The negative is thus uncovered and an exposure of the negative may be made. After the exposure, the light-tight cover is returned to its original position. The lever is moved to the "process" position; this closes the rollers and releases the metal strip. The entire sheet film package may then be pulled out. In this step the rollers burst the developer pod and evenly distribute the developer gel in the entire gap between the negative and the positive.

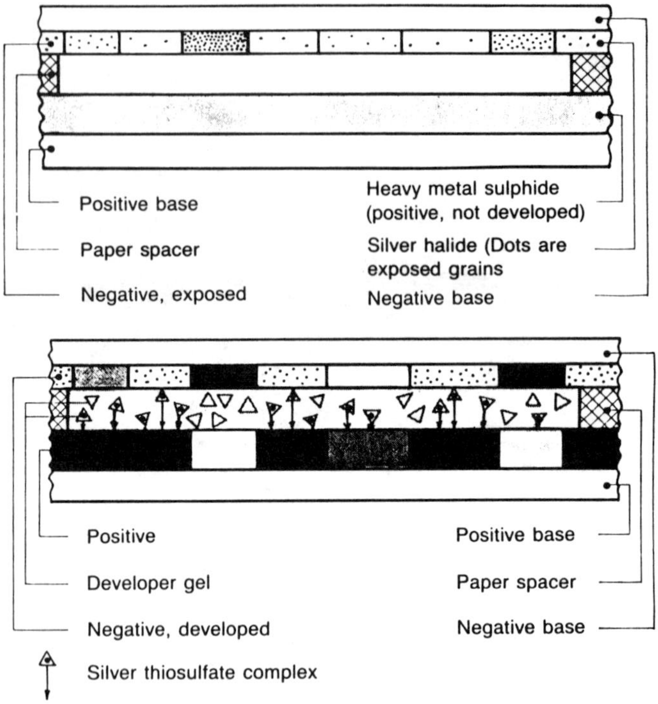

Figure 8-11. Condition of the positive and negative immediately before and after processing. During processing, the developer gel fills the area between the exposed negative and the positive base and develops the exposed silver halide grains to black metallic silver. Wherever unexposed silver halide grains exist they react with the gel to form a soluble silver thiosulfate complex which diffuses to the positive surface where it further reacts with the heavy metal sulfide to form a silver sulfide layer. This layer then further reacts with the developer to form metallic silver (black). Thus, in the areas on the negative where the image was dark, no silver halide was exposed; that silver migrates to the positive and makes a corresponding dark area. Where the image on the negative was bright, the silver halide was developed to form insoluble black silver and was not available to migrate to the positive leaving the latter white.

do not require coating. If the film pack is a P/N type, a usable negative is also produced, but it must be treated and dried before use. The negative in the other instant film types is on a nontransparent backing and cannot be used to make additional prints. The Polaroid 8 in. x 10 in. film is similar to their 4 in. X 5 in. sheet in function except that it requires an external processor, see Figure 8-9C. Kenro Corp.(250 Clearbrook Rd., Elmsford, NY 10523) produces a Model MP1035 camera back for use with the Polaroid 8 in. X 10 in. film.

The Polaroid pack film, not to be confused with the Polaroid sheet film discussed above, is enclosed in a "pack" that fits in a specific camera back, shown in Figure 8-9B. A number of negatives (and their respective prints) are contained within the pack and the first of these is brought into position for exposure by pulling on a paper tab. Once each negative has been exposed, it is brought into contact with the developer and print paper then extracted from the camera back by pulling on another paper tab.

The 35 mm Polaroid B&W process produces a slide film positive but requires a 5 minute processing step outside the camera in a special processor, shown in Figure 8-9D.

The instant process B&W media have the same properties as those described under the wet process B&W media. It must be emphasized that the resolution of the instant process positives is not as great as with the wet process positives. due to the diffusion process described above. However, since the instant process negative has the same general resolution as a wet process negative, a P/N negative, properly processed, can be used to generate high resolution prints in the same way as in the wet process.

Polaroid produces several types of color film yielding either prints or transparencies in several different sizes. These are all corrected for daylight. The color print films, supplied as single sheets or in packs, have a separate negative which must be removed after the prescribed development time.

Polaroid also produces an integral print film with a 3 1/8 in. X 3 1/8 in. format. This film yields a color reversal type print with no negative to remove. Furthermore, the prints are ejected by the camera after exposure and subsequently process in daylight. The integral film pack may contain a battery which is used by certain cameras for powering the ejection motor. This film is used in the SX-70 microscope adaptor mentioned above. A new integral print film, Type 339, is receiving much attention for clean room applications since it does not produce any particulate material in the processing step. Several microscope manufacturers are now producing cameras that use Type 339 film for this purpose.

The 35 mm color film, as with the B&W film described above, requires a 5 minute processing step outside the camera in a special processor.

C. Suggested Photomicrographic Media

It is recognized that many photomicrographers have their own special preferences for specific formats and photomicrographic media and that there are a number of manufacturers of media, both domestic and foreign, however in the interest of brevity, only selected products of the Eastman Kodak Company and the Polaroid Corporation will be listed. Furthermore, only the media for the 16 mm cine, the 35 mm, the 4 in. X 5 in., and the new instant 8 in. X 10 in. cameras is tabulated. Table 8-2 lists the B&W media and Table 8-3 lists the color media. Suitable media for other formats can be inferred from the tables.

V. PREPARATION OF A PHOTOMICROGRAPH

The starting point for a good photomicrograph is a good sample. The characteristics of the sample are dictated by the information needed; beyond that, the sample must be properly prepared, whether it is a microtomed and stained biomedical subject, a polished and etched material sample or an integrated circuit surface. The transmitted light sample is usually capped with a cover glass of the correct thickness and the metallurgical sample surface must be flat and mounted exactly normal to the microscope axis. Cleanliness of the sample is most important. In other words, the sample must be optimal in every respect if a professional photomicrograph is to be obtained.

Selection of the exact area to be photographically recorded depends on the information needed. However, the quality of the resulting photomicrograph may be decidedly improved by thoughtful framing and composition.

A. Framing and Composition

Framing is the process of locating the area of interest within the confines of the film frame as indicated by the reticle in the viewing tube or by the ground glass.

Composition is the arrangement of the sample image within the film frame that satisfies both the technical and aesthetic requirements for the photomicrograph. Aesthetics in this case is both subjective and dependent on the final use of the photomicrograph. It may require no more than alignment of the edge of a square, microelectronic chip with the edge of the film frame or it may demand, in the case of photomicrographic contest entries, the selection of a "showy" or spectacular area and a unique orientation of the sample to provide the most visual impact. Thus the primary requirement for aesthetics in this section is that a photomicrograph provide the intended information in a pleasing manner.

TABLE 8-2
BLACK AND WHITE PHOTOMICROGRAPHIC MEDIA

Format, Manufacturer and Type*	Exposure Index(ASA) Xenon	3200K	γ	Resolving Power	Graini-ness	Spectral Sens.
16 mm						
K. Plus X 7226	50	40	-	High	Ext Fine	Pan
K. Tri X 7278	200	160	-	Med	Fine	Pan
K. 4-X 7277	400	320	-	Med	Fine	Pan
35 mm						
K. Panatomic X	32		0.5-0.8	V High	Ext Fine	Pan
K. Tech Pan	25-320		0.4-3.6	High-Ext High	Ext Fine	Pan
P. Polapan CT	125		2.0	Med	Fine	Pan
K. Tri X	400		0.5-0.8	Med	V Fine	Pan
P. Polagraph HC	400		4.0	Med	Fine	Pan
K. T Max	400-1600		0.6-1.2	V High	Ext Fine	Pan
4 in. X 5 in.						
K. Tech Pan	25-320		0.4-3.6	High-Ext High	Ext Fine	Pan
K. Cont Process	100	50	1.9-3.3	Med-V High	Fine	Ortho
P. 55 P/N (s)	50		1.4	Low**		Pan
P. 665 P/N (p)	75			Low**		Pan
K. T Max	400		0.6-1.2	V High	Ext Fine	Pan
P. 52 (s)	400		1.3-1.4	Low**	Fine	Pan
P. 552 (p)	400		1.3-1.4	Low**		Pan
P. 57 (s)	3000		1.3-1.4	Low**	Med	Pan

*K. Kodak
P. Polaroid
(s) Requires a Polariod model 545 sheet film holder.
(p) Requires a Polaroid model 550 pack film holder.
CT Polaroid continuous tone film.
HC Polaroid high contrast film.

** Negative has very high resolution.

TABLE 8-3
COLOR PHOTOMICROGRAPHIC MEDIA

Format, Mfr., and Type*	Color Bal.	Exposure Index (ASA) Xenon	3200K	Reciprocity Filter/Incr. In 1 sec Exp.	Correction Exp. Time 0 sec Exp.
16 MM					
K. Ektachrome 7275	Tung.	250	400	–	–
K. Ektachrome 7251	Day	400	100	–	–
35 MM					
K. Photomicrography Color Film 2483	Day	16	4	(CC50Y + CC70C) + 1/2 Sec	+ 15 Sec
K. Kodachrome 164-7098	Day	25	-	CC10M + 1 Sec	CC10M +25 Sec
P. Polachrome CS	Day	40		81A + 1 Sec	None + 10 Sec
K. Ektachrome 160-2754	Day	64		CC15B + 1 Sec	CC20B + 15 Sec
K. Ektachrome 177-8471	Day	200	50	CC10R + 1/2 Sec	Not Recom.
K. Ektachrome 102-7341	Day	800/1600		–	–
4 in. X 5 in.					
K. Photomicrography Color Film 2483	Day	16	4	(CC50Y+CC70C) + 1/2 Sec	+ 15 Sec
P. Polacolor 58 (s)	Day	80		80D + 3 Sec	None + 30 Sec
P. Polacolor 59 (s)	Day	80		CC30B + 3 Sec	CC20M + 30 Sec
P. Polacolor 559 (p).	Day	80		Same as Type 59	
K. Ektachrome 164-7098	Day	200	50	CC10R + 1/2 Sec	Not Recom.
8 in, X 10 in.					
P. Polacolor 808 (a)	Day	80		CC10B + 3 Sec	CC5R + 30 Sec
P. Polacolor 809 (a)	Day	80		CC30B + 3 Sec	CC20M + 30 Sec
P. Polacolor 891 (a)	Day	80		CC10R + 3 Sec	–

*K. Kodak
P. Polaroid
(a) Requires a Polaroid model 81-05 or 81-06 film holder and Polaroid 8 in. X 10 in. processor.
(s) Requires a Polaroid model 545 sheet film holder.
(p) Requires a Polaroid model 550 pack film holder.

Composition, for example, involves the consideration of the image orientation as referenced to the frame edges and image position in the frame area. While magnification is not a part of composition, per se, it may affect the final composition of the image.

B. Magnification

The choice of the magnification is somewhat subjective but the magnification chosen must be high enough to resolve the structure of interest, but the magnification must not be so high that the relationship of that structure to the whole is lost. It is helpful for publication purposes to select a magnification that matches certain prescribed values. For example, the American Society for Testing and Materials recommends that the following values be used in preparing a photomicrograph for their publication; 1X, 5X, 25X, 50X, 100X, 150X, 200X, 250X, 500X, 750X, 1000X, 1500X, AND 2000X.

Visual magnification with the compound microscope has been discussed in Chapter 1, Section III and Chapter 2, Section III C. It was pointed out that when visually examining the sample, the enlarged image appears to be about 250 mm (10 in) below the eyepiece; this is a virtual image, i.e., it cannot be observed on a screen placed at that position. However a real image can be observed on a screen at a distance of 250 mm (10 in) above the eyepiece. It is at this position that the photographic film is ideally placed to record the sample structure. The diameter of the image at that distance is almost always large enough to support a full image on a 4"X 5" film format. The 35 mm camera if placed at the 250 mm (10 in) distance would only accept a small portion of the image therefore, it is generally placed much closer to the eyepiece. The image, in this case, is optically corrected by the insertion of a relay lens in the light path as shown in Figure 8-5. In this way the 35 mm image covers roughly the same field of view as the 4" x 5" camera.

Photomicrographic magnification is given by the following:

$$M_P = M_O \cdot M_I \cdot M_E \cdot M_R \cdot B.L./250 \qquad (42)$$

where M_P is the total photographic magnification, M_O is the magnification of the objective lens, M_I is the magnification of any intermediate lens, M_E is the magnification of the eyepiece in the camera path, M_R is the magnification of any relay lens in the camera and B.L. in the total bellows length, i.e. the total distance between the eyepiece and the film plane.

Unfortunately, the magnification of the objective lenses is not always precise and may vary as much as 15% from the nominal value therefore, the most accurate magnifications must be determined using a stage micrometer, preferably traceable to the NIST (National Institute of Standards and Technology, formerly National Bureau of Standards). If a 4" x 5" camera is used, the magnification can be measured on the ground glass focusing screen. Otherwise, a photomicrograph of the stage micrometer can be taken and, if the set-up is not changed, the resulting print can be used for calibration of like photomicrographs. This procedure is described in Table 8-4. The calibration can also be added to each photomicrograph by making a double exposure of the stage micrometer at one edge of the photomicrograph, however for this purpose, the stage micrometer should be black with bright markings.

Having determined the actual magnification of the resulting photomicrograph, let us consider what the magnification should be with respect to the N.A. of the objective lens. A photomicrograph is prepared using a 50X, 0.95 N.A. objective and the total magnification at the film plane is 450 X. The total magnification of 450X is only about one-half of the maximum allowable by the Abbe upper limit of 1000 N.A. as described in Chapter 2, Section III C3. Therefore, it is permissible to photographically enlarge the image 2 times and still expect an increase in resolution. It should be emphasized that the rule of 1000 N.A. applies to photomicrographs viewed at 250 mm (10 in). If an enlargement is made and viewed at a distance of 500 mm (20 in) the enlargement can be 2 times that for normal viewing or a total of 4X enlargement. However, no additional information will be forthcoming from the additional enlargement. In summary, the N.A. of the objective lens must be considered in the selection of magnifications for photomicrographs and their enlargements.

C. Flatness Of Field

When the central portion of the sample image is in sharp focus with a standard achromatic objective lens, the periphery is not, or conversely if the periphery is in focus, the center is not. This is due to the curvature of the object field, as shown in Figure 8-12A and B. During a visual observation the microscopist will constantly focus the microscope and the individual images of the center and the periphery thus obtained are mentally integrated into a composite image that is completely in focus. This is not possible in photomicrography since the film is a single plane that is exposed with a single image. If the periphery of the image is not completely in focus it will be disappointingly evident in the photomicrograph.

TABLE 8-4
MAGNIFICATION CALIBRATION FOR A PHOTOMICROGRAPH

1. Set up the microscope for brightfield observation as described in Chapter 2, Section IV C or Chapter 3, Section IV C.
2. Focus carefully on a stage micrometer and prepare a photomicrograph. If a 4" X 5" camera with ground glass is used, the measurement in Step 3 may be made directly on the ground glass.
3. If the stage micrometer is metric, measure the length of a known spacing on the image with a metric rule. For example, the distance between 0.1 and 0.2 on the print (or ground glass) is found to be 25 mm. The magnification ratio is: 0.2 - 0.1 = 0.1 mm : 25 mm. Multiplying both sides of this ratio by 10; 1 mm (at the sample): 250 mm (at the film), therefore, the magnification is 250X. The procedure is exactly the same for English units except an inch rule is used to measure the image.
4. If a bellows camera or a zoom system is present on the photomicrographic assembly, the magnification can be adjusted to a desired value and confirmed using Steps 2 and 3. Repeat Steps 2, 3, and 4 until the magnification is precisely that desired.

To avoid the problem of field curvature use plano corrected objective lenses; this is a good, but somewhat expensive solution. The improvement over a standard achromatic objective lens is shown in Figure 2-20.

If the standard achromatic objective lenses are to be used, the apparent flatness of field can be improved by increasing the depth of field, as shown in Figure 8-12C. Depth of field and the various methods of increasing depth of field is discussed in Chapter 2, Section III C2. The more commonly used methods to increase depth of field are to use a lower magnification objective lens combined with a higher magnification eyepiece. or decrease the aperture diaphragm setting. It should be emphasized that both of these methods reduce the N.A. and hence the resolution of the image.

D. Focusing

Focusing is that final step before exposing the photomicrograph and if not accomplished with the same care as all the other previous steps, the effort will be wasted. Our experience with visual microscopy indicates that focusing is more critical for the higher magnification objective lenses. This is due to the depth of

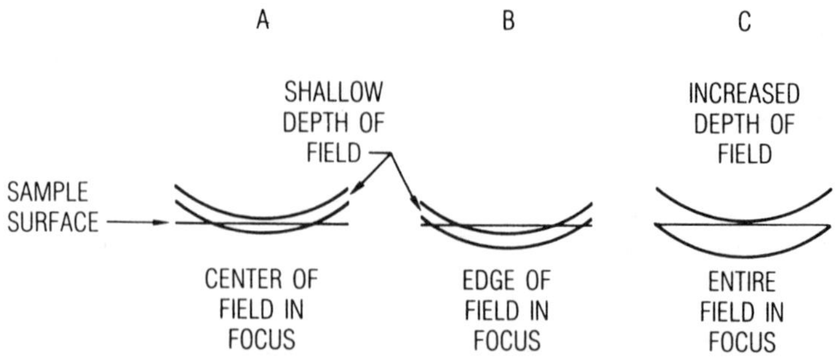

Figure 8-12. A standard achromatic objective lens normally has a curved field of focus. When the aperture diaphragm is set at the usual 0.8 to 0.9 opening, the depth of field is shallow. Interaction of this shallow depth of field with the sample surface, as shown in cross section in (A) and (B), results in either the center or the edge of the field of view being in focus. If the aperture diaphragm is closed sufficiently, the increased depth of field will result in good focus for the entire field of view, as shown in (C). However, this procedure will reduce the resolution of the image and in the extreme case, produce undesirable diffraction effects.

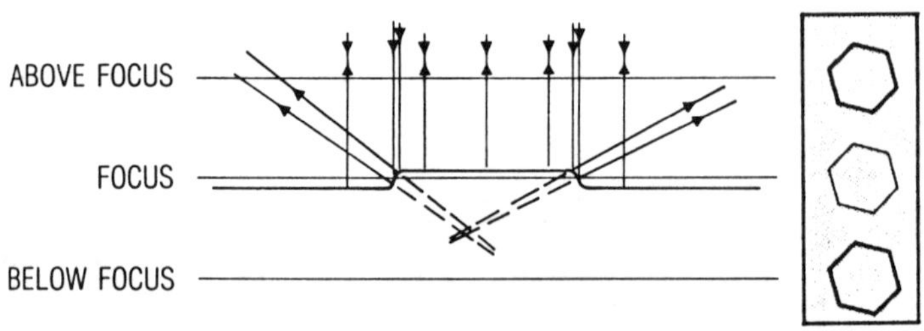

Figure 8-13. Use of Schneiderhöhn lines as a focusing aid. The origin of these bright lines is shown in the cross section at the left. They appear at the boundaries between two grains with slightly different heights; these are the result of polishing or etching. In the representation of the image at the right, the Schneiderhöhn lines move away from the boundary from which they originate when the microscope focus is moved above or below focus. When a Schneiderhöhn line is coincident with the boundary from which it originates, the best focus exists.

field considerations mentioned above. In photomicrography, however, it becomes evident in a short time that it is the low magnification objective lenses (<10X) that require extraordinary care in focusing if a sharp image is to be photographically recorded. This surprising result is due to depth of focus. Depth of focus at the intermediate image plane increases as the square of the objective lens magnification, hence it has a small value for low magnifications. Table 8-5 lists the depth of focus for common sample thicknesses; typical relief for a metallurgical sample is on the order of 2μ while for a histological sample it may be about 8μ.

Another caution with regard to focusing is that metallurgical samples examined on a upright stand must be carefully leveled prior to examination.

The proper mechanics of focusing the microscope for visual observation provides a good basis for photomicrographic practice. For objective lenses less than 10X, focusing is best accomplished with the coarse focus knob. For higher magnifications, where contact between the objective lens and the sample can result in damage to either or both, the following is recommended. Carefully move the objective lens toward the sample using the coarse focus while observing from the side until the objective lens is as near as possible to the sample, then slowly focus upward with the coarse focus while viewing through the eyepiece until the sample image is visible. Final focusing is somewhat different from the purely visual case and is accomplished by starting above focus and slowly adjusting the fine focus knob until the image is just in focus. If that position is passed, repeat the operation as many times as necessary, until it is possible to stop exactly at best focus position.

Having discussed the mechanics of focusing it is necessary to consider the visual criteria that provide the feedback or control for the focusing process. One can say that focus is achieved when a "sharp" image is obtained, however such a definition is subjective. It has been the experience of the author that a definitive measure of correct focus in thin transmitted light samples is the Becke line used by the petrographer. The Becke line is a bright line which is seen to move back and forth across a boundary between two materials of different refractive index as the focal point of the microscope is moved above or below exact focus on that boundary. Since such boundaries always exist in transmitted samples suitable for brightfield investigation, such a technique is useful for the biomedical field as well as for the inorganic fields. The Becke line effect is shown in Figure 5-22 for sample particles which are higher or lower in refractive index than the surrounding medium. The Becke line is brighter and more pronounced as the difference in refractive index between the sample and the surrounding medium increases. Conversely, the Becke line becomes faint and often colored pale orange or blue if the difference in refractive indices is very small. The Becke is seen to move back and forth across the boundary which produces it as the microscope is focused above and below the position of correct focus. The criterion for correct focus is when the Becke line is coincident with the boundary which produces it.

A similar effect, known as the Schneiderhohn line is observed in incident light microscopy. This effect, shown in Figure 8-13, is the result of slightly different elevations between adjacent grains in a sample. The Schneiderhohn line, as with the Becke line, is seen to move back and forth across the boundary which produces it as the microscope is focused above and below the position of correct focus. The criterion for correct focus is when the Schneiderhohn line is coincident with the boundary which produces it.

TABLE 8-5
DEPTH OF FOCUS FOR TYPICAL SAMPLE THICKNESSES AS A FUNCTION OF OBJECTIVE LENS MAGNIFICATION

OBJECTIVE LENS MAGNIFICATION	DEPTH OF FOCUS (μm) 2 μm SAMPLE THICKNESS	8 μm SAMPLE THICKNESS
2.5	0.01	0.05
4	0.03	0.1
6.3	0.08	0.3
10	0.2	0.8
16	0.5	2
25	1	5
40	3	13
63	8	32
100	20	80

Now that we have a set of criteria for focus, it remains to make the position of correct focus coincident with the film plane. If the camera is a flexible bellows, 4"x 5" or larger camera, focusing is normally accomplished by examining a ground glass screen in a darkened room. Focusing is easily achieved with a ground glass screen, unless the structure is too fine. In this latter case, a ground glass screen with a central, clear area and reticle can be used with a focusing magnifier. The magnifier must be carefully focused on the reticle in the clear area, otherwise it may be focused on other than the film plane. Use of the magnifier is also recommended for low magnification photomicrographs where the objective lens magnification is less than 10X.

If a fixed bellows camera is used, focusing is usually accomplished via a right angle viewing and focusing (v&f) tube. The more modern microscopes may incorporate the v&f capability into one of the binocular eyepieces thus eliminating the need for an additional v&f tube on the camera assembly; all that follows may be applied to this configuration as well.

The 35mm camera is most commonly a fixed bellows camera, therefore, it will serve as the model for the following discussion. The first step in using the v&f tube is to critically focus its reticle using the knurled ring at the eye lens. For a transmitted light microscope, obtain a sample image then move off sample to a clear area. If an incident light microscope is used, focus on a clean first surface mirror. Turn the knurled ring of the eye lens out as far as possible. Allow the eye to relax and while gazing into the v&f tube turn the knurled ring until the focusing reticle is sharply in focus. If the position of best focus is passed repeat the process, stopping at the exact point of sharpest focus. Once this step has been accomplished, the sample image can be accurately focused at the film plane by critically focusing it at the reticle.

Some photomicrographers recommend as a check on the focus to move the eye back and forth a short distance in front of the v&f tube after focusing the image. If the image is exactly in focus, the image and the reticle image will appear stationary. If the image is not exactly in focus, the image will appear to move to one side or the other relative to the reticle image.

It was indicated above that photomicrography with objective lenses of less than 10X is difficult because of the very small depth of field. For this purpose, Hinsch (1979) has recommended the use of an eyepiece telescope, see Chapter 7, Section VII, which is held manually in front of the v&f tube to provide additional magnification of the reticle and sample images. The eyepiece telescope is first focused on some remote object, then it it placed in front of the v&f tube. The reticle in the v&f tube is then focused in exactly the same way as described above. Focusing of the sample image is accomplished with the coarse focus knob while viewing through the eyepiece telescope to indicate the correct point of focus. This method will yield results that are reproducible from operator to operator.

A few 35 mm cameras are equipped with a v&f tube that has a ground glass focusing plate that is interchangeable with the clear plate and reticle. The ground glass very accurately defines the image plane but the image is not as sharp as with the clear glass.

As a final note, it is occasionally worthwhile to spend a few minutes after focusing the image to see if the image remains in focus. If the image drifts out of focus within a minute of focusing, the instrument should be referred to professional service personnel for repair.

E. Determination Of Exposure Time

The final step to taking a photomicrograph is exposing the film, however to get the desired results, the correct exposure time must be used. If an automatic camera or a photomicroscope is used, a sensor monitors the light level from the sample and provides the correct exposure time, based on inputs from the microscopist such as film speed and illumination type. These units are available from all major microscope manufacturers. Certain automatic cameras can base exposures on either a small spot or the entire field intensity and some can even provide information for correcting the color temperature of the illumination. In general, the modern automatic cameras provide very good results and are recommended for their speed and ease of operation as well as their overall accuracy. Care must be exercised in matching the camera set-up parameters as carefully as possible to the actual image conditions. Since the automatic units are so specific in their capabilities, no standard procedure can be given for their operation.

The exposure times for a manually operated camera can be determined in either of two ways; with the two-step method or with a external exposure meter.

The two step method of exposure determination involves the taking of a number of test exposures with a given photographic set-up and from this selecting the correct exposure time for taking the actual exposure. This procedure is given in Table 8-6 for 35mm film, Polaroid 4"X 5" pack film, 4"X 5" wet process sheet film, and 4"X 5" Polaroid sheet film. It is imperative to record the illumination settings as well as the other conditions of the photomicrographic set-up if there is to be any delay between the exposure calibration and the photomicrography of the sample.

When the test exposures have been processed, the correct exposure time is determined by inspection of the processed film or in the case of the Polaroid films the finished prints. Figure 8-14 illustrates the results of the procedure in Table 8-6 for the four types of film. The B&W negatives should be evaluated in the darker areas of the sample, that is, the lighter parts of the negative. It is important to select the test negative which shows the greatest detail in these areas without being too dark. High contrast films, such as contrast process ortho, have a narrower exposure latitude that the continuous tone films therefore require more care in selection of the proper exposures.

If the test strip is too dark, make another test strip and decrease each of the original exposure times by a factor of 16. If this is not possible using the shutter settings, use neutral density filters. On the contrary, if the test strip is too light increase each of the exposure times on a new test strip by a factor of 16.

The reversal films and Polaroid prints should prove easier to evaluate than the negatives; here the criterion for optimum exposure is good detail in all areas where they were present in the sample. In this case, if the test strip is too dark, make another test strip and increase each of the exposure times by a factor of 16; if it is too light, decrease each of the exposure times on a new test strip by a factor of 16.

Figure 8-14. The two step exposure method. (A) is a two step test print for 4 X 5 Polaroid film. (B) is a series of test negatives for 35 mm film. The procedure for this is given in Table Figure 8-6.

Once the correct exposure time has been established for a particular set-up, the sample may be photographed. It is also possible to use the exposure time for that particular set-up to determine the exposure for other set-ups if the illumination remains constant or is at least reproducible in intensity. the exposure for these new set-ups is given by:

$$\frac{\text{NEW EXP}}{\text{OLD EXP}} = \frac{\text{NEW B.E.}^2}{\text{OLD B.E.}^2} \cdot \frac{\text{NEW MAG}^2}{\text{OLD MAG}^2} \cdot \frac{\text{OLD ASA}}{\text{NEW ASA}} \cdot \frac{\text{OLD N.A.}^2}{\text{NEW N.A.}^2} \qquad (43)$$

In this expression, EXP is the exposure time in seconds, MAG is the total magnification, B.E. is the bellows extension, ASA is the film speed in ASA units, and N.A. is the effective N.A. of the objective lens/condenser system. Although this expression is universally applicable, it is definitely not suitable for rapid or production photomicrography.

The second method for determination of exposure times is the exposure meter method. The exposure meters available vary widely in price and capability. These range from automatic, precalibrated units that determine exposure times and use that to control the shutter to those hand held units that are designed for general photography. The exposure meters designed for photomicrography that monitor the intensity via the binocular or trinocular eyepiece tube must be calibrated using the two step method described above. Once calibrated these instruments will take into consideration changes in illumination intensity, magnification, and N.A.; ASA changes are made with a switch. Bellows extension effects must be calculated using equation (43). It is important with these instruments to use the same eyepiece tube or same position in front of the focusing tube for every measurement.

TABLE 8-6
THE TWO STEP METHOD FOR EXPOSURE CALIBRATION

1. Set up the microscope as described in Chapter 2, Chapter 3, or Chapter 4 and focus the microscope for the sample of interest. After set up do not change the illumination or any part of the optical system, including, the filters, the field diaphragm and aperture diaphragm.

2a. For 35 mm film or Polaroid 4" X 5" pack film, load a roll (or pack) of film into the camera back and advance film to the first frame.

2b. For 4" X 5" wet process sheet film, insert a film carrier into the camera and withdraw the dark shield until it is almost completely removed from the film pack.

2c. For Polaroid 4" X 5" sheet film, set lever on the film back to "insert" position, insert a film pack into the film back and pull out the light-tight cover to the stop.

3. Expose the film at a time anticipated to be too short for a correct exposure (for example 1/60 sec).
4a. For 35 mm film, advance the film one frame. For Polaroid 4" X 5" pack film, pull the tab and extract the print. Label the Polaroid print with the exposure time.
4b. For the 4" X 5" wet process sheet film; reinsert the dark shield 1/2", repeat Step 3 and reinsert the dark shield an additional 1/2".
4c. For the Polaroid 4" X 5" sheet film, reinsert the light-tight cover 1/2", repeat Step 3 and reinsert the light-tight cover an additional 1/2".
5. Increase the exposure time by one increment and make another exposure. (in the example of Step 3 the exposure time would now be 1/30 sec)
6a. Advance the 35 mm film one frame or pull the tab for the Polaroid 4" X 5" pack film and extract the print. (Label Polaroid print with the exposure time)
6b. For the 4" X 5" wet process sheet film reinsert the dark shield an additional 1/2".
6c. For the Polaroid 4" X 5" sheet film reinsert the light-tight cover an additional 1/2".
7. Increase the exposure time by one increment and make another exposure. (in the example of Step 3 the exposure time would now be 1/15 sec)
8. Repeat steps 6 and 7 two more times for 35 mm film or Polaroid 4" X 5" pack film. (in the example of step 3 the successive exposure times would be 1/8 sec and 1/4 sec.) For the wet process and Polaroid 4"x 5" sheet films repeat Steps 6 and 7 once more.
9a. Process the 35mm film. If Polaroid 4" X 5" B&W pack film was used, coat the prints.
9b. For 4" X 5" wet process sheet film, completely close the dark shield and process sheet of film.
9c. For Polaroid 4" X 5" sheet film, completely reinsert the light-tight cover, set lever on the film back to "process" position, and pull out the film pack to initiate the processing. Allow to process for the required time, remove print from pack and coat.

The Horseman exposure meters (available from Calumet Photographic Inc. 890 Supreme Dr., Bensenville, IL 60106), are ideal for the 4" X 5" flexible bellows camera as they slide into the camera just like a film pack. An integrated exposure meter is precalibrated and automatically compensates for bellows extension and filters. They have an ASA range from 3 to 12,000 and a shutter speed range from 1/500 sec to 30 min.

Hurtgen and Massios (1979) have described the use of a hand held, general purpose, exposure meter for photomicrography. They suggest that the meter be placed reproducibly at the exit pupil, or Ramsden Disk of either the eyepiece or the focusing tube each time a reading is taken. The room should be darkened to minimize the effect of ambient lighting on the reading.

To calibrate the hand held exposure meter a two step calibration is performed and a reading taken with the exposure meter. The meter is set at the ASA value for the film and the f-stop value corresponding to the optimum exposure time determined in the two step calibration is used as a reference point. For each sample exposure the exposure meter held in the same point; the exposure time is found opposite the reference point value. This instrument will correct for differences in illumination intensity, magnification, and N.A. ASA changes can be made on the scale but the same reference value is used. Again, bellows extension effects must be calculated using equation (43).

Vetter (1973) has described three problems encountered when determining exposures for fluorescence photomicrography. The first problem is the extremely low light levels; these can be measured with the usual meters only if the fluorescent material occupies >50% of the field. The second problem is that the fluorescence is scattered randomly over a dark field making full field metering of intensity necessary. Thirdly, a number of the common fluorochromes used in biological work, such as fluorescein isothiocyanate (FITC), quenches or fades rapidly on exposure to the exciting radiation. This last problem makes it necessary to make a measurement on one field of view, determine the exposure, then change to a new field of view when making the actual exposure.

F. Correction of Illumination for Color Film

Various sources of illumination have different color temperatures, as described in Chapter 2, Section III A. Furthermore, the color temperature of the illumination as it exposes the film can be altered by the the microscope optics, mounting media, stains, etc. The problem of color temperature correction is handled in several ways by the major microscope manufacturers, for example, Reichert and Zeiss provide a control circuit in the power to the lamp which regulates the color temperature of the lamp to 3200 K. These and others also rely on the relatively uniform, daylight (6000 K) output of the xenon arc lamp. Ideally no filters would be required for these illuminators if the appropriate film type isused. Other manufacturers such as Olympus have modules as a part of the camera which can be used to monitor the illumination color temperature. Correction of the illumination is then achieved manually with filters, see Chapter 7, Section VI B.

Lacking any of the features described above, the illumination may be corrected by preparing a series of test films under carefully defined conditions using as necessary, color correction and color balancing filters. This procedure is described in Table 8-7.

TABLE 8-7
COLOR CORRECTION OF ILLUMINATION FROM INCANDESCENT AND HALOGEN LAMPS

1. Set up the microscope as described in Chapter 2, Chapter 3, or Chapter 4 and focus the microscope for the sample of interest. Record the position of the control for the illumination intensity and do not change the illumination during this procedure.

2a. For 35mm film or Polaroid 4" X 5" pack film, load a roll (or pack) of film into the camera back and advance film to the first frame.

2b. For 4" X 5" wet process sheet film, insert a film carrier into the camera and withdraw the dark shield.

2c. For Polaroid 4" X 5" sheet film, set lever on the film back to "insert" position, insert a film pack into the film back and pull out the light-tight cover to the stop.

3. Determine the exposure time. Insert the first filter from the appropriate chart below into the light path, correct the exposure using the associated multiplier, and expose the film.

FILTER SEQUENCE FOR TUNGSTEN FILM

#	WRATTEN FILTER	EXPOSURE TIME MULTIPLIER
1	80D	1
2	80C	1
3	80B	2
4	80C	2

FILTER SEQUENCE FOR DAYLIGHT FILM

#	WRATTEN FILTERS	EXPOSURE TIME MULTIPLIER
1	80A	4
2	80A + 82	4
3	80A + 82A	4
4	80A + 82B	8
5	80A + 82C	8

4a. Advance the 35 mm film one frame or pull the tab for the Polaroid 4" X 5" pack film and extract the print. (Label Polaroid print with the filter number.)

4b. For the 4" X 5" wet process sheet film reinsert the dark shield and remove carrier. Repeat Step 2b.

4c. For the Polaroid 4" X 5" sheet film reinsert the light-tight cover, set lever on the film back to "process" position, and pull out the film pack to initiate the processing. Allow to process for the required time, remove print from pack and coat. (Label print with filter number) Repeat Step 2c.

5. Repeat Steps 3 and 4 with successive filters until the series is completed.

6a. Process the 35mm film. (label each frame with filter number)

6b. process the 4" X 5" sheet film. (label each sheet with filter number)

7. Select the film frame most closely matching the visual image of the sample and use the respective filter(s) in conjunction with the illumination intensity setting established in step 1.

Slight variations in tint between the photomicrograph and the visual image may be eliminated using color correction filters, see Chapter 7, Section VI B. A procedure similar to that described in Table 8-7 can be used to determine the best of these filters for the specific application.

G. Contrast Enhancement for Black and White Film

If a differentially stained sample is imaged on B&W film, it is quite possible that the gray levels of the two different color stains may be imaged as the same shade of gray. One of the simplest means of increasing contrast between different colored areas of a sample is to use ortho film, see Section IV A. If the two contrasting colors in the sample fall into two different groups (1) violet, blue, green and (2) yellow, orange, red, then the ortho film, alone, may effect a contrast difference.

Panchromatic B&W film that is sensitive to the complete visible spectrum allows the use of a wide range of color filters to provide color contrast. The selection of color filters for use in B&W photomicrography is covered in Chapter 6, Section VI B.

H. Photomicrographic Procedures

The various considerations necessary for the production of a high quality photomicrograph have been discussed, all that remains is to integrate these into a procedure. The great diversity of photomicrographic equipment will make it impossible to cover every operation for each instrument therefore, a generalized procedure will be described for both fixed bellows cameras and flexible bellows cameras. The special topics of cine, time lapse and stereo photomicrography will also be addressed.

1. Fixed Bellows Camera Procedure

The fixed bellows cameras commonly include the 35 mm format, the 3-1/4" X 4-1/4" format and occasionally the 4" X 5" format. For reference, the arrangement of the fixed bellows camera is shown in Figure 8-15; the procedure given below in Table 8-8 is based on this arrangement.

TABLE 8-8
PHOTOMICROGRAPHIC PROCEDURE FOR FIXED BELLOWS CAMERA

1. Set up the microscope using the illumination method of choice as described in Chapters 2, 3, and 4. Critically focus the microscope on the sample using the visual optics. Make sure that the camera shutter is closed.

2a. For 35 mm film or Polaroid 4" X 5" pack film, load a roll (or pack) of film into the camera back and advance film to the first frame.

2b. For 4" X 5" wet process sheet film, insert a film carrier into the camera and withdraw the dark shield.

2c. For Polaroid 4" X 5" sheet film, set lever on the film back to "insert" position, insert a film pack into the film back and pull out the light-tight cover to the stop.

3. If necessary, redirect the image to the camera and refocus the image at the same plane as the v&f tube reticle. (The v&f tube reticle should have been previously focused as described in Section V D, above.)

4. Frame the image as necessary by translating and/or rotating the sample on the stage. Adjust the magnification by changing objective lens, zoom lens (if present) and camera eyepiece with due consideration for factors such as resolution and field of view. Evaluate the flat ness of field and make the necessary adjustments to optimize the flatness as described in Section V C, above.

5. Add filters to the illumination for color correction in color photomicrography or for contrast in B&W photomicrography. If necessary, adjust illumination intensity to the value that matches the color correction filters, see Section V F, above.

6. If the camera has a manual system, determine the exposure time and set the shutter for that time. If the camera is automatic, set the appropriate controls such as film ASA.

7. Critically refocus the image in the v&f tube.

8. Wait 5 to 10 seconds to allow vibrations to settle then make the exposure by tripping the shutter.

9a. Advance the 35 mm film one frame or pull the tab for the Polaroid 4" X 5" pack film and extract the print.

9b. For the 4" X 5" wet process sheet film reinsert the dark shield and remove carrier.

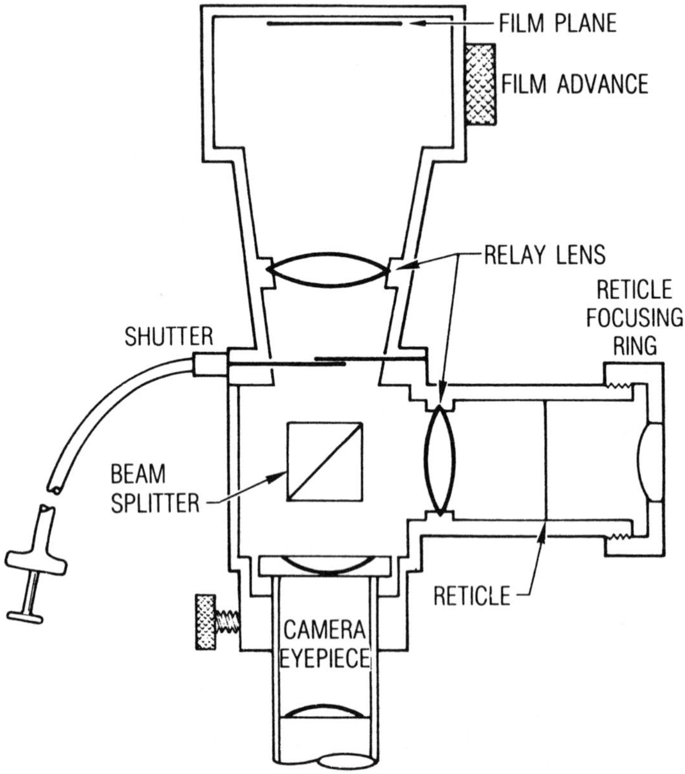

Figure 8-15. The fixed bellows camera.

9c. For the Polaroid 4" X 5" sheet film reinsert the light-tight cover, set lever on the film back to "process" position, and pull out the film pack to initiate the processing. Allow to process for the required time, remove print from light-tight cover and coat, if required.

10. Redirect the image to the visual optics, if necessary.

---

2. Flexible Bellows Camera Procedure

The flexible bellows cameras commonly include the 4" X 5" format and in some cases the 8" X 10" Polariod format. For refrence, the arrangement of the bellows camera is shown in Figure 8-16; the procedure given below in Table 8-9 is based on this arrangement.

TABLE 8-9
PHOTOMICROGRAPHIC PROCEDURE FOR FLEXIBLE BELLOWS CAMERA

1. Set up the microscope using the illumination method of choice as described in Chapters 2, 3, and 4. Critically focus the microscope on sample using the visual optics.
2. If necessary, redirect the image to the camera then open the shutter.
3. Dim the lights in the room and refocus the image on the ground glass screen.
4. Frame the image as necessary by translating and/or rotating the sample on the stage. Adjust the magnification by changing objective lens, zoom lens (if present), camera eyepiece and bellows extension with due consider ation for factors such as resolution and field of view. Evaluate the flatness of field and make the necessary adjustments to optimize the flatness as described in Section V C, above.
5. Add filters to the illumination for color correction in color photomicrography or for contrast in B&W photomicrography. If necessary, adjust illumination intensity to the value that matchs the color correction filters, see Section V F, above.
6. Critically refocus the image on the ground glass screen.
7. Determine the exposure time, close the shutter and set the shutter for that exposure time. Use great care not to disturb the microscope/camera arrangement until the exposure has been made.

8a. For 4" X 5" wet process sheet film, insert a film carrier into the camera and withdraw the dark shield.

8b. For Polaroid 4" X 5" sheet film, set lever on the film back to "insert" position, insert a film pack into the film back and pull out the light-tight cover to stop.

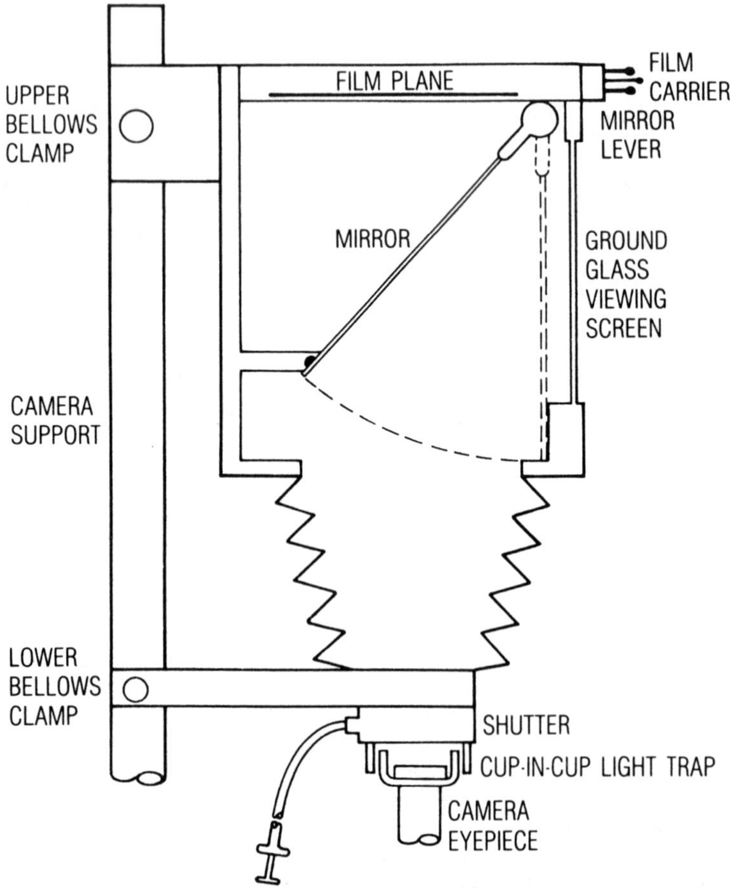

Figure 8-16. The flexible bellows camera.

8c. For Polaroid 4" X 5" pack film, load a pack of film into the camera back and advance film to the first frame by pulling on the tab.

9. Wait 5 to 10 seconds to allow vibrations to settle then make the exposure by tripping the shutter.

10a. For the 4" X 5" wet process sheet film reinsert the dark shield and remove carrier.

10b. For the Polaroid 4" X 5" sheet film reinsert the light-tight cover, set lever on the film back to process" position, and pull out the film pack to initiate the processing. Allow to process for the required time, remove print from pack and coat.

10c. For the Polaroid 4" X 5" pack film pull the tab and extract the print.

---

3. Cine Photomicrography

Cine photomicrography is simply photomicrography in which a motion-picture camera is used to record the image of the sample in motion, be it single celled animals, mitosis of cells or grain growth in metals. It has the capability of showing action in real time, or greatly increasing the speed of a slow process (time lapse) or slowing a process (slow motion) depending on the relative film speed in the camera and the projector. Both B&W and color films are available with reasonable resolution. Few manufacturers now provide cine camera attachments because of the relative ease and convenience of video recording, even though, at this time, the video systems lacks the resolution of the photographic film. Therefore, since used cameras and a few new instruments are still available a brief discussion of cine photomicrography is given. Nikon and Olympus offer complete cine camera assemblies for use with their microscopes; these systems also have a time-lapse capability.

A number of differences exist between cine photomicrography and still photomicrography, among these are: (1) the sample field and the focus must be monitored continuously during the exposure especially if there is significant motion in the sample; (2) the exposure time is dependent on the film speed or frames per second that are exposed (except for time-lapse); (3) the camera records time as well as images; and (4) the considerable vibrations of the operating camera must be completely isolated from the microscope.

A typical cine photomicrographic arrangement is shown in Figure 8-17. The microscope is mounted on a vibration isolation plate, see Chapter 7, Section II A. Framing of the sample image can be greatly facilitated if the microscope has a revolving stage and a mechanical stage.

If a standard cinecamera is used, it must be optically coupled to the microscope monocular or trinocular tube with a

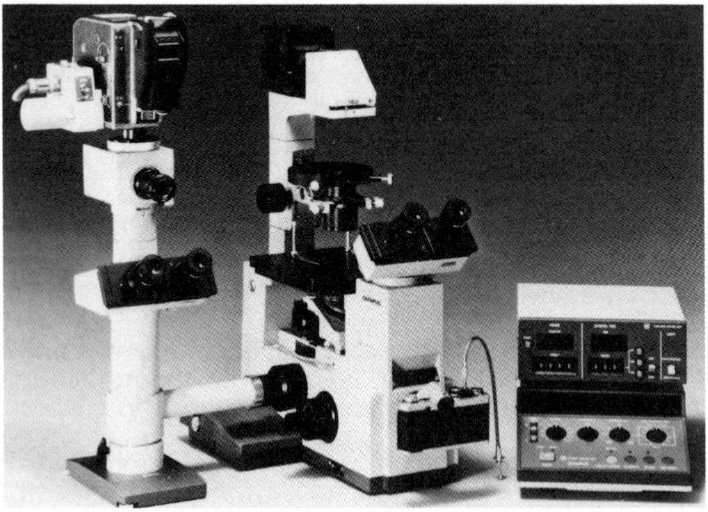

Figure 8-17. A cine camera (upper left) with time lapse time capability. The timer system is at the upper right. Courtesy of Olympus Corporation.

focusing assembly. This assembly should: (1) contain a reticle that provides a definable focusing plane that can be made optically coincident with the film plane; (2) have a focusing tube that permits focusing the reticle and the image of the sample; (3) have an exposure monitoring photocell; and (4) possibly have facilities for superimposing time and caption information. The focusing assembly clamps rigidly to the microscope but no solid attachment is made to the camera to minimize the transmission of vibration to the microscope. Light leakage at the camera attachment is eliminated with a pair of nested cups.

The cine camera is mounted on a rigid camera support. For this purpose a 16 mm camera is recommended as a good compromise between cost and capability. The camera should have: (1) frame rate of 8 to 64 frames per second (fps); (2) an adjustable rotary shutter; (3) the facility for instantaneous and time exposures in single frame mode; (4) motorized film transport; (5) 100 foot film capacity, minimum; and (6) ideally, a frame counter. The camera should be equipped with a 50 mm or 75 mm lens and in this sense may be considered similar to an integral lens, still camera as described in Section III above. The camera lens must be set at infinity and the lens diaphragm must be wide open. The lens diaphragm has no function in the cinemicrographic set-up.

If an exposure meter is present, it is used to set the value of the adjustable rotary shutter, otherwise, the correct shutter setting can be determined with a short test strip of film. Table 8-10 gives the shutter openings corresponding to various shutter speeds for a number of film speeds. If the exposure times are out of the range of the shutter openings it is necessary to change the illumination level, however if color film is used, color temperature of the illumination must be considered.

TABLE 8-10
SHUTTER OPENINGS (IN DEGREES)
FOR VARIOUS EXPOSURE TIMES

FILM SPEED IN FPS	EXPOSURE TIMES 1/250"	1/125"	1/60"	1/30"	1/15"
8	-	-	48	96	192
18	-	52	108	216	-
24	35	70	144	-	-
48	69	138	-	-	-
64	92	184	-	-	-

Time-lapse is a category of cine photomicrography in which the sample is photographed at a slower frame rate than the film is subsequently projected. The process under observation is thus accelerated. A three times acceleration can be achieved by photographing at 8 fps and projecting at a normal rate of 24 fps. For further acceleration of the process, special equipment is required to provide a designated time interval alternated with single exposures. Table 8-11 provides a listing of the designated time intervals for a total action time when using a 100 ft roll of 16 mm film having about 4000 frames. Total projection time for the 100 ft roll is 4.2 minutes at 16 fps and 2.8 minutes at 24 fps.

During the action time the sample must not move either laterally or axially. If the area of interest, such as a living organism, does move out of the field of view the microscopist may gradually move the mechanical stage to recenter the sample, however this is not practical unless the time-lapse rate is fairly rapid. A sample, for example, a polished section in a high temperature furnace, will expand axially when heated. This latter problem points out the need to check the focus often during a time-lapse study. Finally, the illumination must remain constant for the exposures. For those sample which are light sensitive, it may be necessary to add an electromagnetic shutter between the illuminator and the microscope that can be opened only during the exposure time. This shutter may be activated by the time-lapse controller.

Chang, Mohapatra and Johnson (1977) have described a simple time-lapse control circuit that is used with a motorized 35 mm camera. This unit was not designed to compete with a 16 mm system that can provide 4000 images, but it can serve for slow events where quality slides or photographs are needed.

TABLE 8-11
TIME-LAPSE INTERVALS AND ACCELERATION FACTORS

Total Time	Designated Time Interval	Acceleration Upon Projection At: 16 FPS	24 FPS
15 min	4.4 FPS	3.6	5.5
30 min	2.2 FPS	7.2	11
1 hr	1.1 FPS	14.5	22
2 hr	1.8 Sec/frame	28.8	43
4 hr	3.6 Sec/frame	57	86
8 hr	7.2 Sec/frame	114	172
1 day	21.6 Sec/frame	320	480

Slow motion cine photomicrography can be accomplished in a simple manner by using the highest speeds available on the camera and projecting the image at the slowest speeds available on the projector. For rapid events it is necessary to use a high speed camera. High intensity illumination, such as the xenon arc lamp must be used to provide reasonable exposures and even then, the photographic emulsions must be very sensitive. The generally greater vibration of the high speed cameras must be carefully isolated from the microscope.

4. Problems

Photomicrography is a complicated process involving the precise interaction of illuminator, microscope, adapter, and camera either as individual items or in combination as a photomicroscope. In this assemblage there are enough variables that even the most experienced operator occasionally has a non ideal photomicrograph. Table 8-12 is a list gleaned from many sources, including personal experiences, that should provide help in case of difficulty. The table addresses film problems occurring with both wet process and instant films.

This table assumes a suitable sample, examined with an appropriate type of illumination. In those cases where a problem is related to the sample, the microscope or uniformity of illumination, the reader is referred to the Chapters 2, 3, or 4 for help.

TABLE 8-12
PHOTOMICROGRAPHY PROBLEMS

PROBLEM	CAUSE	CORECTION
No Image	Roll film not properly loaded.	Load roll film per the manufacturer's instructions.
	Image not exposing film.	Remove light shield before making exposure. Verify that shutter is functioning and that image light is present by removing the film and looking into the camera while tripping shutter.
Partial Image (Image cut off on one end)	Light shield not fully removed.	For wet process sheet pull film light shield completely out of film carrier. For Polaroid sheet film pull the light-tight cover out until it stops.
Partial Image (Image cut off on 1 or 2 sides)	Polaroid sheet film only.	Film pulled out too rapidly during processing step. Finger pressure on external part of roller assembly for model 500 film holder only.

Partial Imaage (Circular outline)	Small field eyepiece used.	Use a wide angle eyepiece, a photographic eyepiece, or extend bellows, if present.
	Shutter not located at the camera eyepiece exit pupil.	Reposition the adapter so that the shutter coincides with the exit pupil of the camera eyepiece.
Partial Image (Elliptical or off center circular outline)	Optical axes of microscope and flexible bellows camera not coincident.	Position microscope so that its optical axis coincides with that of the camera.
Unsharp Image	Vibration: Look for evidence of vibration with a petri dish 1/2 full of water. Ripples in the water surface indicate vibration.	If vibration is continuous, use a vibration isolation platform, see Chapter 7, Section II A. If vibration is due to focusing, wait a few seconds after focusing before releasing shutter. Also check to see that nesting cups are not touching in a flexible bellows camera. Do not use a focal plane shutter.
	Poor focus; refocus and take another micrograph, if still unsharp, try remaining corrections.	Refocus image on ground glass or at the focusing reticle. Refocus focusing reticle. Parfocalize focusing reticle in binocular eyepiece with the camera eyepiece, if present. Use an eyepiece telescope in front of v&f tube to focus low power (<100X) images. Observe image for drift; if present, refer microscope to service organization for repair.
	V&F tube reticle not optically coincident with film plane.	Refer camera to service organi-zation for repair.
	Empty magnification.	Total magnification too high for effective N.A. of objective lens.

Unsharp image around edge.	Objective lens not plano corrected.	Use planachromatic objective lenses if available, or, reduce aperture diaphragm opening slightly and refocus.
Contrast too high or too low	Microscope incorrectly adjusted or wrong type film.	Readjust microscope per Chapters 2, 3, or 4. If microscope is not at fault, evaluate film for specific application. If B&W film is used, color contrast can be modified with filters.
Colors not correct.	Filters not installed, wrong filters used, or wrong setting of illumination for existing filters.	Use correct filters for film and illumination types. Verify that illumination power setting is correct for the filters.
	Reciprocity failure of film; color shift in dark areas.	Use longer exposure time. An additional filter may be needed (with additional exposure time).
Light patches in print.	Light leak between eyepiece and camera.	Use nested cup light trap.
	Holes in flexible bellows.	Replace bellows.
	Defective light seal between camera and film carrier.	Refer camera to service organization for repair.
	Defective light seal in film carrier.	Replace film carrier.
Out-of-focus black spots on print.	Dust on camera eyepiece lens.	Clean camera eyepiece lens often.

REFERENCES

Chang, C. S., S. C. Mohapatra, and W. H. Johnson, Functional Photography, 18, Nov. 1977.

Hinsch, J., American Laboratory, 11, #4, 35, April 1979.

Hurtgen, T. P. and G. A. Massios, American Laboratory, 109, April 1979.

Loveland, R. P., "Photomicrography; a Comprehensive Treatise", 2 Vols., John Wiley and Sons, NY, 1971.

Vetter,J. P., American Laboratory, 8, April 1973.

9

Photomacrography

I. INTRODUCTION

Photomacrography is the process of photographically documenting a sample at magnifications at >1X to 50X. This magnification range, which is coincident with that of the stereo microscope, provides a crucial bridge between the sample and the higher magnification images afforded by a compound optical microscope or an electron microscope. Furthermore, the photomacrograph can provide documentation on both the location and the orientation of micro samples removed from a larger sample. Samples for photomacrography may range from an unprepared object, such as a geological sample, to large biological thin sections and macroetched metallurgical samples.

II. PHOTOMACROGRAPHIC INSTRUMENTATION

Instruments for photomacrography range from a single lens camera to the stereo microscope with camera and the very sophisticated photomicroscopes. Acharia (1986) has also described the use of a standard photographic enlarger as a photomacrographic system in which the sample is placed in the negative carrier and the recording negative placed in the print frame. The relative simplicity of photomacrographic systems belies the care necessary to obtain a high quality image. Hyzer (1975) expressed this fact in these words; "The constraints imposed by limits of diffraction, depth of field, failure of the reciprocity law, empty magnification, lens aberrations, lighting and exposure are more restricting in photomacrographic work than in almost any other field of photography."

A typical single lens, flexible bellows camera is shown in Figure 9-1. These cameras consist of a base support, lens carrier with shutter, and bellows with film back. The interchangeable lenses with an integral, adjustable aperture mount on the lens carrier. The film back contains a ground glass viewing screen which rests at the film plane when the film holder is not in place. The distance between the lens and the film plane is established with the lens focus knob. On this instrument, the distance between the sample and the entire camera is also adjustable by means of the camera focus knob. This latter feature makes it possible to focus the camera without altering the magnification. The bellows cameras usually support a 4" x 5" format however some with a 35 mm format are available.

The Wild M400 Photomakroskop, a superior fixed bellows macrocamera, is shown in Figure 9-2. This instrument has a zoom system that compensates for the lack of a flexible bellows and provides a total magnification range from 3.1X to 63.6X on 4" X 5" film. The Photomakroskop has the appearance of a stereo microscope but differs in that it has an adjustable aperture diaphragm and a binocular viewing and focusing system that gives only a two dimensional image. This instrument supports an automatic advance 35 mm camera, a Polaroid 3-1/4" X 4-1/4" film pack or a 4" X 5" camera back. The exposure console accommodates films with an ASA of 2.5 to 16,000 with reciprocity correction, if necessary. It also corrects exposure for darkfield situations. This controller also has provisions for flash triggering and advancing the 35 mm film.

The compound microscope can also be used for photomacrography. Some of the research or so-called universal instruments are equipped with macro accessories, including special condenser systems. The Reichert MeF-3 has a unique photomacrographic system for incident light photomacrography. Their Macrodual system, Figure 9-3, mounts on the side of the metallograph and images samples beneath it at magnifications from 1X to 9X.

Stereo microscopes are dual compound microscopes that have magnifications in the range of the photomacrograph, therefore, they are well suited as a basis for photomacrography. A typical photomacrographic arrangement is shown in Figure 9-4. Since the stereo microscope has two distinct optical paths, the camera can be attached to either eyepiece. In fact, stereo pair photomacrographs can be taken by attaching a stereo camera and recording both image paths. It must be recognized that the image from a single eyepiece is not a normal incidence image as each optical path is tilted about 3° from the normal so that they can focus on the same location; this is shown in Figure 9-5A. Wild has designed their M7S stereo microscope with an objective lens that may be displaced enough that the photo optical path has completely centered optics, this is illustrated in Figure 9-5B.

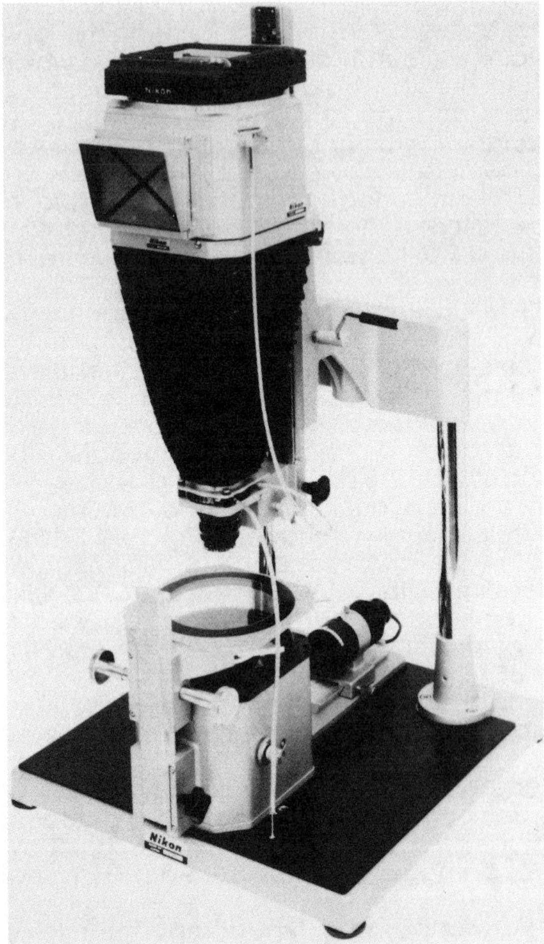

Figure 9-1. The flexible bellows camera equipped with several lenses and a variety of illuminators is a very useful instrument for photomacrography over a wide range of magnifications. Courtesy of Nikon, Inc.

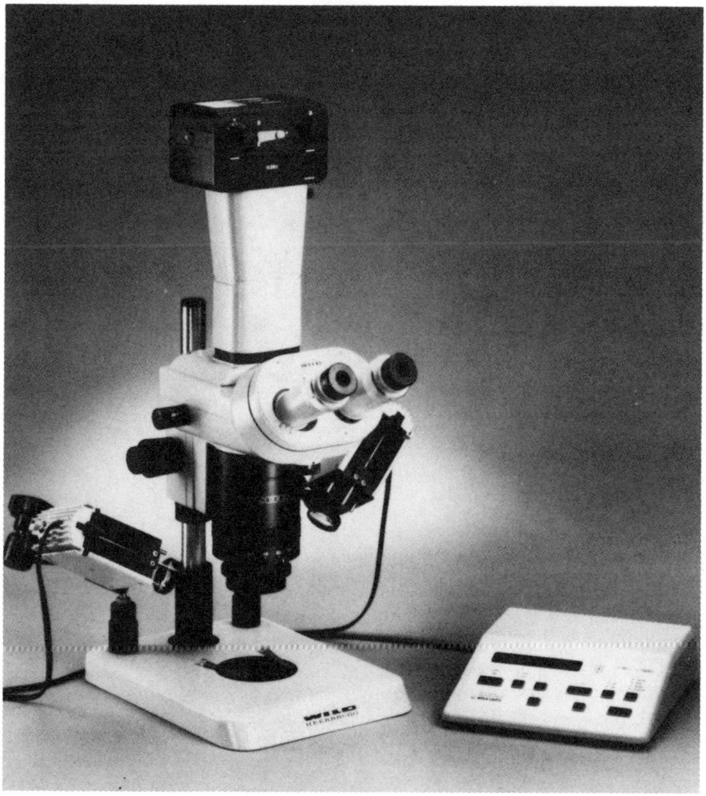

Figure 9-2. The Wild Photomakroskop has a high quality optical system equipped with a zoom system and is capable of magnifications from 3X to 63X with the 4 X 5 format. Courtesy of Wild Leitz USA, Inc.

A

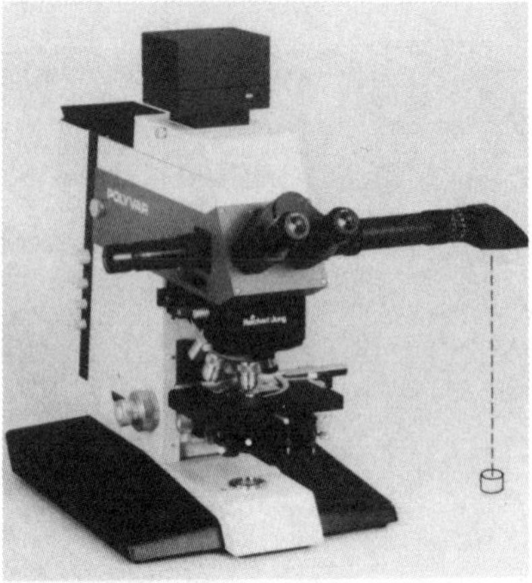

B

Figure 9-3. A. The Reichert Macrodual attachment affixed to the right side of their Polyvar-Met microscope provides magnifications of 1 to 10X for macro samples. B. An incident brightfield illuminator is also available for use with the Macrodual attachment. Courtesy of Cambridge Instruments Inc., Scientific Instruments Division.

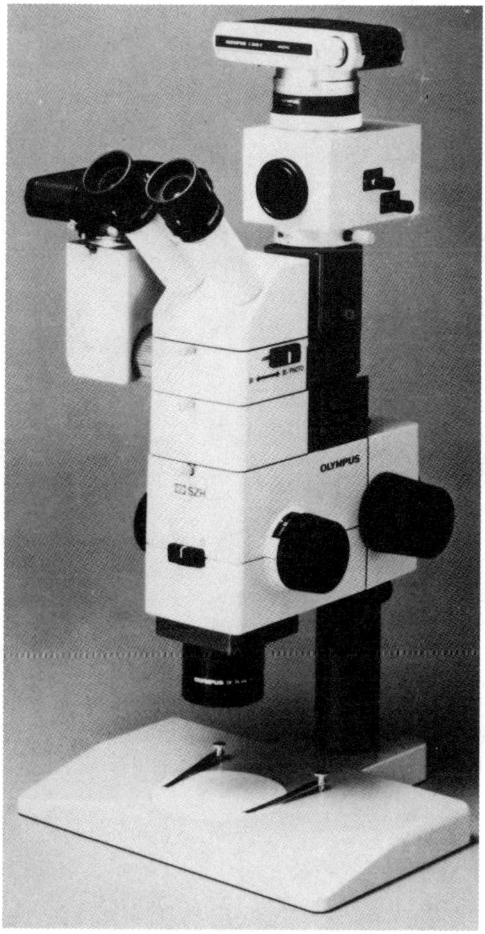

Figure 9-4. The stereo microscope with its low magnification range is ideally suited for photomacrography when equipped with a camera. This instrument is equipped with two camera ports. Courtesy of Olympus Corporation.

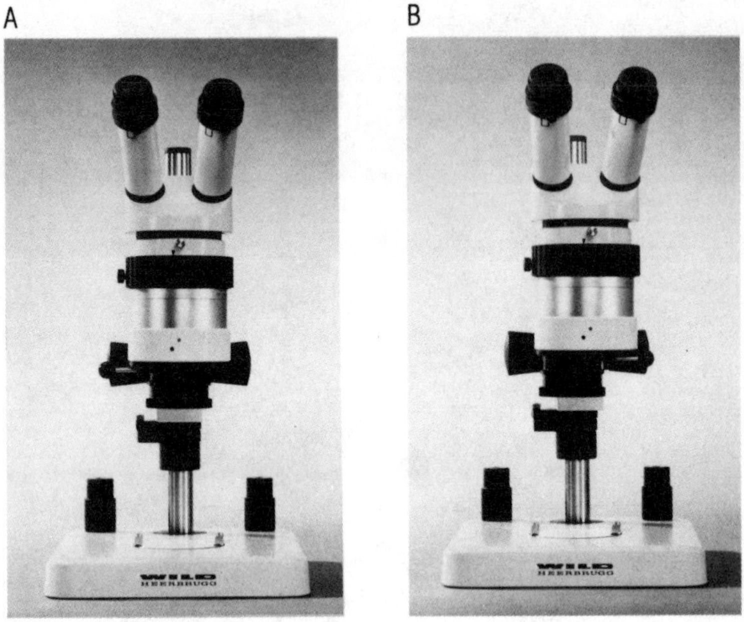

Figure 9-5. The image axis of each eye's viewing tube in a stereomicroscope, (A), is inclined at 3° to the vertical so that each eye may view the same area and thereby acquire a stereo image. As a result of this 3° tilt, photomicrographs taken with a stereomicroscope will have a slight difference in magnification from left to right. In the Wild M7S, CMO stereomicroscope (A) this problem can be avoided by translating the objective lens, as shown in (B). In this case, a camera on the left eyepiece will make use of the objective lens in a centered position. Courtesy of Wild Leitz USA, Inc.

III. PREPARATION OF A PHOTOMACROGRAPH

A. General

Photomacrography can seem deceptively easy to perform with the ready availability of highly automated instruments and this perception may be true if one must record the images of a large number of similar samples. But when a wide variety of samples are to be recorded, there is considerable challenge in the framing and illumination of each sample as well as in establishing the appropriate depth of field to produce a useful photomacrograph.

Since photomacrographic samples are quite diverse, it is perhaps worthwhile to discuss them, prior to describing the process of photographically recording them. Objects with a significant three-dimensionality are usually examined with incident illumination. If a sample has multiple surfaces such as a geological sample or a manufactured item, the three dimensionality can be accentuated with incident darkfield illumination. If the sample has a unique specular or reflective surface such as an integrated circuit wafer, that sample can be best examined in incident bright-field illumination. Transparent and thin translucent materials with planar or relatively planar surfaces can be examined effectively in transmitted brightfield or darkfield illumination. Large thin sections of biomedical subjects are excellent examples of transmitted light samples.

B. Framing and Composition

Framing and composition are analogous to that described in Chapter 8, namely the arrangement of the sample image within the film frame that satisfies both the technical and aesthetic requirements for the photomacrograph. Composition for photomacrography not only requires a consideration of the orientation of the image in reference to the frame edges and general position in the frame area but also the direction of view. To achieve a desired viewing direction the sample may be oriented on the stage with a small block of modeling clay, judiciously placed so that it will not be seen by the camera. Typically, a square or rectangular sample is oriented so that its edges are parallel to the print edges.

C. Magnification

Magnification, as such, is not usually the goal when preparing a photomacrograph, it is often merely increased until the sample, or that part of the sample that is of interest, nicely fills the film frame. The sample areas for several camera lenses is given for both the 35 mm and the 4" X 5" formats in Table 9-1. The actual value of the magnification is, nevertheless, important and is often documented simply by placing a scale adjacent to the sample, as shown in Figure 9-6.

TABLE 9-1
SAMPLE DIMENSIONS TO FILL FRAME FOR 35 MM AND 4" X 5" FORMATS AT VARIOUS MAGNIFICATIONS

MAGNIFICATION RATIO	35 MM FORMAT IN MM (INCHES)	4" X 5" FORMAT IN MM (INCHES)
1:4 (reduction)	96 X 144 (3.8 X 5.7)	360 X 460 (14 X 18)
1:3	72 X 108 (2.8 X 4.3)	255 X 335 (10.5 X 13.5)
1:2	48 X 72 (1.9 X 2.3)	180 X 230 (7 X 9)
1:1	24 X 36 (0.9 X 1.4)	90 X 115 (3.5 X 4.5)
2:1 (mag.)	12 X 18 (0.5 X 0.7)	45 X 58 (1.8 X 2.3)
3:1	8 X 12 (0.3 X 0.5)	30 X 38 (1.3 X 1.5)
4:1	6 X 9 (0.2 X 0.4)	23 X 28 (0.9 X 1.1)
5:1	4.8 X 7.2 (0.2 X 0.3)	18 X 22 (0.7 X 0.9)
10:1	2.4 X 3.6 (0.1 X 0.2)	9 X 11 (0.35 X 0.45)
20:1	1.2 X 1.8(0.05 X 0.07)	4.5 X 4.8 (0.18 X 0.22)
50:1	0.5 X 0.7(0.02 X 0.03)	1.8 X 2.3 (0.07 X 0.09)

The magnification for the single lens, flexible bellows macrocamera depends both on the focal length of the lens and the bellows length; it can be calculated as follows:

$$M = \frac{B.L. - FL}{FL} \tag{44}$$

where B.L. is the distance from the lens to the film plane and FL is the focal length of the lens. For example, if the bellows length is 250 mm and the focal length of the lens is 25 mm, the magnification is 9X. Figure 9-7 shows the total magnification for a number of different camera lenses of different focal lengths and various bellows extensions. Magnification for the stereo microscope case is given by equation (41) in Chapter 8.

In the case of the bellows camera, the same magnification may be maintained for a number of photomacrographs, by focusing the entire camera, that is, the lens, the bellows and the film carrier. In this case focusing must not be accomplished by moving the lens carrier as this will change the magnification as well.

If a bellows camera or a compound microscope with a zoom system is used, the magnification can be set to a desired value by adjusting the enlargement of the scale's image at the viewing screen to a given value. This procedure is described in Table 9-2. For example, a 5X magnification is obtained when a 1 cm scale is enlarged to 5 cm on the ground glass screen.

Figure 9-6. While it is generally easy to relate a photomacrograph to the actual object, it is always desirable to include a scale in the picture as shown here to indicate the actual magnification in the picture.

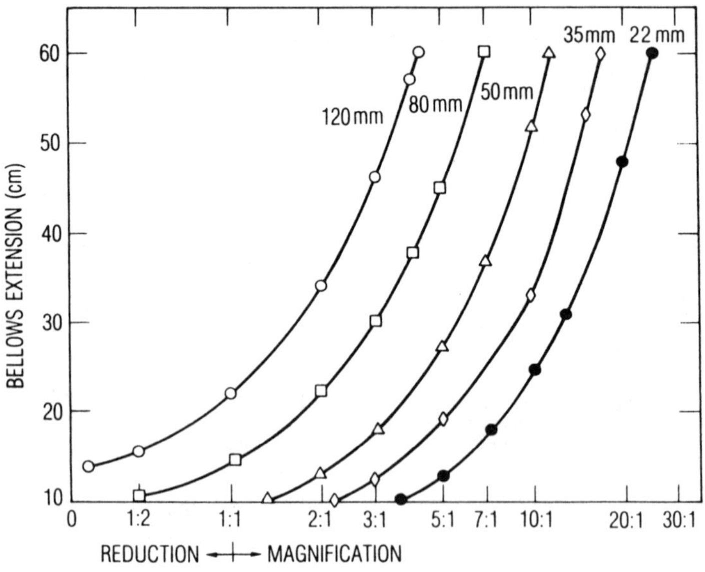

Figure 9-7. Relationship between the bellows extension and magnification (or reduction) for several focal length camera lenses. Redrawn, courtesy of Wild Leitz USA, Inc.

TABLE 9-2
ESTABLISHING THE MAGNIFICATION FOR A PHOTOMACROGRAPH

1. Place an inch or metric rule on the stage of the photo macroscope and illuminate it.
2. Open the shutter and the adjustable aperture then focus the image of the rule.
3. Using a second, similar rule, measure the size of as large a unit as possible in the image. For example, 1" in the image is found to measure 4.7" with the second rule. The magnification is then 4.7X.
4. Continuing the example, let us assume that the magnification we desire is 5.0X, then increase the bellows length or the zoom value slightly and refocus the photomacroscope. If 4X is the desired value, then reduce the bellows length or the zoom value slightly.
5. Repeat steps 3 and 4 until the desired magnification is obtained.

D. Depth Of Field

The major concern in the photomacrography of bulk samples is that of sufficient depth of field. The depth of field is dependent on the magnification and the aperture of the system. The depth of field, as with a microscope objective decreases as the magnification is increased and as the aperture of the system is increased. By way of illustration, the depth of field is given in Table 9-3 for several magnifications and two aperture values. The criterion for acceptable focus in this table is the circle of confusion which is the maximum permissible diameter of the image of a point source at the film plane. In practice, the lowest magnification camera lens that will give the desired depth of field, with appropriate bellow extension should be used.

Reducing the aperture opening will also increase the depth of field, however it simultaneously reduces the image brightness and sharpness. Therefore, the best aperture opening balances the depth of field against the exposure time and overall image detail. If a large depth of field is required, it may be necessary to photograph the sample at a magnification lower than that which would fill the film frame and make a subsequent photographic enlargement of the B&W negative or color transparency to achieve the desire image size.

TABLE 9-3
DEPTH OF FIELD* AT SEVERAL MAGNIFICATIONS
FOR TWO APERTURE VALUES IN MM (0.001 INCH)

MAGNIFICATION	35 MM FORMAT f/8		35 MM FORMAT f/16		4" X 5" FORMAT f/8		4" X 5" FORMAT f/16	
1:5 (REDUCTION)	14	(550)	29	(1100)	LARGE		LARGE	
1:2	3	(120)	6	(240)	9.6	(400)	19.2	(800)
1:1	1	(40)	2	(80)	3.2	(125)	6.4	(250)
5:1 (MAG)	0.1	(4)	0.2	(8)	0.4	(16)	0.8	(32)
10:1	0.06	(2)	0.11	(4)	0.08	(3)	0.2	(8)
30:1	SMALL		SMALL		0.06	(2)	0.11	(4)

*Assumes a circle of confusion of 0.03 mm for the 35mm format and 0.01 mm for the 4" X 5" format.

E. Exposure Determination

Exposure is determined in exactly the same manner as is described in Chapter 8, Section V E.

F. Filters

The use of filters for photomacrography is exactly the same as for photomicrography; the applications for color and B&W films are given in Chapter 8, Sections V F and G, respectively. The uses of polarizing filters will be addressed in the following section in the discussion of illumination techniques.

IV. PHOTOMACROGRAPHIC PROCEDURES

The procedures for the various types of photomacrography will be given in tabular form in this section. Since illumination is such an important aspect of successful photomacrography, each type will be preceded by a discussion of lighting techniques.

A. Transmitted, Brightfield Photomacrography

Photomacrography in transmitted, brightfield illumination is most commonly used for large cross sections, of naturally colored or stained biomedical or geological samples. The optimal illumination in this case is Koehler illumination in which the source is sharply imaged on the aperture diaphragm in the camera lens. In this way the illumination is very even and intense over the entire area of the photographic negative.

A typical transmitted, brightfield assembly is shown in combination with a flexible bellows camera in Figure 9-8. This

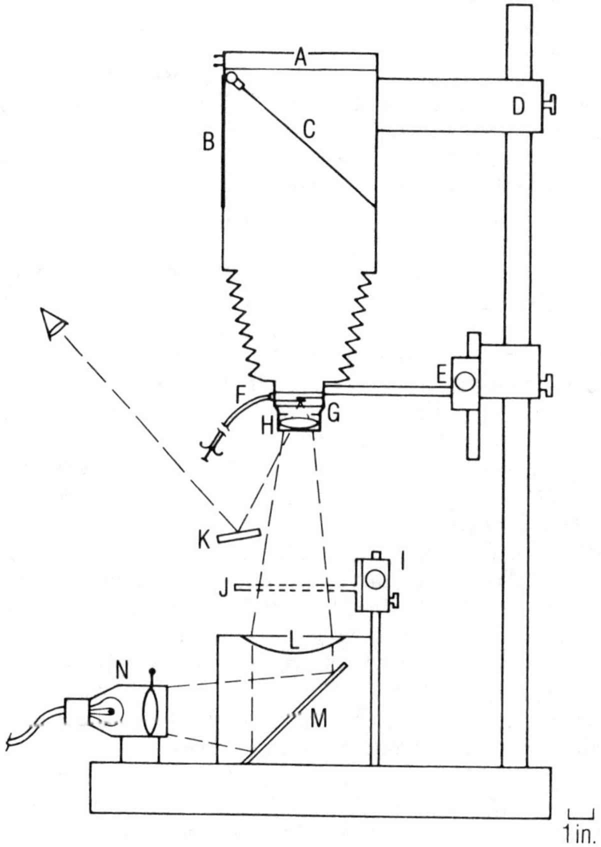

Figure 9-8. Arrangement of a flexible bellows camera for transmitted light, brightfield illumination. The mirror, K, is hand held and used to view the image of the filament of the lamp, N, as it is focused on the camera aperture diaphragm, G.

A	- Film carrier	H	- Camera lens
B	- Viewing screen *	I	- Stage focus knob
C	- Reflex screen *	J	- Stage
D	- Upper camera support **	K	- Hand held mirror
E	- Lower camera support **	L	- Condenser lens
F	- Shutter and release	M	- Substage mirror
G	- Aperture diaphragm	N	- Focusable illuminator
*	- Optional	**	- May be combined into one

assembly consists of a lamp housing, mirror and condenser with a focusable sample stage. The assembly will accept interchangeable condenser lenses, each having a focal length that is matched to a specific camera lens. For the longer focal length (lower magnification) camera lenses it may be necessary to use an additional negative lens in the lamp housing to adequately illuminate the matching, large diameter condenser lenses. Since the light levels are relatively high in transmitted brightfield illumination, a simple incandescent lamp is usually adequate for the purpose.

A similar transmitted, brightfield arrangement is available for most modern stereo microscopes. To provide proper Koehler illumination most manufacturers provide a spectacle or condenser lens to match each of their objective lenses. Leitz/Wild has a very flexible optical arrangement in their Transmitted-Light Stand EB in which any one of several condenser lenses may be placed into the optical path to match the requirements of the objective lens. This stand may also be used with the Wild M400 Photomakroskop. A typical transmitted, brightfield arrangement for a stereo microscope is shown in Figure 9-9.

Transmitted light combined with a polarizing capability is very useful for examination of birefringent crystals, fibers and geological thin sections as well as the effects of stress on transparent materials. The addition of a revolving stage and a 1st order plate, see Chapter 7, Section VI G, will permit the evaluation of preferred orientation of crystals or fibers in thin sections of materials.

Most manufacturers produce, at least, a simple polarized light capability for their stereo microscopes. These accessories consist of a polarizer that is mounted in a fixed position in or on the transmitted light stage and a rotatable polarizer that mounts on the objective lens of the microscope. Accessories that compliment the polarizers are a revolving stage and a 1st order plate, see Chapter 7, Section VI G.

The polarizing capability is established by placing the polarizing filter in the illuminator and affixing the analyzer to the objective lens. With the illuminator operating and no sample in place, rotate the analyzer in its housing until it is crossed with respect to the polarizer, i.e., no light can be see through the instrument. A birefringent sample introduced onto the stage will be bright, often in brilliant colors, against a dark background. For materials of very low birefringence, the addition of a 1st order plate between the polarizer and analyzer will produce color effects. The 1st order plate should be rotated around the optical axis until a uniform pink field is produced. A clear sheet of cellophane can be substituted for a 1st order compensator plate to enhance the polarizing effects.

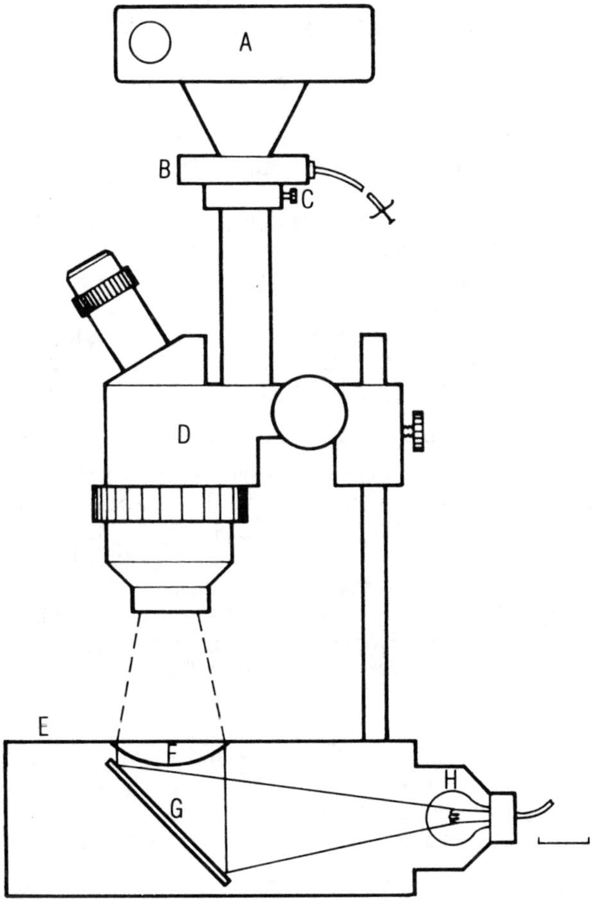

Figure 9-9. Arrangement of a stereo microscope for transmitted light, brightfield illumination.

A - Camera
B - Shutter and release
C - Camera attachment
D - Stereo microscope
E - Transmitted light base
F - Condenser lens
G - Substage mirror
H - Lamp

PROCEDURE FOR TRANSMITTED, BRIGHTFIELD PHOTOMACROGRAPHY WITH A SINGLE LENS, FLEXIBLE BELLOWS CAMERA

1. Refer to Figure 9-8 for the identification of the various items described in this procedure.
2. Select an approximate magnification from Table 9-1 based on the film format and the approximate sample size.
3. Select the longest focal length camera lens that will give this magnification from Figure 9-7.
4. Adjust the bellows length to give the desired magnification from Figure 9-7.
5. Insert the correct condenser (spectacle) lens in the illuminator base that matches the selected objective lens.
6. Turn on the illuminator, close the camera lens diaphragm, and open the shutter.
7. Adjust the position of the lamp, the focus of the lamp condenser or the camera focus (using the camera focus knob) to obtain an image of the lamp filament on the leaves of the camera lens diaphragm. This image should be sharp, centered and slightly larger that the diameter of the diaphragm. A mirror, as shown in Figure 9-8, can be used to view this image on the diaphragm. Once this focus has been obtained do not change the position of the camera relative to the illuminator.
8. Open the camera lens diaphragm to one stop less than full open.
9. Place the sample slide on the stage, focus, and frame it. (use the stage focus knob only). The image is viewed on the ground glass screen (if a reflex housing screen is used, rotate the mirror so that the image may be seen on the front screen). The room must be darkened for this operation.
10. Determine the exposure, close the shutter, and rotate the reflex mirror, if used, out of the light path.

11a. For 35 mm film or Polaroid 4" X 5" pack film, load a roll (or pack) of film into the camera back and advance film to the first frame. Withdraw the dark shield.

11b. For 4" X 5" wet process sheet film, insert a film carrier into the camera and withdraw the dark shield.

11c. For Polaroid 4" X 5" sheet film, set lever on the film back to "insert" position, insert a film pack into the film back and pull out the light-tight cover to the stop.

12. Press the shutter release to expose the film.

13a. Advance the 35mm film one frame or pull the tab for the Polaroid 4" X 5" pack film and extract the print.

13b. For the 4" X 5" wet process sheet film reinsert the darkshield and remove carrier.

13c. For the Polaroid 4" X 5" sheet film reinsert the light-tight cover, set lever on the film back to "process" position, and pull out the film pack to initiate the processing. Allow to process for the required time, remove print from light-tight cover and coat, if required.

PROCEDURE FOR TRANSMITTED, BRIGHTFIELD PHOTOMACROGRAPHY USING A STEREO MICROSCOPE WITH CAMERA

1. Refer to Figure 9-9 for the identification of the various items described in this procedure.
2. Turn on the illumination. Place the sample slide on the stage, focus, and frame it using the microscope eyepieces. Select the objective lens, accessory lens and the zoom setting to give the desired magnification.
3. Insert the condenser or spectacle lens, if available, in the illuminator base that matches the selected objective lens.

4a. For 35 mm film or Polaroid 4" X 5" pack film, load a roll (or pack) of film into the camera back and advance film to the first frame. Withdraw the dark shield.

4b. For 4" X 5" wet process sheet film, insert a film carrier into the camera and withdraw the dark shield.

4c. For Polaroid 4" X 5" sheet film, set lever on the film back to "insert" position, insert a film pack into the film back and pull out the light-tight cover to the stop.

5. Transfer the image to the camera path. Focus the reticle in the camera v&f tube as described in Chapter 8, Section V D. Adjust the magnification and framing as necessary then critically refocus the image using the camera v&f tube.
6. Determine the exposure and press the shutter release to expose the film.

7a. Advance the 35mm film one frame or pull the tab for the Polaroid 4" X 5" pack film and extract the print. Reinsert the dark shield.

7b. For the 4" X 5" wet process sheet film reinsert the dark shield and remove carrier.

7c. For the Polaroid 4" X 5" sheet film reinsert the light-tight cover, set lever on the film back to "process" position, and pull out the film pack to initiate the processing. Allow to process for the required time, remove print from light-tight cover and coat, if required.

8. Return the image to the microscope eyepieces for examination of the next sample.

B. Transmitted Darkfield Illumination

Darkfield illumination exists when the light from below the sample does not intercept the objective lens, this is described in Chapter 4, Section III. The light that does reach the camera is diffracted, refracted, or otherwise scattered by the sample, thus, the sample appears bright with a dark background. This type of illumination is particularly useful for thin, transparent samples which are essentially colorless, that is, unstained or with no natural color.

The flexible bellows camera may be used for transmitted darkfield illumination by placing the sample on the stage of the transmitted brightfield assembly, if available. In this case, the sample may be focused using the stage focus knob, as shown in Figure 9-10. Otherwise, a suitable, fixed stage may be assembled from a glass plate and appropriate supports as shown in Figure 9-11. For the latter arrangement, the sample is focused by moving the entire camera with the camera focus knob.

The transmitted light stand is ideal when using the stereo microscope for transmitted darkfield illumination. Normally there is little working space under these stands, therefore it may be necessary to use an external mirror to redirect the light from the illuminator to the sample as shown in Figure 9-12. The fixed glass plate as shown in Figure 9-11 may also be used with the stereo microscope.

Almost any lamp located so as to illuminate the sample at about a 45° angle can be used for most transmitted, darkfield applications, this is the oblique illumination described in Chapter 4, Section II. Wild offers an elegant brightfield/darkfield stand for their macro and stereo instruments which gives a full azimuth darkfield illumination.

PROCEDURE FOR TRANSMITTED, DARKFIELD PHOTOMACROGRAPHY WITH A SINGLE LENS, FLEXIBLE BELLOWS CAMERA

1. Refer to Figure 9-10 for the identification of the various items described in this procedure.
2. Select an approximate magnification from Table 9-1 based on the film format and the approximate sample size.
3. Select the longest focal length camera lens that will give this magnification from Figure 9-7.
4. Adjust the bellows length to give the desired magnification from Figure 9-7.
5. Place the sample slide on the stage, turn on the illuminator and open the shutter.
6. Adjust the position of the lamp to illuminate the sample from beneath at about 45° to the axis of the camera. Focus the lamp condenser lens so that the sample is completely illuminated.

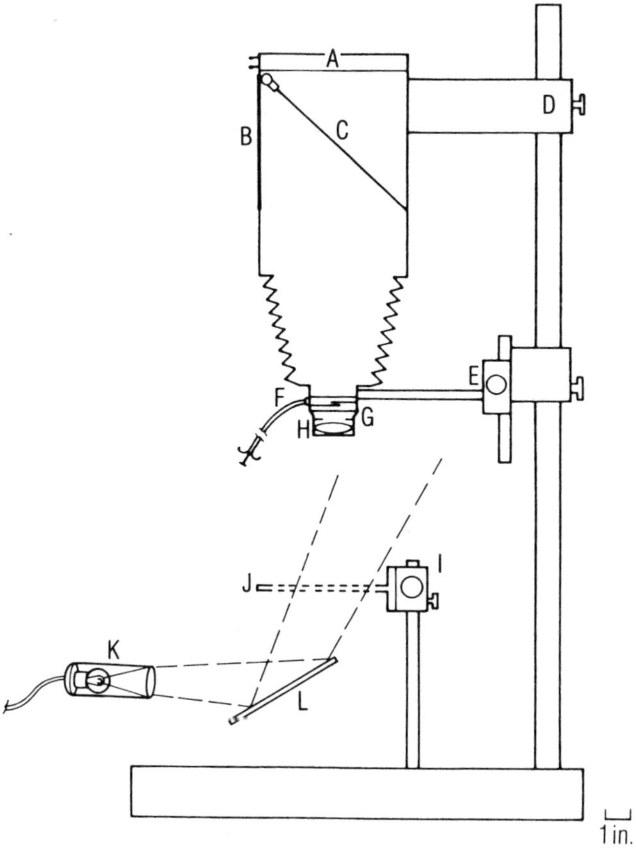

Figure 9-10. Arrangement of a flexible bellows camera for transmitted light, darkfield illumination.

A - Film carrier
B - Viewing screen*
C - Reflex screen*
D - Upper camera support**
E - Lower camera support**
F - Shutter and release
* - Optional
G - Aperture diaphragm
H - Camera lens
I - Stage focus knob
J - Stage
K - Focusable illuminator
L - Substage mirror
** - May be combined into one

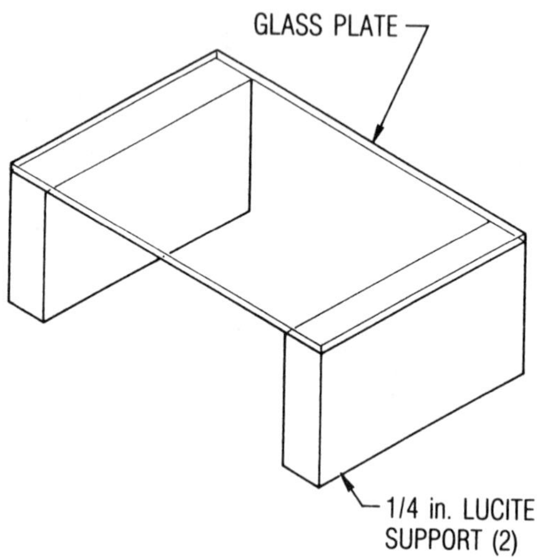

Figure 9-11. A fixed glass plate stage assembly for use with the flexible bellows camera or stereo microscope with an incident light stand for transmitted light darkfield illumination, replacing I and J in Figure 9-10 or the transmitted light base, E in Figure 9-12. The height of the Lucite (Reg.TM) supports should be at least 2 inches for greatest convenience.

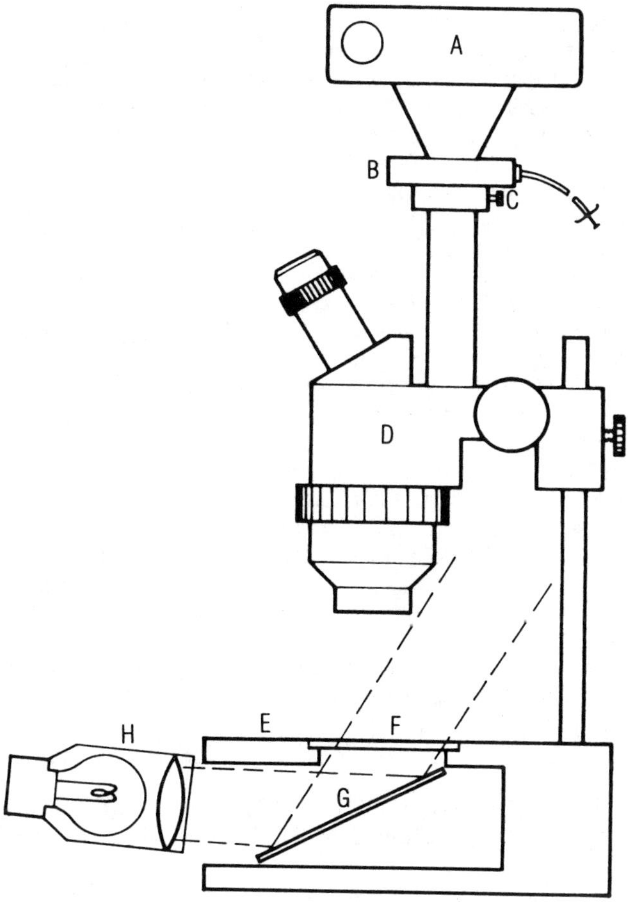

Figure 9-12. Arrangement of a stereo microscope for transmitted light, darkfield illumination.

A - Camera
B - Shutter and release
C - Camera attachment
D - Stereo microscope
E - Transmitted light base
F - Stage support plate
G - Substage mirror
H - Lamp

7. Open the camera lens diaphragm to one stop less than full open.
8. Focus and frame the sample image. Focusing is accomplished with the stage focus knob if a transmitted brightfield assembly is used. If a fixed, glass plate stage is used, focusing is accomplished with the camera focus knob. The image is viewed on the ground glass screen (if a reflex housing screen is used, rotate the mirror so that the image may be seen on the front screen). The room must be darkened for this operation.
9. Determine the exposure, close the shutter, and rotate the reflex mirror out of the light path, if used.

10a. For 35 mm film or Polaroid 4" X 5" pack film, load a roll (or pack) of film into the camera back and advance film to the first frame. Withdraw the dark shield.

10b. For 4" X 5" wet process sheet film, insert a film carrier into the camera and withdraw the dark shield.

10c. For Polaroid 4" X 5" sheet film, set lever on the film back to "insert" position, insert a film pack into the film back and pull out light-tight cover to the stop.

11. Press the shutter release to expose the film.

12a. Advance the 35 mm film one frame or pull the tab for the Polaroid 4" X 5" pack film and extract the print.

12b. For the 4" X 5" wet process sheet film reinsert the dark shield and remove carrier.

12c. For the Polaroid 4" X 5" sheet film reinsert the light-tight cover, set lever on the film back to "process" position, and pull out the film pack to initiate the processing. Allow to process for the required time, remove print from light-tight cover and coat, if required.

PROCEDURE FOR TRANSMITTED, DARKFIELD PHOTOMACROGRAPHY USING A STEREO MICROSCOPE WITH CAMERA

1. Refer to Figure 9-12 for the identification of the various items described in this procedure.
2. Adjust the position of the lamp and accessory mirror to illuminate the sample from beneath at about 45° to the axis of the camera. Focus the lamp condenser lens so that the sample is completely illuminated.
3. Place the sample slide on the stage, focus, and frame it using the microscope eyepieces. Select the objective lens, accessory lens and the zoom setting to give the desired magnification.

4a. For 35 mm film or Polaroid 4" X 5" pack film, load a roll (or pack) of film into the camera back and advance film to the first frame. Withdraw the dark shield.

4b. For 4" X 5" wet process sheet film, insert a film carrier into the camera and withdraw the dark shield.

4c. For Polaroid 4" X 5" sheet film, set lever on the film back to "insert" position, insert a film pack into the film back and pull out the light-tight cover to the stop.

5. Transfer the image to the camera path. Focus the reticle in the camera v&f tube as described in Chapter 8, Section V D. Adjust the magnification and framing as necessary then critically refocus the image using the camera v&f tube.

6. Determine the exposure and press the shutter release to expose the film.

7a. Advance the 35 mm film one frame or pull the tab for the Polaroid 4" X 5" pack film and extract the print. Reinsert the dark shield.

7b. For the 4" X 5" wet process sheet film reinsert the dark shield and remove carrier.

7c. For the Polaroid 4" X 5" sheet film reinsert the light-tight cover, set lever on the film back to "process" position, and pull out the film pack to initiate the processing. Allow to process for the required time, remove print from light-tight cover and coat, if required.

8. Return the image to the microscope eyepieces for examination of the next sample.

C. Incident Brightfield Illumination

Photomacrography with incident brightfield or normal incident illumination finds much application in the documentation of integrated circuit chips and wafers as well as macropolished and etched metallurgical samples. Thus, the technique works well for the recording of any planar, specular surface.

Only one manufacturer, Nikon produces incident brightfield attachments for their single lens bellows cameras, and these are limited to their 65 mm and 35 mm camera lenses. For all other single lens bellows cameras the illumination system described below must be used for incident brightfield applications.

The newer stereo microscopes provide a much better prospect for incident brightfield illumination. Leitz/Wild and Olympus both have incident brightfield illuminator modules with a built-in lamp that fit most of their respective stereo microscopes; these are often referred to as coaxial illuminators. Leitz/Wild also has an external attachment that produces either brightfield or darkfield illumination; this unit also has a built-in lamp.

Zeiss has an external attachment for their stereo microscopes that is illuminated from an external lamp using a fiber optic link. Nikon also offers a half-reflecting mirror attachment for their stereo microscopes which will give brightfield illumination. It should be emphasized that these two attachments do require an external illuminator.

For those instruments that are not equipped with an incident brightfield illuminator, suitable devices can be assembled easily and inexpensively that will provide a uniform, bright, Koehler illumination. Figures 9-13 and 9-14 show one such arrangement for a bellows camera and a stereo microscope, respectively. This system has been used to properly illuminate a flat, specular sample as large as 64 mm (2-1/2") in diameter. The illumination is supplied by a standard, external, focusable microscope illuminator. The condenser lens has a focal length of about 230 mm (9 in.) and a diameter of 162 mm (6-1/2"). It should be noted that for a sample to be fully illuminated, the diameter of the condenser lens must be somewhat more than twice the diameter of the sample. The partially reflecting mirror is ideally a thin glass plate with a coating that reflects 50% of the incident illumination. However, an uncoated piece of glass may be used with somewhat lower efficiency. The mirror is mounted at a 45° to the axis of the camera; this may be easily accomplished by gluing a triangular piece of clear plastic on either side of the mirror as shown in Figure 9-15a. A piece of black matte paper should also be placed as shown in Figures 9-13 and 9-14 to minimize undesirable reflections. A laboratory type, scissor jack makes a very suitable stage for incident light photomacrography, the height adjustment in this case becomes the stage focus knob. For samples on the order of 1/2 inch diameter or less, a focusable microscope illuminator can be used without a separate condenser lens. A standard cover glass can conveniently be used as the partially reflecting mirror. The cover glass is mounted in a slot cut in the eraser of a pencil, as shown in Figure 9-15B. The pencil serves as a pivot rod and support that may be rotatea for adjustment. A small mound of modeling clay can be used as a temporary support for the pencil.

Two alternate methods should be mentioned for obtaining incident brightfield images without the use of a partially reflecting mirror; these are illustrated in Figure 9-16. In the first alternate method, the illuminator is aimed at the sample at a very high angle and the sample is tilted toward it until the light from the illuminator is reflected from the sample into the camera. This technique is less than ideal since there is some variation in focus and magnification across the sample. To remedy the focus, the lens aperture must be closed somewhat to increase the depth of field. The minor differences in the magnification can be minimized if the camera is the press type that has swing and shift capabilities.

The second alternate method for brightfield illumination also requires a press camera with shift and swing capabilities. Here again, the illuminator is aimed at the sample at a very high angle. The sample surface is not tilted but translated in the direction of the illuminator and the film plane is shifted away from the direction of the illuminator so that the light reflected from the sample is

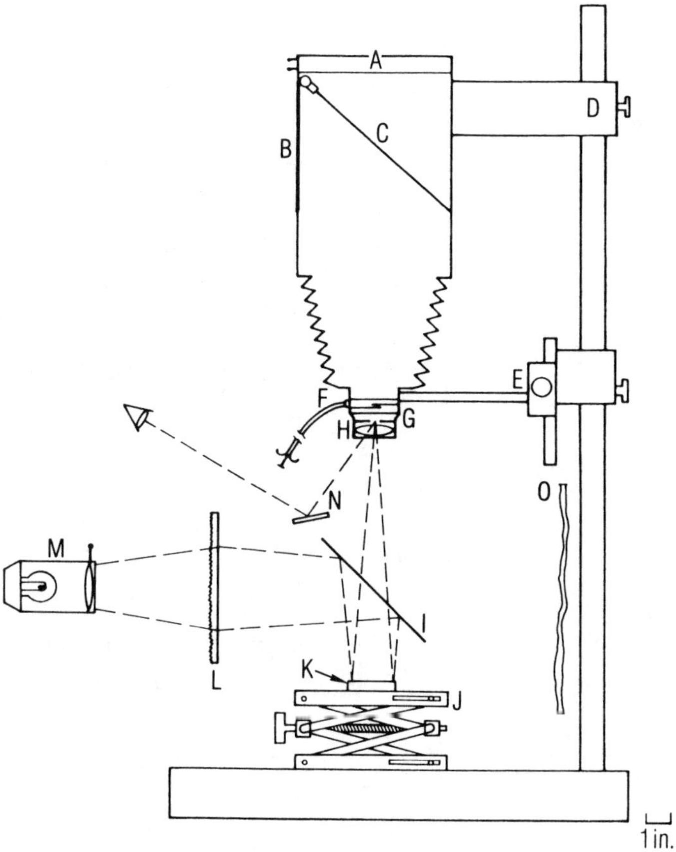

Figure 9-13 Arrangement of a flexible bellows camera for incident light, brightfield illumination. The mirror, N, is hand held and used to view the image of the lamp, M, as it is focused on the camera aperture diaphragm, G.

A - Film carrier
B - Viewing screen*
C - Reflex screen*
D - Upper camera support**
E - Lower camera support**
F - Shutter and release
G - Aperture diaphragm
H - Camera lens
I - Partially reflecting mirror
J - Stage (laboratory jack)
K - Sample
L - Condenser lens
M - Focusable illuminator
N - Hand held mirror
O - Sheet of black paper with matte or dull surface
* - Optional
** - May be combined into one

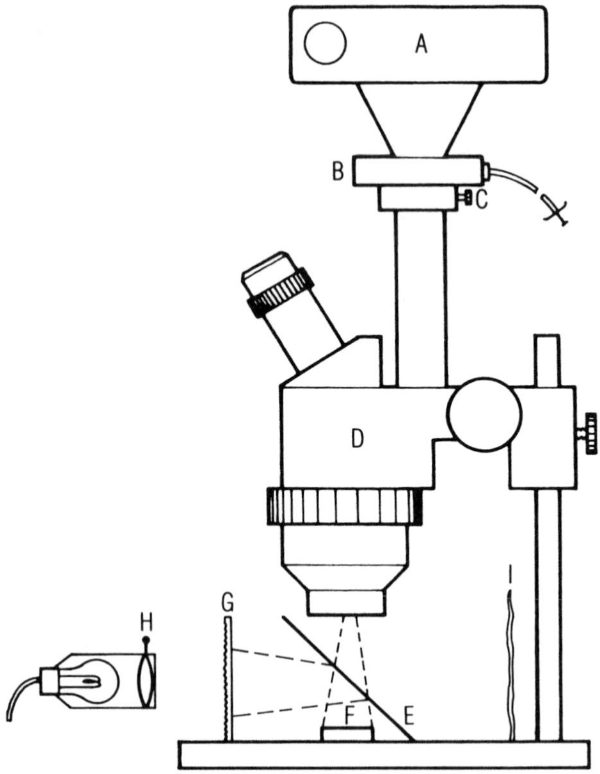

Figure 9-14. Arrangement of a stereo microscope for incident light, brightfield illumination.

A - Camera
B - Shutter and release
C - Camera attachment
D - Stereo microscope
E - Partially reflecting mirror
F - Sample
G - Fresnel condenser lens
H - Focusable illuminator
I - Sheet of black paper with matte or dull surface

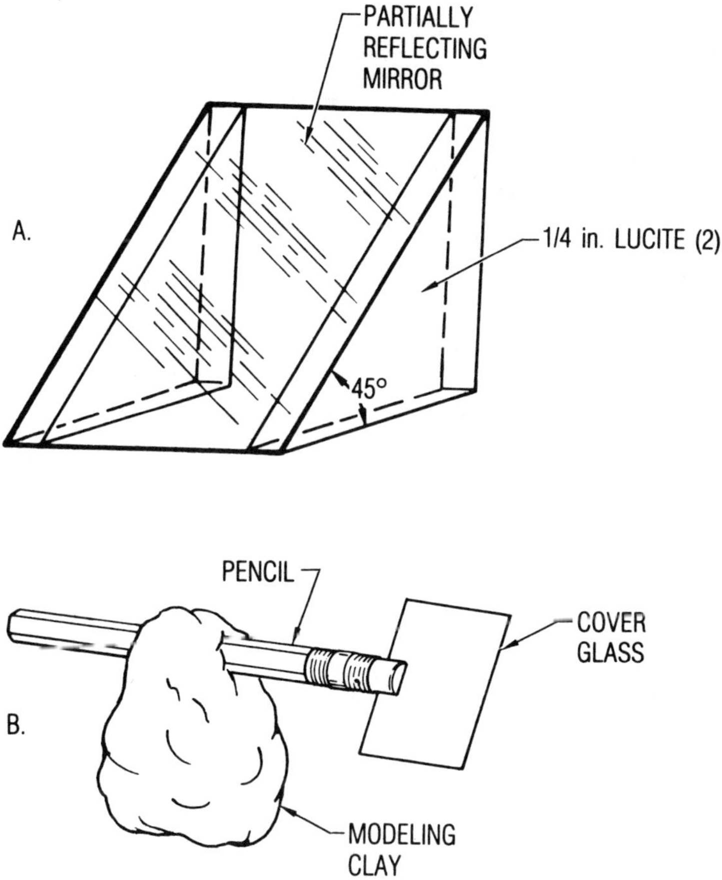

Figure 9-15. Two forms of user constructed partially reflecting mirrors suitable for the arrangements in Figures Figure 9-13 and Figure 9-14. (A) is constructed of a partially reflecting mirror, ideally about 4" X 4" and two, 45° angle supports of 1/4" thick, clear Lucite (Reg.TM). The mirror, ideally, should have about 50% transmission, however, a plain sheet of glass may be used. (B) is a temporary, partially reflecting mirror for small samples. The 45° angle must be established manually in this case.

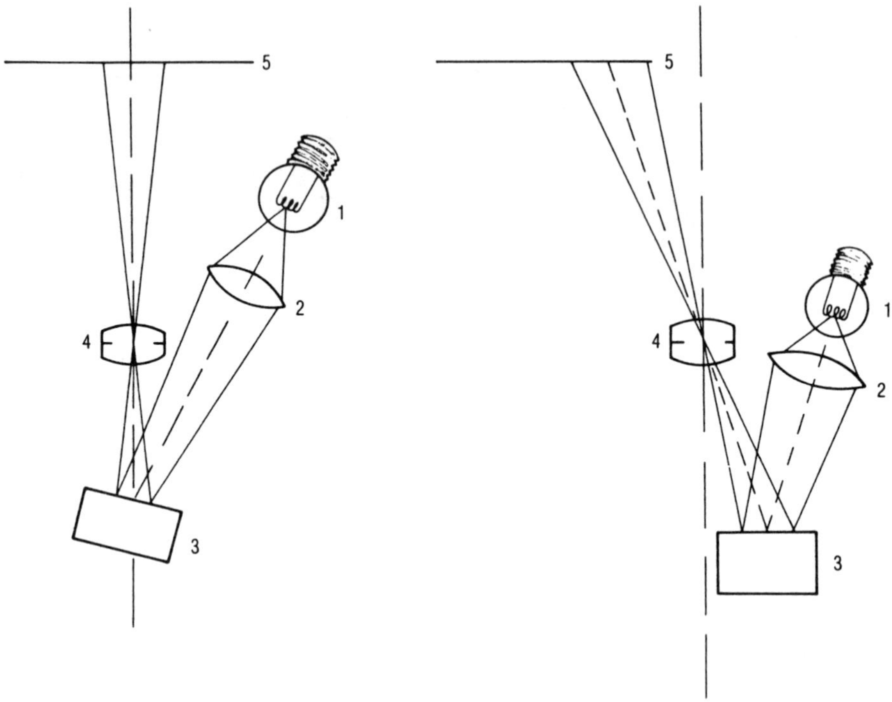

Figure 9-16. Two alternate methods for obtaining incident light, brightfield illumination. In the left illustration, the sample, 3, must be tilted so that the light from the illuminator will reflect into the lens of a bellows camera or stereo microscope. In the right illustration, a press camera with shift and swing capability is required. The film plane is shifted one direction and the sample is displaced in the opposite direction to provide the desired brightfield illumination.

1 - Lamp
2 - Condenser lens
3 - Sample
4 - Camera lens
5 - Film plane

intercepted by the film. This method suffers to some extent because the shift of the film plane introduces some minor distortion of the image.

PROCEDURE FOR INCIDENT, BRIGHTFIELD PHOTOMACROGRAPHY WITH A SINGLE LENS, FLEXIBLE BELLOWS CAMERA

1. Refer to Figure 9-13 for the identification of the various items described in this procedure.
2. Select an approximate magnification from Table 9-1 based on the film format and the approximate sample size.
3. Select the longest focal length camera lens that will give this magnification from Figure 9-7.
4. Adjust the bellows length to give the desired magnification from Figure 9-7.
5. Level the sample if necessary and place it on the stage.
6. Turn on the illuminator, close the camera lens diaphragm,and open the shutter. Place the partially reflecting mirror over the sample. Position the condenser lens as close as possible to the partially reflecting mirror. If a cover glass is used as a partially reflecting mirror, as described above, rotate it about the axis of the pencil as well as azimuthally to illuminate the sample.
7. Focus the sample. The sample image is viewed on the ground glass screen (if a reflex housing screen is used, rotate the mirror so that the image may be seen on the front screen). The room must be darkened for viewing.
8. Adjust the position of the lamp, the focus of the lamp condenser, the partially reflecting mirror and, if necessary, the camera focus (using the camera focus knob) to obtain an image of the lamp filament on the leaves of the camera lens diaphragm. This image should be sharp, centered and slightly larger that the diameter of the diaphragm. The back side of the partially reflecting mirror can be used to view this image on the diaphragm. Once this focus has been obtained do not change the position of the camera relative to the illuminator.
9. Open the camera lens diaphragm to one stop less than full open.
10. Critically focus the sample and frame it (use the stage focus knob only).
11. Determine the exposure, close the shutter, and rotate the reflex mirror out of the light path, if used.

12a. For 35 mm film or Polaroid 4 in. X 5 in. pack film, load a roll (or pack) of film into the camera back and advance film to the first frame. Withdraw the dark shield.

12b. For 4 in. X 5 in. wet process sheet film, insert a film carrier into the camera and withdraw the dark shield.

12c. For Polaroid 4 in. X 5 in. sheet film, set lever on the film back to "insert" position, insert a film pack into the film back and pull out the light-tight cover to the stop.

13. Press the shutter release to expose the film.

14a. Advance the 35 mm film one frame or pull the tab for the Polaroid 4 in. X 5 in. pack film and extract the print.

14b. For the 4 in. X 5 in. wet process sheet film reinsert the dark shield and remove carrier.

14c. For the Polaroid 4 in. X 5 in. sheet film reinsert the light-tight cover, set lever on the film back to "process" position, and pull out the film pack to initiate the processing. Allow to process for the required time, remove print from light-tight cover and coat, if required.

PROCEDURE FOR INCIDENT, BRIGHTFIELD PHOTOMACROGRAPHY USING A STEREO MICROSCOPE WITH CAMERA

1. Refer to Figure 9-14 for the identification of the various items described in this procedure.
2. Level the sample if necessary and place it on the stage.
3. If a coaxial or an attached brightfield incident illuminator is used, turn on the illumination and proceed to step 5.
4. If external illumination is used, place the partially reflecting mirror over the sample. Position the conden ser lens as close as possible to the partially reflecting mirror. Turn on the lamp and illuminate the sample. Focus the illumination so that an image of the lamp filament is formed on a sheet of paper held directly in front of the microscope objective lens. If a cover glass is used as the partially reflecting mirror it may be necessary to rotate it about the axis of the pencil as well as azimuthally to illuminate the sample.
5. Focus the sample and frame it using the microscope eyepieces. Select the objective lens, accessory lens and the zoom setting to give the desired magnification.

6a. For 35 mm film or Polaroid 4 in. X 5 in. pack film, load a roll (or pack) of film into the camera back and advance film to the first frame. Withdraw the dark shield.

6b. For 4 in. X 5 in. wet process sheet film, insert a film carrier into the camera and withdraw the dark shield.

6c. For Polaroid 4 in. X 5 in. sheet film, set lever on the film back to "insert" position, insert a film pack into the film back and pull out light-tight cover to the stop.

7. Transfer the image to the camera path. Focus the reticle in the camera v&f tube as described in Chapter 8, Section V D. Adjust the magnification and framing as necessary then refocus image using the camera v&f tube.

8. If the sample is not evenly illuminated with the attached illuminator, check the leveling of the sample. If the sample is not evenly illuminated with the external illuminator, adjust the position of the lamp, the focus of the lamp condenser and the position of the partially reflecting mirror.
9. Critically focus and frame the sample.
10. Determine the exposure and press the shutter release to expose the film.
11a. Advance the 35 mm film one frame or pull the tab for the Polaroid 4 in. X 5 in. pack film and extract the print. Reinsert the dark shield.
11b. For the 4 in. X 5 in. wet process sheet film reinsert the dark shield and remove carrier.
11c. For the Polaroid 4 in. X 5 in. sheet film reinsert the light-tight cover, set lever on the film back to "process" position, and pull out the film pack to initiate the processing. Allow to process for the required time, remove print from light-tight cover and coat, if required.
12. Return the image to the microscope eyepieces for examination of the next sample.

D. Incident Darkfield Illumination

Incident darkfield photomacrography is the most common illumination for a large variety of objects, ranging from opaque to transparent where the external shape must be documented. Illumination techiques very widely and in some cases, considerable experimentation may be required to achieve the desired results. It is safe to say that no single type of illumination will suffice for a wide range of samples.

Among the illuminators used for incident, darkfield photomacrography are fixed focus or focusable, external microscope illuminators, ring lamps, flash lamps, fiber optic illuminators and fluorescent lamps. Selected examples of these illuminators are shown in Figure 1-8.

The Bowens Texturelite, represented a truely unique and versatile ring type illuminator that mounted around the camera lens. The illuminator had four quadrants each containing both an incandescent "modeling" lamp and a flash lamp. Any or all of these lamps could be used with or without aperture masks to achieve a wide variety of illuminations. Unfortunately, this lamp is no longer available.

Fluorescence of a sample may be recorded by illuminating it with an ultraviolet lamp and attaching an appropriate barrier filter, see Chapter 7, Section VI E, to the camera or objective lens.

The various illuminators mentioned above may be used separately or in any combination and with light tents to give a very uniform illumination that is particularly useful for shiny or specular samples. Several suitable light tent designs are shown in Figure 9-17.

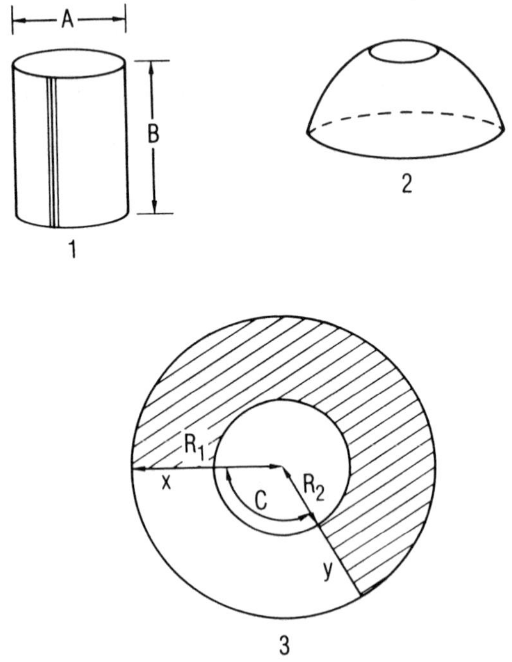

Figure 9-17. Three light tents suitable for photomacrography. (1) is a cylinder with diameter A and height B which may be constructed of white paper joined with clear tape. The diameter should be somewhat larger than the sample and the height should permit placement between the stage and the camera lens. (2) is a section of a ping-pong ball and is suitable only for small samples, Albertson (1964). (3) is constructed of white paper. When joined with clear tape at X and Y, a truncated cone is formed that serves as an excellent light tent. The values of R1, R2 and the angle C determine the diameter and height of the tent, McComb (1981).

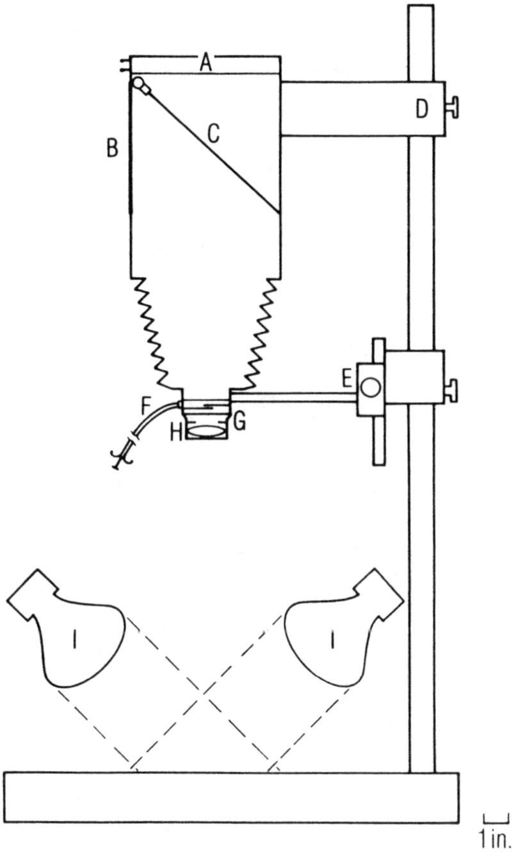

Figure 9-18. Arrangement of a flexible bellows camera for incident light, darkfield illumination.

A - Film carrier
B - Viewing screen*
C - Reflex screen*
D - Upper camera support**
E - Lower camera support**
F - Shutter and release
G - Aperture diaphragm
H - Camera lens
I - Illuminator(s)
* - Optional
** - May be combined into one

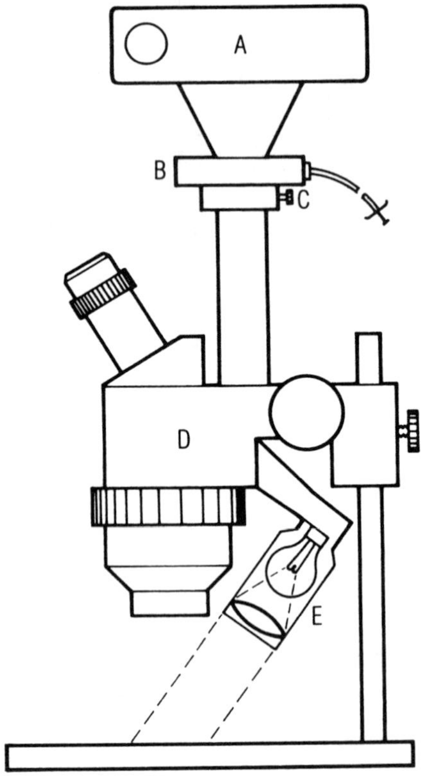

Figure 9-19. Arrangement of a stereo microscope for incident light, darkfield illumination.

A - Camera
B - Shutter and release
C - Camera attachment
D - Stereo microscope
E - Illuminator(s)

The sample, to be properly documented, must be clearly contrasted from the stage on which it rests. A wide range of samples can be contrasted using a selection of white, black and colored craft papers as covers for the sample stage. Harsh shadows on the sample resulting from a single illuminator can be softened by a second, weaker source of illumination at a higher angle and aimed in the opposite direction. Strong shadows cast by the sample on the stage can be very effectively eliminated by mounting the sample on a glass plate suspended several centimeters above the stage. In this latter case the paper sheet remains on the stage and provides contrasting color and a featureless background.

Polarizing filters may be used to minimize severe reflections from sample surfaces. One filter, the polarizer is mounted on the illuminator and the second, the analyzer is mounted on the camera or objective lens. Each filter should be rotated about its optic axis until the offending reflection has been reduced to an acceptable level.

Immersion in water or other fluids offer another means of reducing unwanted reflections from the surfaces of transparent materials. This method has been used successfully for the documentation of inclusions in gemstones. The various factors guiding the selection of the fluid are; (1) the sample must be insoluble in the fluid, (2) the fluid should have a refractive index different from the sample if the sample edges are to be accentuated, (3) the fluid must have a similar refractive index if internal details are to be seen most distinctly. Proper care must be given to avoid open flame and to provide adequate ventilation when organic fluids are used.

Typical bellows camera and stereo microscope arrangements for incident, darkfield illumination are shown in Figures 9-18 and 9-19, respectively. Table 9-4 describes suitable incident, darkfield illuminators to be used for various types of samples. It is usually necessary to make a number of adjustments to the illumination to achieve the most desirable lighting of the sample; experimentation and innovation are often the keys to successful lighting. As mentioned above, it may be necessary to use more than one type of illuminator to obtain the optimum effect.

TABLE 9-4
SUGGESTED ILLUMINATION OF VARIOUS SAMPLE TYPES FOR INCIDENT, DARKFIELD PHOTOMACROGRAPHY

SAMPLE TYPE	ILLUMINATION
Flat, specular	Use single illuminator at low angle.
Blocky, translucent	Use one illuminator at 45° and a second weaker lamp opposite the first to soften shadows, as

	necessary. Some transmitted light may be useful.
Blocky, transparent with planar surfaces	Use one illuminator at 45° and a second weaker lamp opposite the first to soften shadows, as necessary Use polarizers on lamps and analyzer on camera lens. Rotate polarizers to minimize bright reflections. Some transmitted light may be useful. Immersion in a fluid of different refractive index may be useful.
Blocky, opaque	Use one illuminator at 45° and a second weaker lamp opposite the first to soften shadows, as necessary.
Blocky, opaque with specular surfaces	Use one illuminator at 45° and a second weaker lamp opposite the first to soften shadows, as necessary. Use polarizers on lamps and analyzer on camera lens. Rotate polarizers to to minimize bright reflections. Darken room or use a light tent to eliminate unwanted reflections.
Blocky, opaque with deeply fissured surface	Use a ring light mounted around the camera lens.
Spherical, specular	Use a light tent and illuminate it uniformly around its periphery. The image of the camera lens cannot be eliminated from the surface of the sample.
Inclusions in glass or plastic or in a vial.	Immerse in water or mineral oil and use light tent.

PROCEDURE FOR INCIDENT, DARKFIELD PHOTOMACROGRAPHY WITH A SINGLE LENS, FLEXIBLE BELLOWS CAMERA

1. Refer to Figure 9-18 for the identification of the various items described in this procedure.
2. Select an approximate magnification from Table 9-1 based on the film format and the approximate sample size.
3. Select the longest focal length camera lens that will give this magnification from Figure 9-7.
4. Adjust the bellows length to give the desired magnification from Figure 9-7.
5. Place the sample on the stage.

6. Turn on the illuminator, open the lens diaphragm, and open the shutter.
7. Focus at a point about the midlevel of the sample using the stage or camera focus knob. The sample image is viewed on the ground glass screen (if a reflex housing screen is used, rotate the mirror so that the image may be seen on the front screen). The room must be darkened for viewing.
8. Position the lamp(s) to achieve the desired illumination.
9. Adjust the camera lens diaphragm to give the necessary depth of field, i.e., the entire area of interest is in sharp focus.
10. Determine the exposure, close the shutter, and rotate the reflex mirror out of the light path, if used.

11a. For 35 mm film or Polaroid 4" X 5" pack film, load a roll (or pack) of film into the camera back and advance film to the first frame. Withdraw the dark shield.

11b. For 4" X 5" wet process sheet film, insert a film carrier into the camera and withdraw the dark shield.

11c. For Polaroid 4" X 5" sheet film, set lever on the film back to "insert" position, insert a film pack into the film back and pull out the light-tight cover to the stop.

12. Press the shutter release to expose the film.

13a. Advance the 35 mm film one frame or pull the tab for the Polaroid 4" X 5" pack film and extract the print.

13b. For the 4" X 5" wet process sheet film reinsert the dark shield and remove carrier.

13c. For the Polaroid 4" X 5" sheet film reinsert the light-tight cover, set lever on the film back to "process" position, and pull out the film pack to initiate the pro cessing. Allow to process for the required time, remove print from light-tight cover and coat, if required.

PROCEDURE FOR INCIDENT, DARKFIELD PHOTOMACROGRAPHY USING A STEREO MICROSCOPE WITH CAMERA

1. Refer to Figure 9-19 for the identification of the various items described in this procedure.
2. Turn on the illumination and place the sample on the stage.
3. Position the lamp(s) to achieve the desired illumination.
4. Focus the sample and frame it using the microscope eyepieces. Select the objective lens, accessory lens and the zoom setting to give the desired magnification.

5a. For 35 mm film or Polaroid 4" X 5" pack film, load a roll (or pack) of film into the camera back and advance film to the first frame. Withdraw the dark shield.

5b. For 4" X 5" wet process sheet film, insert a film carrier into the camera and withdraw the dark shield.

5c. For Polaroid 4" X 5" sheet film, set lever on the film back to "insert" position, insert a film pack into the film back and pull out the light-tight cover to the stop.
6. Transfer the image to the camera path. Focus the reticle in the camera v&f tube as described in Chapter 8, Section V D. Adjust the magnification and framing as necessary.
7. Critically focus using the camera v&f tube.
8. Determine the exposure and press the shutter release to expose the film.
9a. Advance the 35 mm film one frame or pull the tab for the Polaroid 4" X 5" pack film and extract the print. Reinsert the dark shield.
9b. For the 4" X 5" wet process sheet film reinsert the dark shield and remove carrier.
9c. For the Polaroid 4" X 5" sheet film reinsert the light-tight cover, set lever on the film back to "process" position, and pull out the film pack to initiate the processing. Allow to process for the required time, remove print from light-tight cover and coat, if required.
10. Return the image to the microscope eyepieces for examination of the next sample.

E. Stereoscopic Photomacrography

When a person with normal vision redirects his attention from a remote object to a nearby object, two autonomic responses occur, the eyes accommodate and converge, i.e., the eyes rotate inward and refocus on the new object. If the object is between 0.25 meters and 200 meters distant, the images produced by the two eyes with a specific accommodation and focus for the specific distance, give rise to an impression of 3 dimensionality for the viewed object. The near limit of stereoscopic vision is further reduced by using the stereo microscope, see Chapter 1, Section III. If the images in each eyepiece of a stereo microscope are recorded photographically, the resulting two prints or tranparencies, known as a stereo pair, will recreate the stereo image when examined in an appropriate viewer. It is regrettable, however, that about 10% of the population does not have stereo vision, therefore they will not see a 3 dimensional image from the stereo pair.

This section discusses stereoscopy or the mechanics of producing stereo photomacrographs. Magnification for this purpose is usually from 2X to about 40X; the upper limit is due to insufficient depth of field beyond this magnification. The angle between the two views ranges from about 3° at 2X to about 7° at 5X. Although, higher magnifications would seem to require larger angles based on geometric considerations, 7° seems to be a satisfactory tilt angle for stereo photomacrographs up to 40X.

Since stereoscopy is important for samples with a three dimensional or blocky character, stereoscopic photomacrography makes use mainly of incident, darkfield illumination. To be sure, some 3-dimensionality may be perceived in biomedical sections as well, so the discussions here will also apply to transmitted brightfield and darkfield illumination.

Perhaps the most straightforward arrangement, shown in Figure 9-20, for taking stereo photomacrographs was offered by the American Optical Company for their Cycloptic stereomicroscopes. A Graflex stereo camera was attached to the stereo microscope with a hinged support so that the image could be framed and focused visually then the camera was swung into position directly over the eyepieces to record a stereo pair of images simultaneously. This attachment is no longer available.

A simple tilting stage for incident illumination giving a fixed 7° tilt can be made by gluing a short section of the wooden stem of a cotton swab across the center of the short dimension of a 1" X 3" glass slide. This is illustrated in Figure 9-21. Hyzer (1972) has described a tilting stage that will accommodate larger samples.

The procedure to prepare a stereo pair of a sample is given in Table 9-5. For optimum results, various areas of the sample should have height differences in the macro range, that is, they should be observable in a stereo microscope. For a typical arrangement of the simple tilting stage in position under a macro camera, refer to Figure 9-22.

Stereo pairs in the form of prints are easily viewed with a stereoscope. A variety of stereoscopes are available from Ward's Natural Science Establishment, Inc. (5100 W. Henrietta Road, Rochester, NY 14692). The right and left images are placed on the table below the stereoscope on their respective sides with their tilt, "Y" axes, at right angles to the line between the viewing lenses or the "X" direction as shown in Figure 9-23. Center the image for the right eye then translate the left image until the two images tend to fuse; it is at that point that the stereo effect will be obvious. Minor rotation of one of the images may be necessary to optimize the effect. When in the correct position, the stereo pair may be glued or taped to a larger piece of art board so that subsequent viewing may be made without realignment of the two pictures.

TABLE 9-5
PREPARATION OF A STEREOSCOPIC PHOTOMACROGRAPH USING A SIMPLE TILTING STAGE

1. Select the type of illumination suitable for the sample.
2. Set-up the photomacrographic system and define the direction "X" as parallel to a line between the eyepieces. If possible, rotate the camera so that the long axis of the film, as shown in Figure 9-22 is parallel to X.

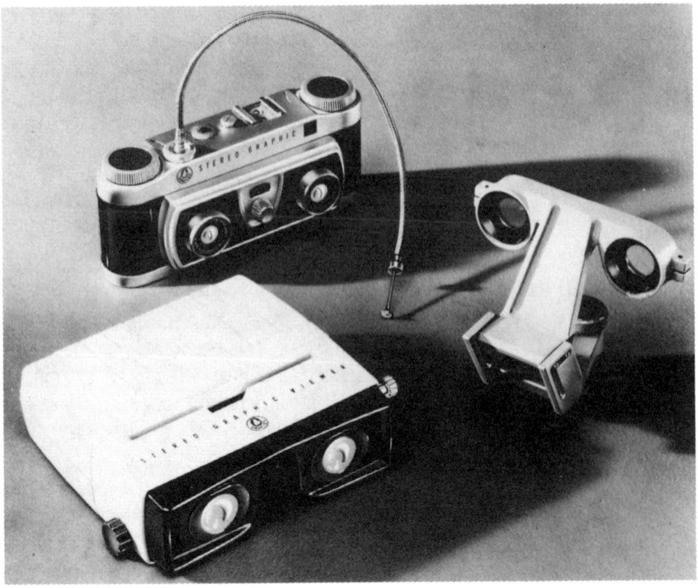

Figure 9-20. A novel stereo attachment (right) that positioned a Graflex stereo camera (top) over the eyepieces of a stereo microscope. The resulting stereo pair transparencies could be examined in the stereo viewer (lower). This attachment is, however, no longer available. Courtesy of Cambridge Instruments Inc., Optical Systems Division.

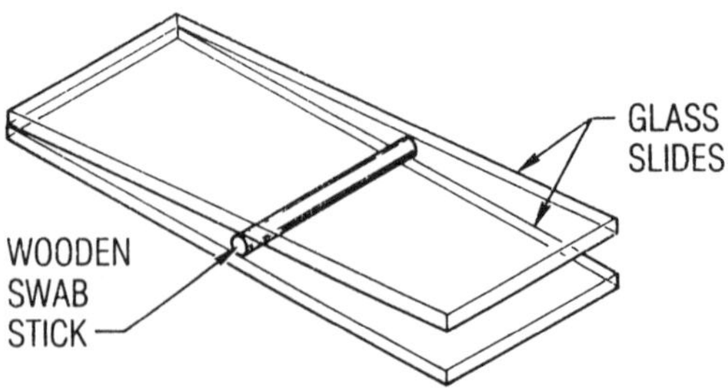

Figure 9-21. A simple, user constructed tilt stage for production of stereo pair photomacrographs. The tilt angle is about 7°.

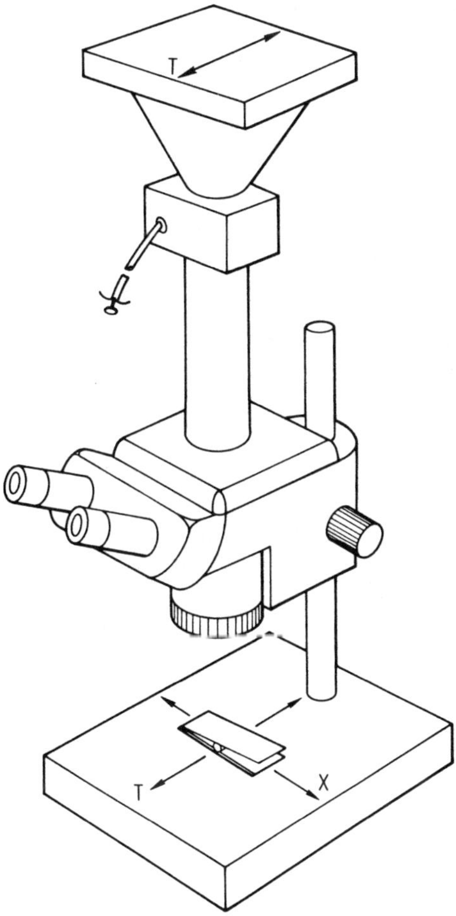

Figure 9-22. The tilt stage of Figure 9-22 positioned on the stage of a photomacrographic instrument (either a stereo microscope or a flexible bellows camera may be used). The tilt axis of the stage, T, is parallel to the wooden swab stick. The tilt axis, T, of the photomacrograph is parallel to that of the stage. The sample is placed on the stage and tilted so that the right side of the stage is low. The photomacrograph taken in this position is labeled "R". The stage is then tilted so that the left side of the stage is low. The photomacrograph taken in this position is labeled "L". The tilt axis is marked on each photomacrograph. These photomacrographs constitute a "stereo pair".

3. Mount the sample on the tilting stage. Orient the axis of the tilting stage, T, at right angles to the X axis in Figure 9-22.
4. Tilt the stage to one side. Focus and frame the sample. If a stereo microscope is used, select a reference point on the sample with a crosswire eyepiece. If a single lens bellows camera is used select a reference point on the sample and make a pencil mark on the ground glass screen at that point.
5. Expose the photomacrograph.
6. Tilt the stage to the opposite position. Focus and frame the sample. If necessary, translate the sample in the "X" direction to make the reference point on the sample coincide once again with the crosswire or pencil mark.
7. Expose the second photomacrograph.
8. Process to produce stereo prints or transparencies and view with stereoscope or stereo projector, respectively.

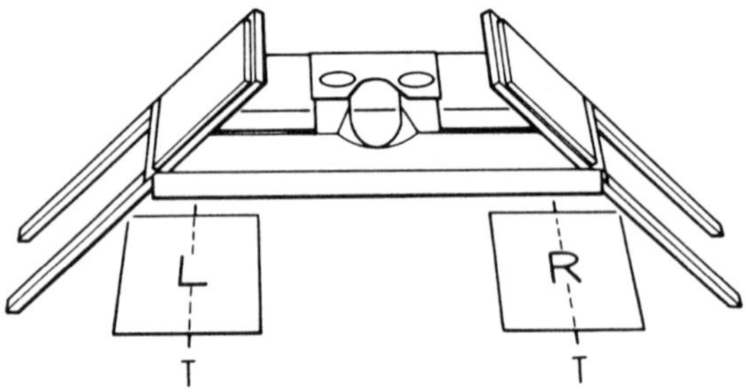

Figure 9-23. The stereo pair photomacrographs may be examined with a stereoscope by placing them in the position shown. Place one photomacrograph under the stereoscope so that it is properly centered in the field of one eye then adjust the other photomacrograph by translation and slight rotation (if necessary) until a single fused, stereo image is observed.

Elaborate schemes have been devised by others to identify the right and left images, i.e., the images to be viewed by the right and left eyes, respectively. However, it is relatively easy for the observer to correctly identify the two images correctly by inspection of the finished prints or transparencies. If the images are reversed the stereo image is confused; actually the 3 dimensionality is reversed and objects which protrude outward from the sample seem to protrude inward. If this is observed, merely reverse the positions of the two prints.

Stereo transparencies may be mounted as a pair and examined with either a hand-held viewer or with a stereo projector.

F. Problems

The problems encountered in photomacrography of two general types. The instrumentation problems are similar to those encountered in photomicrography and, as such, are covered in Chapter 8, Section V H 3. The illumination is, undoubtedly, the major problem in photomacrography. As mentioned above, the illumination may be defined by sample type to some extent, but in many cases it requires a certain amount of experimentation to achieve the optimum results.

REFERENCES

Acharia, K. M., Functional Photography, 23, May/June 1986.

Albertson, C. E., The Microscope, 14, #7, 253, 1964.

Hyzer, W. G., Photomethods, 25, February 1975.

Hyzer, W. G., Research/Development, 51, December 1972.

McComb, G., Functional Photography, 18, July/August 1981.

10

Selection, Care and Maintenance of the Optical Microscope

I. INTRODUCTION

This chapter is an attempt to make microscopy better for the microscopist in terms of acquiring the correct instrument, placing it in an ideal environment, both for the instrument and the operator, and assuring that the instrument continues to perform in an optimal fashion. Also in this chapter is a collection of techniques for evaluating the objective lens, the most important part of the microscope with respect to image quality.

II. SELECTION OF A MICROSCOPE

The selection of a microscope may be considered a threefold process which includes: (1) the definition of the basic requirements for the instrument predicated on its application; (2) the selection of the accessories, commensurate with the allowed budget; and (3) the choice of the specific microscope which incorporates the above information but moderates it in the context of various subjective preferences of personal or corporate origin.

The basic question in the definition of requirements is whether the instrument is to be used for low power investigations (<50X) or at higher magnifications. This will quickly determine if we are concerned with a stereo microscope or if we must consider a compound microscope. The further requirements for a compound microscope are strongly a function of the application. Table 10-1 lists the items that must be considered as a function of the various major technical disciplines. For a discipline not listed, the table nevertheless provides an essential check list.

Once we have established the basic configuration of the microscope, we can acquire or develop the appropriate stand. Long life and sturdiness is usually associated with a metal stand, however with the advent of some of the newer polymers, a reasonably sturdy stand may incorporate a considerable number of plastic parts. Delrin plastic in bearings and focusing mechanisms can improve their ease of operation and extend operating life. For many single applications a simple stand is all that is required, however, if a number of illumination types are required a research grade stand is a wise choice. The options are great, for example, one may consider an instrument for a special application which has only one fixed objective lens while at the other extreme, a quintuple nosepiece is available with a push button activated, motorized changer.

Most other additions to the basic stand are in the form of illumination, additional contrasting capabilities, measuring devices and conveniences. These can be relatively inexpensive or can tax the most generous budget. Chapters 2 and 3 discuss these additions to the transmitted and incident light microscopes, respectively and Chapter 7 lists various specialized accessories.

An incandescent source such as the halogen lamp is quite suitable for illumination if only visual and modest photomicrographic work is performed. If much darkfield, polarized light, phase contrast, or interference contrast photomicrography is anticipated, a xenon lamp may be a wise choice. For most users a binocular head reduces fatigue during protracted viewing.

The preceding information has be slanted toward the selection of a new instrument, however a used microscope can be an economical alternative if: (1) it fully matches the basic configuration requirements as determined above, (2) it is equipped with or can be equipped with the appropriate accessories and, (3) it is in good mechanical and optical condition. In this latter regard, the instrument should be thoroughly evaluated by a professional service organization. An instrument without anti-reflection coated optics, see Chapter 1, Section III, should not be considered for serious microscopical work.

III. THE MICROSCOPE'S ENVIRONMENT

The environmental requirements for the light microscope should not be underestimated. These requirements are of two types, those which deal with the physical environment necessary for obtaining maximum performance from the instrument and and those which affect the comfort of the microscopist. The latter requirements involve ergonomics or the science of human engineering.

The room space required for a single microscope will vary to some extent with the complexity of the instrument. For a simple biological microscope the floor space should not be less than 2 M

TABLE 10-1
MICROSCOPE CONFIGURATION AND ACCESSORIES FOR VARIOUS TECHNICAL DISCIPLINES

ITEM	BIOLOGIC RESEARCH	CERAM-OGRAPHY	MEDICINE AND LIFE SCIENCES
STAND			
Upright	3	2	3
Inverted	2	3	0
W/Bertrand Lens	0	2	0
W/35 MM Camera	3	1	3
W/Large Format Camera	1	3	0
W/Video Camera (a)	2	1	0
ILLUMINATION			
Transmitted Light	3	2	3
Incident Light	0	3	0
Brightfield	3	3	3
Darkfield	1	1	2
Phase Contrast	1	0	1
DIC (b)	2	2	2
Polarized Light (b)	1	3	2
Modulation Contrast	1	2	1
Incident Fluorescence (c)	1	1	1
Trans. Fluorescence	1	0	1
Electronic Enhancement(a)	1	1	1
OBJECTIVE LENS			
Achromat	2	1	2
Planachromat	2	1	3
Fluorite/Apochromat(c)	1	3	2
Strain Free (b)	1	1	0
Long Working Distance	1	1	0
STAGE			
Standard Stage	3	0	3
Large Stage	0	0	0
Revolving Stage	1	3	0
Motorized Stage	1	0	0
Computerized Stage	1	0	0
MEASUREMENT			
Eyepiece Reticle	2	2	3
Filar Eyepiece	1	2	0
Computer Line Width (a)	1	1	0
Computer Image Analysis(a)	1	1	0

LEGEND: 1 = May be desirable under certain conditions and/or not important.
2 = Secondary importance and/or desirable.
3 = Necessary and/or highly desirable.
a = Video camera required for electronic enhancement, computer line width and computer image analysis.

TABLE 10-1
MICROSCOPE CONFIGURATION AND ACCESSORIES FOR VARIOUS TECHNICAL DISCIPLINES (Continued)

METALLURGY	OPTICAL CRYSTALLOGRAPHY AND PETROGRAPHY	SEMICONDUCTOR MANUFACTURING MASK	WAFER	ASSEMBLY
0	3	3	3	3
3	2	0	0	0
0	3	0	0	0
1	1	0	0	1
3	1	3	3	3
2	0	3	3	1
0	3	3	0	0
3	2	0	3	0
3	3	3	3	3
2	0	0	2	3
0	0	0	0	0
2	0	0	3	1
2	3	0	0	0
2	0	0	2	0
0	1	0	1	1
0	0	0	0	0
2	0	2	2	2
1	1	0	0	1
1	1	3	3	2
2	0	0	0	1
3	3	0	3	1
1	1	0	0	3
2	0	0	0	3
1	0	3	3	1
3	3	0	0	0
1	0	2	2	0
2	0	2	2	0
3	3	3	0	1
2	0	2	2	0
2	0	3	3	0
2	0	0	0	0

b = Strain free objective lenses required for polarized light and most DIC disciplines.

c = Newer fluorite objective lenses desirable for incident light fluorescence microscopy because of their higher UV transmission.

by 2 M or approximately 6 ft by 6 ft. The walls, ceiling, and floor should be of a nondusting material. For example, concrete walls should painted with a semi-gloss enamel, preferably an off-white color, and concrete floors should be covered with a vinyl flooring. If a "drop ceiling" is used, the tiles also should be of a nondusting variety.

The environment for the instrumentation is listed in Table 10-2. The mean temperature adopted for the laboratory is often dependent on the climate or even the custom or culture in the laboratory's location but good temperature regulation is important. Humidity that is below the recommended values give rise to problems of static discharge while values above the recommended range may permit the growth of molds and mildew. Also the microscopical laboratory should be free from corrosive gases, particularly hydrofluoric acid which will etch the glass of the lenses.

If central air heating and/or cooling is required and especially if outside air is used, some form of filtration should be used to preclude particulate material from the laboratory atmosphere. This filtration becomes very extensive and culminates in a HEPA type absolute filter if clean room conditions must prevail. In the latter case the entire room may receive filtered air or the microscope may be placed in a laminar flow hood with the filtered air flowing over the table and its contents only.

It is also important to include a small vent to the outside with an exhaust blower for either a xenon lamp, if it is one of the ozone producing variety, or for a carbon arc source. In this way, any toxic or noxious gases produced by the lamp will not cause health problems.

The bench for the microscope must be sturdy and may be either of wooden or metal construction. The ideal bench top is 25 to 30 mm (1 to 1-1/4 in.) thick wood or plywood with a bonded composition surface (such as Formica™). The laboratory furniture with a natural serpentine rock or equivalent top is also excellent.

Vibration is a serious problem for critical microscopy and especially for photomicrography. The sources of vibration are many and include nearby rail and street traffic, heavy machinery, and foot traffic in the building. The obviously desirable situation is to place the microscopy laboratory in a location free from vibration, this can seldom be realized, although a ground floor is usually the best for a given site.

For that laboratory site plagued with vibration, techniques and materials have been developed to minimize vibration at the instrument. One of the simplest approaches is an instrument support with vibration isolators built into its base; these set on the bench and support a single microscope. These supports increase the height of the instrument about 75 mm (3 in).

TABLE 10-2
MICROSCOPY LAB ENVIRONMENT

Temperature Range	19 - 24°C (66 - 75°F) (Note 1)
Temperature Tolerance	± 3°C (5°F)
Humidity	35 - 55% R.H. (Note 2, 3)

Note 1: For critical work, refractive index oil should be selected for use at the ambient temperature.
Note 2: If static sensitive materials or potentially explosive materials are used, the R.H. should be in the range of 50 - 60%.
Note 3: If the instrument is to be used in the tropics, it should be treated to prevent fungus attack.

For more severe vibration, benches are available which incorporate either passive or active pneumatic vibration isolation systems. With these systems it is possible to design a bench to eliminate, or, at least, markedly reduce almost any type of vibration.

IV. ERGONOMICS

While the above considerations are important to a well functioning optical microscopy laboratory they are not sufficient. The missing ingredient is ergonomics, that is, the design of a work area that considers the needs and comfort of the microscopist. The height of the bench must be correct if the microscope is to be used for extended periods of time. If the microscope is not ergonomically designed it may be necessary to provide elevated arm rests. The chair must be comfortable and fully adjustable for each microscopist. Furthermore the chair should be designed to provide proper back and leg support.

Other ergonomic considerations include the temperature, humidity, and illumination of the laboratory. The temperature and humidity values listed in Table 10-2 are quite suitable for the microscopist.

Lighting is an important consideration in an optical microscopy laboratory, if it is either too dark or too bright it can result in eye strain for the microscopist. Typical illumination levels in the fully illuminated room should be about 125 foot-candles at the table surface. If detailed work is to be performed with the unaided eye, the illumination should be about 200 foot-candles. Incandescent illumination is ideal since these lamps respond smoothly to frequent on-off cycles without producing the flicker that the common fluorescent lamps exhibit upon starting. Instant start fluorescent lamps are an acceptable second choice.

If photomacrography and photomicrography are to be performed that require the viewing of a focusing screen, it is important that the room lighting be controllable from the microscope work station. An excellent arrangement involves the use of low voltage switches at each work station as well as at the room entrance(s) that control a relay which in turn switches the lighting. This is shown schematically in Figure 10-1.

The above considerations should provide the basis for an ideal laboratory environment for the operation of one instrument. By appropriate scaling of the floor space the laboratory may accommodate a number of instruments, however if photomacrography and photomicrography are to be performed it is desirable to have a separate room for each instrument to avoid lighting conflicts.

V. TESTING OF A MICROSCOPE

Modern microscopes have been carefully designed, often with the aid of computers to achieve the highest performance commensurate with the specific objective lens type. For this reason objective lens testing is little used today. However, if an objective lens has been subjected to misuse or abuse, it may be wise to perform some testing to verify that it is satisfactory for further use. These tests should also provide a basis for optical evaluation of used instruments.

The various objective lens properties that can readily be evaluated by the microscopist are listed in Table 10-3 along with the various procedures for their evaluation.

The star test was developed for telescope objective lens evaluation and the test basically consisted of studying the diffraction rings on either side of correct focus for a star image. Coles(1926) published a procedure for applying the star test to the objective lenses for transmitted light microscopes. He utilized the very fine pinholes in a silvered microscope slide as "stars". A very suitable star test plate for transmitted light objective lenses can be fabricated from a clean glass microscope slide, one side of which has been coated with an opaque layer of aluminum in a vacuum evaporator. The coating should be lightly abraded with lens tissue until very small holes are visible in the layer. Attach a cover glass of 0.017 mm thickness for European instruments or 0.18mm for American or Japanese instruments using Canada Balsam. Coles also describes a procedure for making the star test plate with a chemically deposited silver film. Incident light objective lenses can be tested on a transmit light instrument; in this case, use no coverglass on the star test plate.

The star test procedure is described in Table 10-4. Zeiss had, at one time, offered an Abbe Test Plate, which was an adaptation of the star test plate.

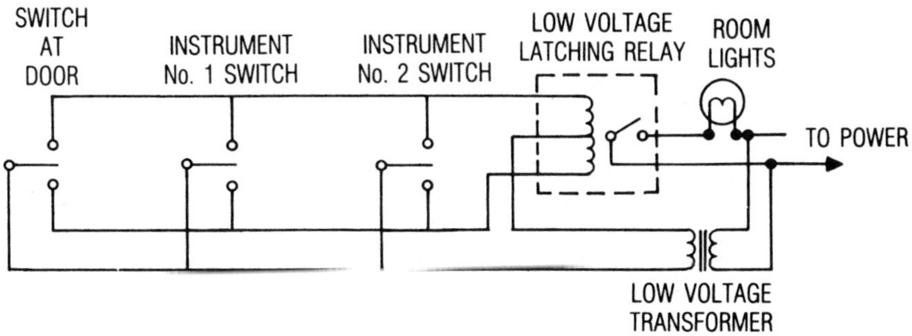

Figure 10-1. Suggested switching arrangement for the room lighting of a photomicrographical laboratory with two microscopes. As many instrument switches as needed may be added for a larger laboratory.

TABLE 10-3
METHODS FOR TESTING OBJECTIVE LENSES

OBJECTIVE LENS PROPERTIES	TESTING METHODS TRANSMITTED LIGHT	INCIDENT LIGHT
Spherical aberration	Star test	Star test*
Chromatic aberration	Star test	Grating
Distortion	Stage micrometer	Stage micrometer
Astigmatism	Star test	Star test*
Centering	Star test	Star test*
Flatness of field	Stage micrometer	Stage micrometer
Resolution	Diatom slide or Aperometer	Grating or Aperometer
Contrast	Well-stained blood smear	Grating

*Must be performed with transmitted light illumination using a star test plate without cover glass.

The stage micrometer is a useful test sample to determine both distortion and flatness of field for the objective lenses of either transmitted or incident light microscopes; the procedures for these tests are given in Table 10-5.

The Olympus microscope company has offered a test plate for the testing of incident light microscopes that consists of a 600 line per millimeter aluminized glass grating. Using this test plate it is possible to evaluate incident light objective lenses for a number of objective lens properties as indicated in Table 10-6. 600 line per

TABLE 10-4
STAR TEST FOR MICROSCOPE OBJECTIVE LENSES

1. Carefully set-up a transmitted light microscope per Chapter 2, Section IV C.
2. Place the star test slide on the stage, select a small circular hole in the silvering, and center this hole in field of view.
3. Observe the diffraction rings which appear on either side of correct focus and which disappear at correct focus. Typical diffraction rings are shown in Figure 10-2. The more nearly the diffraction rings are alike in intensity and spacing, above and below focus, the better the objective lens is corrected for spherical aberration. Excessive inequality in the spacings of the rings suggests zonal spherical aberration; this should be evaluated by a professional service person.
4. If the rings below focus are more pronounced than those above focus, the objective lens has some undercorrected spherical aberration. If the microscope is equipped with a draw tube, extend the draw tube slightly and repeat steps 3 and 4 until the diffraction rings are equally sharp above and below focus. If this effect is pronounced with a fixed tube microscope, the suitability of the objective lens should be evaluated by a professional service person.
5. If the rings are more pronounced above focus than below focus, the objective lens has some overcorrected spherical aberration. If the microscope is equipped with a draw tube, shorten the draw tube slightly and repeat steps 3 and 4 until the diffraction rings are equally sharp above and below focus. If this effect is pronounced with a fixed tube microscope the suitability of the objective lens should be evaluated by a professional service person.
6. If the rings are elliptical below focus in one direction and elliptical above focus but at 90° to the first direction, the objective lens has astigmatism.
7. If the rings are circular but not concentric this suggests poor centering of one or more lenses in the objective lens.
8. Select a somewhat larger hole or tear in the silvering of the star plate and examine the marginal fringes above and below exact focus. An achromatic objective lens will show reddish-purple margin below focus and yellowish-green margin above focus. A fluorite or semiapochromatic objective lens will exhibit a narrow red margin below focus and a narrow green margin above focus. An apochromatic objective lens should have no color in the marginal fringes. Note: the appropriate compensating eyepieces should be used for the fluorite and apochromatic objective lenses.

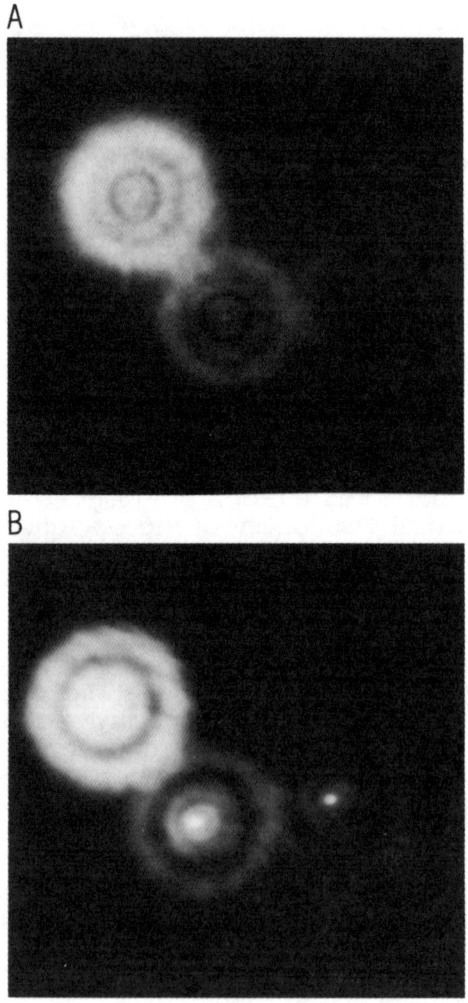

Figure 10-2. Star test image. The uniform intensity and spacing of these diffraction rings for both above (A) and below (B) focus conditions are characteristic of a lens well corrected for spherical aberration.

TABLE 10-5
TESTS FOR FLATNESS OF FIELD AND DISTORTION USING THE STAGE MICROMETER

1. Carefully set-up the transmitted light microscope per Chapter 2, Section IV C or the incident light microscope per Chapter 3, Section IV C.
2. Center the origin of a stage micrometer in the field of view and focus critically on the center of the field of view. The sharpness of the stage micrometer scale as it approaches the edge of the field of view is a measure of the flatness of field, i.e., the farther toward the edge of the field of view that the scale may be read without refocusing, the flatter the field of the objective lens. NOTE: the setting of the aperture diaphragm will affect this measurement. The depth of field is increased by reducing the aperture diaphragm setting thus yielding an apparent improvement in the flatness of field; this is at the expense of resolution. This effect is shown in Figure 10-3.
3. Focus on a stage micrometer line near the edge of the field of view, as shown in Figure 10-4. The line must be reasonably long. If it is straight, the objective lens has little or no distortion. If the ends of the line curve away from the center of the field of view, the objective lens has pin cushion distortion, if they curve toward the center of the field of view the objective lens has barrel distortion. NOTE: these results can be affected by the eyepiece type, therefore be sure that the eyepiece is correct for the objective lens under test.

---

millimeter (1.7 micron spacing) and 1200 line per millimeter (0.8 micron spacing) gratings, suitable for this test, are available from Edmund Scientific (101 E. Gloucester Pike, Barrington, NJ 08007)

The resolution of a given objective lens can be determined by direct or indirect measurement. An example of the direct test is given in Table 10-6 where the ability of an incident light objective lens to resolve the rulings of a grating is a measure of its resolution. The frustule or siliceous shell of certain diatoms, as shown in Figure 10-5, have very fine periodic structures and thus make excellent direct test subjects for transmitted light objective lenses. A prepared slide of recent marine diatoms, is available from Ward's Natural Science Establishment, Inc. (5100 W. Henrietta Rd. Rochester, NY 14692). A search of such a slide for the finest structure resolvable, followed by a measurement of the spacings in that structure, gives a value for the resolution of the objective lens used.

The resolution of an objective lens is mathematically related to its numerical aperture, see Chapter 2, Section III C., therefore if we know the N.A. of an objective lens we can determine its resolution. Normally the N.A. of the objective engraved on its body

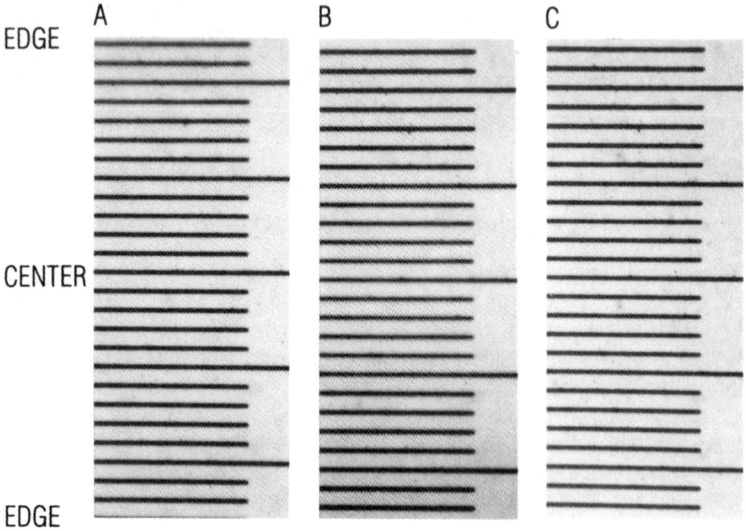

Figure 10-3. Test for flatness of field using a stage micrometer. (A) show the image obtained with a standard achromatic objective lens focused for the center of the field of view; note that the edge is out of focus. (B) with same objective lens as (A) but focused at the edge of the field of view. (C) is taken with a plano objective lens.

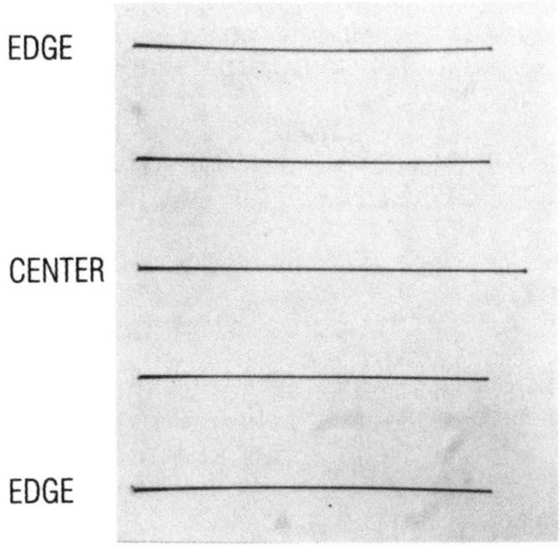

Figure 10-4. Examination of the lines of a stage micrometer (or other object with straight line structure) at the edge of the field of view gives an indication of distortion. The slight outward curvature at the ends of the lines on either edge of the field of view in this photomicrograph is the result of pincushion distortion.

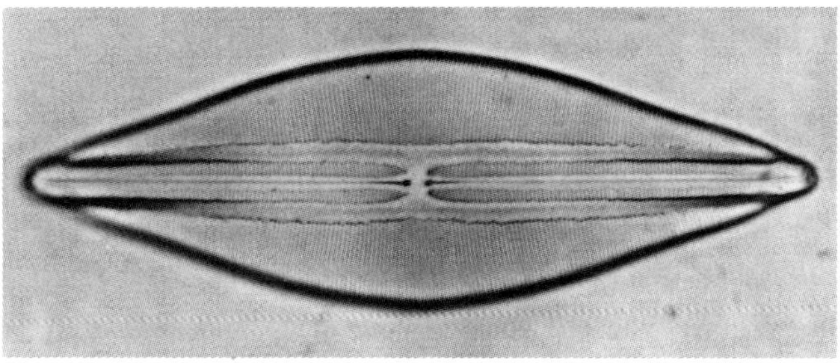

Figure 10-5. The striation spacings of various diatoms may be used to confirm the resolution capabilities of an objective lens. This diatom, Navicula Lyra, has striation spacings of 1.7 m and is nicely resolved with a 25X, 0.50 N.A. objective lens. Sample courtesy of H. Adler and photomicrograph courtesy of Alan DeHaas. 300X

is sufficiently accurate for most purposes, however if the N.A. of a certain objective lens is in question it is measurable with an apertometer. Thus the apertometer is a means of indirectly measuring the resolution of an objective lens.

TABLE 10-6
TESTING THE OBJECTIVE LENS OF AN INCIDENT LIGHT MICROSCOPE USING A DIFFRACTION GRATING*

1. Set-up the incident light microscope as described in Chapter 3, Section IV C.
2. Focus carefully on the center of the grating surface.
3. The lines should be black on a white background. If faint color is visible along the edges of the lines it is indicative of chromatic aberration.
4. The lines should be in focus from end to end for a plano objective. A lack of flatness of field is indicated when the lines in the periphery are out of focus.
5. The lines should be straight, if not this indicates distortion. If the ends of the lines are curved toward the center of the field, barrel distortion is present. If the ends of the lines are curved away from the center of the field, pin cushion distortion is present.
6. The lines should be clearly separated with edges clearly visible, this indicates that the lines are properly resolved.

*NOTE: The spacings on these gratings are not sufficiently small to adequately test the resolution of the higher N.A. objective lenses.

Apertometers are no longer available commercially, however both Spencer and Welford (1961) and Hewlett (1965) have described their construction and use.

Strain must be minimal in the glass elements of an objective lens used for polarized light applications such as petrography and Nomarski interference microscopy. If complete extinction is not obtained, as in the case of petrography, or if anomalous interference effects are observed which remain in a certain area of the DIC image it is likely that the objective lens is strained. Strain effects are commonly observed if nonpolarizing objective lenses are used or if a polarizing objective has be subjected to severe impacts, such as dropping it.

To evaluate an objective lens for strain, remove an objective lens from a transmitted light, polarizing microscope and critically adjust the polarizer and analyzer for maximum extinction. Then install the subject objective lens and note any increase in the transmitted light. If no increase in transmitted light is observed, the lens is acceptable for polarizing applications. If some light is observed, slowly loosen the objective lens in its threaded mount while observing the image. If no maximum in extinction is observed in the first complete rotation of the objective lens, it is defective and must not be used for polarizing observations. However, if a specific position shows a normal extinction, that objective may be used if fixed in that position; normally a shim is used to seat the lens permanently in the optimum position.

A common defect in objective lenses is decementing. Decementing occurs in achromatic lenses from use of incorrect cleaning solvents or perhaps as a result of age. It will be recalled that achromatic lenses are made of two types of glass that are joined with an organic cement. If the cement bond is degraded in some way, air gaps are introduced that produce undesirable reflections and loss of contrast. The decementing can be readily seen in an objective lens in a bright field arrangement using the Bertrand Lens or an eyepiece telescope. A typical decemented lens is shown in Figure 10-6. If such a defect is observed, the objective lens must be replaced.

VI. OPERATOR CARE AND MAINTENANCE

This section addresses those maintenance procedures that can and should be performed by the microscopist to keep the microscope in good operating condition. This includes such items as correct handling of the instrument, replacement of lamps and alignment of the illuminator. Cleaning of the instrument is discussed in considerable detail and includes the removal of unavoidable dust and dirt as well as the occasional residues of samples, mounting media, etc.

A. The Microscope Stand

The microscope should be handled with great care. It is best carried in an upright position with an arm under the base and steadied by grasping the limb of the instrument with the other hand. Remember that the eyepieces are held in position by gravity, thus they will likely fall out if the instrument is tilted. Avoid any sudden mechanical shocks to the instrument.

Avoid forcing any adjustment of the microscope, this may be indicative of misadjustment, a defective mechanism, or, at the very least, solidified lubricant. It is unwise to attempt lubrication of a microscope because mixing of various lubricants can produce uncertain and sometimes undesired results. For this reason, it is normally recommended that an instrument with a stiff or frozen adjustment be professionally serviced. If the instrument is in a facility that is in a remote area, some hints for operator lubrication can be obtained from Bennett(1986).

Forcing of the focusing mechanism is a common problem that results from the improper focusing of an incorrectly positioned

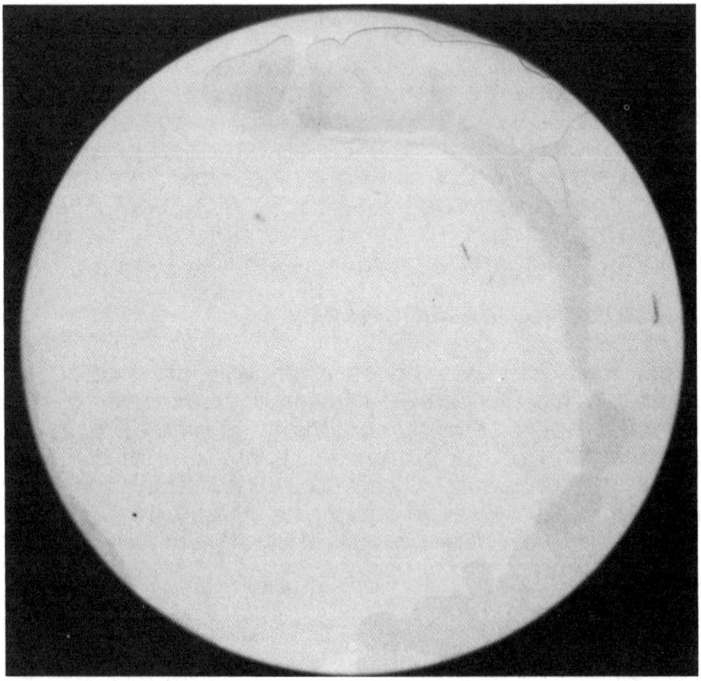

Figure 10-6. A decemented objective lens as seen with an eyepiece telescope or Bertrand lens. This condition will introduce some scattered light into the image and will gradually become worse with age. This condition can also be observed by removing the objective lens and examining it with a stereo microscope.

stage. In this case, the stage height will allow the objective lens to impact the sample during focusing, possibly causing damage to the objective lens, the sample and the fine focus mechanism. It should be emphasized that this type of damage can be avoided by using the procedure described in Chapter 2, Section IV C. However, the first step to avoiding this damage is to adjust the stage height, if possible; this is given in Table 10-7.

The need for cleaning of the microscope stand can be greatly reduced by covering the instrument when not in use. The manufacturer supplied plastic cover is recommended if it is not sticky or brittle from loss of plasticizer. A good cover for any instrument is a thin, tight weave, Nylon cloth. If an eyepiece or objective lens is to be removed from the instrument, for even a short period of time, the opening should be covered with one of the shipping caps.

TABLE 10-7
ADJUSTMENT OF STAGE HEIGHT

1. Loosen the clamp that locks the stage to its dovetail and lower the stage.
2. Place the shortest working distance objective lens into viewing position.
3. Move the stage to its highest position with the coarse focusing knob. Adjust the fine focus to to the center of its travel.
4. Place a typical sample on the stage and cover it with a piece of lens tissue.
5. Gently raise the stage on the dovetail until the filter paper is lightly compressed between the sample surface and the objective lens. Tighten the clamp to fix the stage in that location.

When cleaning of the microscope stand is required, use a clean, lintless cloth lightly moistened with water containing a small amount of mild detergent. Quickly follow the cleaning by wiping with a dry lintless cloth. For minor cleaning, some microscopists recommend wiping the instrument with a soft chamois.

Any residue of mounting medium or immersion oil on the stand or stage should be removed immediately after examinations are completed, using a cloth lightly moistened with xylene (Certified ACS reagent grade xylene is suitable). Following this solvent cleaning, the xylene should be removed as quickly as possible using a clean, dry cloth. It is wise to follow the solvent removal with the above detergent cleaning.

B. Care Of Objective Lenses and Eyepieces

Since objective lenses and eyepieces are removable from the microscope it is wise to consider their care as separate from the instrument. Eyepieces should be handled with reasonable care and should be stored in a desiccator if not on the instrument. The objective lenses are delicate optical systems, some with lenses the size of a pinhead, therefore they should be handled with great care. The strain-free objective lenses are particularly sensitive and should be protected from any violent contact or mechanical shock. If an objective lens is not mounted on a microscope it should be stored in its shipping container; if it is mounted on a quick change clutch, that assembly should be stored in a desiccator. In the various manipulations of objective lenses and eyepieces, avoid touching the lens surfaces with the fingers.

A good practice to follow in changing objective lenses is to remove or install the lens with one hand while cupping the other hand under the lens to catch it, should it fall.

Oil should be cleaned from the immersion objective lens (and transmitted light condenser lens) immediately after use. To accomplish this, two cotton tipped swabs are selected and the tip of each is covered with one half of a 4"x 6" sheet of lens tissue. Two suitable lens tissues are: Pelco Lens Paper (supplied by Ted Pella, Inc., P.O. Box 2318, Redding, CA 96099) and FISHER brand Lens Paper (available from Fisher Scientific, 711 Forbes Ave., Pittsburgh, PA 15219). Lightly moistened the tip of one swab with xylene. Since the exterior lens element of the oil immersion objective is quite small, the swab is gently rotated between the thumb and forefinger while it contacts the lens; follow this by cleaning the oil from the support around the lens. After the cleaning, immediately dry the objective lens with the second swab. If the condenser lens has been "oiled" to the sample slide for optimum N.A., it should be cleaned in the same manner. Examine the surface of the lens in reflected light for any evidence of smearing; if the surface is not completely clean repeat the process. Very oily lenses may require additional cleaning with a second set of swabs.

In addition to the special case of immersion oil removal, all externally exposed lenses require general cleaning from time to time. The eye lenses of eyepieces are natural collection sites for atmospheric dust, eyelashes, dandruff and mascara, these can be most easily observed by noting that they move as an eyepiece is rotated in its tube during the viewing of a sample image. Dust and dirt on the objective lens generally manifests itself only as a degradation of the image quality; it is possible to observe this dust or dirt with the Bertrand lens or the eyepiece telescope.

Cleaning of the external lens surfaces should be accomplished using the following procedures. Before any physical contact is made with the lens surfaces, any loose dust or debris should be blown off using a clean rubber ear syringe. An aerosol canister containing compressed gas can also be used for dusting. However, the aerosol container must be held upright when dusting to prevent the spraying of liquid onto the lens; this may cause spotting or possibly thermal shock damage of the lens. NOTE: This is the only operator-performed cleaning treatment recommended for glass surfaces which are not normally on the exterior of the microscope. This limitation is imperative because the lens coating used in the interior of the microscope are generally softer and more prone to scratching than the exterior lens coatings (see Chapter 1, Section III). If interior lens surfaces or mirrors are not adequately cleaned with this procedure, the cleaning should be referred to professional service personnel. Under no circumstances should either the objective lens or the eyepiece be disassembled to clean the inner surfaces of the individual lens elements.

Further cleansing of exterior lens surfaces may be accomplished using modified swabs and a water solution of a mild detergent. Two suitable products for the cleaning of lenses are: Kodak Lens Cleaner (available from Eastman Kodak Suppliers) and Cleaner for Plastic-Glass-Metal, Catalog # MS-260 (manufactured by Miller Stephenson Chemical Co., Inc., Danbury, CT 06810). Two swabs are selected and the tip of each covered with one half of a 4"x6" sheet of lens tissue; this latter process must be accomplished without touching either the swab tip or the tissue immediately covering the tip with the fingers. Prepare a foam of the Kodak Lens Cleaner by vigorously shaking the bottle and then squirt a small amount of the resulting foam on one covered swab tip. If the Miller-Stephenson cleaner is used, a light spray of the one swab tip is sufficient for this purpose. Proceed to clean the lens with the moistened swab by placing the moistened tip at the center of the lens and working with light pressure toward the outside of the lens in a spiral motion. Immediately repeat this process using the dry swab. For very small objective lenses, the swab may be gently rotated between the thumb and forefinger while it contacts the lens; the drying step is accomplished in the same manner. Examine the surface of the lens in reflected light for any evidence of smearing; if the surface is not completely clean repeat the process. When clean, a coated lens will have a uniform bluish (or rarely, a yellowish) color of the anti-reflection coating.

Very oily lenses may require additional cleaning with a second set of swabs for a complete cleansing. If this procedure fails to clean the lens, the cleaning should be referred to professional service personnel. DO NOT use alcohol or strong solvents for cleaning lenses as these may attack the cements used to hold the lens elements together. For the same reason an eyepiece or objective should never be submerged or soaked in any solvent.

The objective lenses also must be protected from corrosive vapors, especially in chemical microscopy where chemical reactions are carried out on the microscope stage and in metallography where samples which have been treated with very strong etchants are in close proximity. A number of approaches to protect the objective lens have been suggested; these are listed in Table 10-8.

C. Installation Of Lamps

The quality and uniformity of the image is very dependent on the alignment of the optical system; this includes in a very significant way, the lamp. Therefore lamp installation must be performed in a way that will also maintain correct alignment. Since each microscope type will have slight variations in the procedure for alignment of the lamp, the manufacturer's instructions will provide the best and most rapid procedure. If that information is not available, a generalized, installation procedure will be described based on typical lamp housings. The two more common lamps, the incandescent filament lamp, including the halogen lamp, and the xenon arc lamp, are described along with their typical failure modes. The installation procedure is given for the two lamp types followed by alignment procedures for the several illuminators described below.

The common, incandescent filament lamp and the halogen lamp may be used in any of the illuminators described below. These lamps may have a threaded, bayonet or bipin base as shown in Figure 10-7; these may operate at voltages from 6 to 110 volts. The filament of these lamps is either a circular or a somewhat elongate coil of tungsten wire. The broad surface presented by the elongate or flattened coil filament yields a significantly larger source of light for the condenser system, however it must be used in a specific orientation for this feature to be effective, as is shown in Figure 10-8.

These lamps usually fail when the tungsten filament evaporates to the point that a section overheats and melts, producing an open circuit. Another cause of an incandescent filament lamp failing to operate is weakened electrical contacts in the lamp socket; this may cause the solder termination of the lamp to melt and lose contact. This latter condition must be corrected before a new lamp is installed.

To replace an incandescent filament lamp or a quartz halogen lamp, first disconnect the power from the lamp, and allow the housing to cool to room temperature if the lamp has been operating. Remove the lamp support from the housing and then remove the old lamp. The glass envelope of the incandescent lamps is usually strong enough to withstand the stresses of removal and installation however, they should be treated with reasonable care. Prepare the new lamp by cleaning the glass envelope thoroughly with glass cleaner and allowing it to dry. Handle the cleaned lamp only with clean cotton or nylon gloves as fingerprints

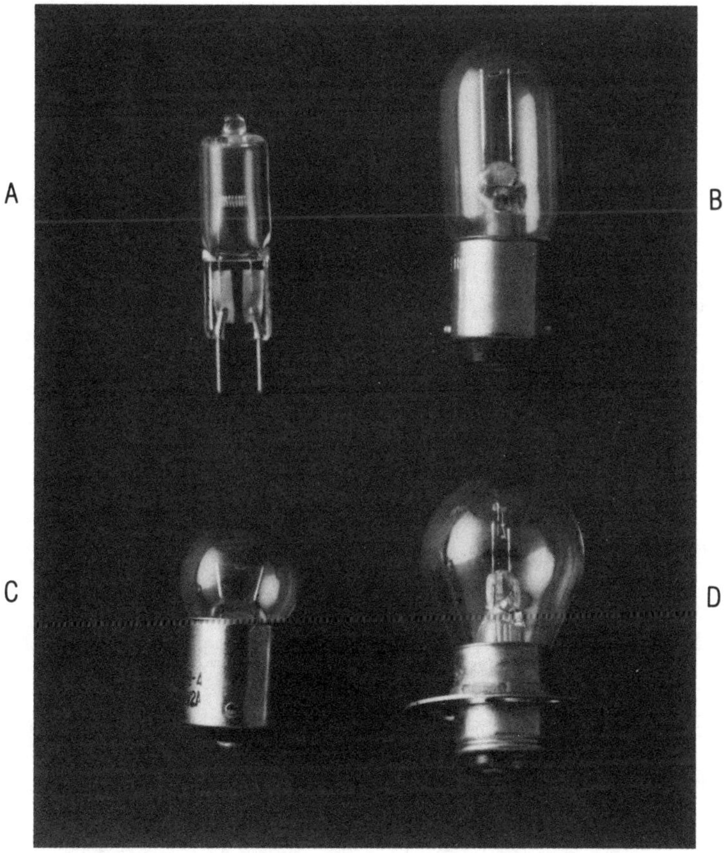

Figure 10-7. Typical incandescent illuminator lamps.

(A) A bipin quartz-halogen lamp. 12 V, 100 W.
(B) A tungsten filament lamp with two base contacts. 6 V, 15 W.
(C) A tungsten filament lamp with single base contact. 6 V, 20 W.
(D) A prefocused tungsten filament lamp. 6.5 V, 18 W.

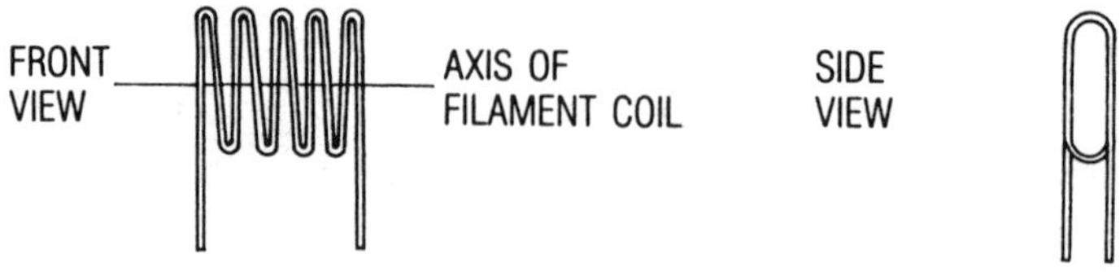

Figure 10-8. Two views of the filament of a typical, modern incandescent or halogen microscope lamp. Correct alignment of the lamp is with the axis of the filament coil at right angles to the optical axis of the illuminator.

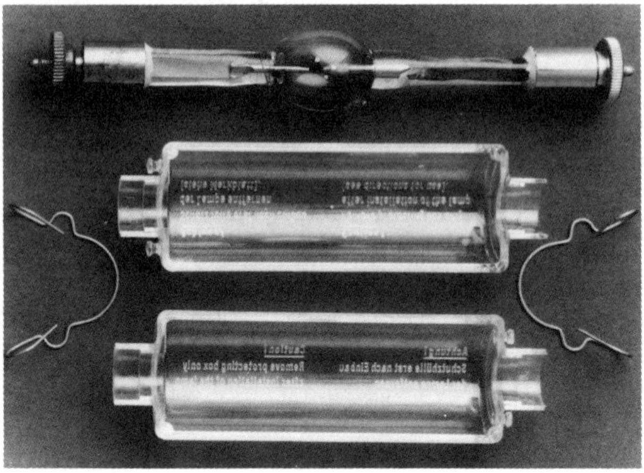

Figure 10-9. A 150 W, high pressure, xenon arc lamp shown with the two halves of its protective shipping cover and the cover locking clips. Since these lamps are filled with xenon under high pressure they must be handled only while enclosed in the cover. Courtesy of McBain And Associates.

become burned into the glass and cause uneven illumination. Install the lamp in its support and then return the support to the lamp housing. If the incandescent lamp is a "precentered" type, the housing may have no adjustments except the lamp condenser focus. In that case, adjust the condenser to give a uniformly bright field on a "reference" brightfield sample as described in Table 10-9.

TABLE 10-8
METHODS FOR PROTECTION OF OBJECTIVE LENSES FROM CORROSIVE VAPORS

REFERENCE	PROCEDURE
Chamot and Mason (1960)	Construct a gas-tight sleeve that is sealed with a #1 cover glass. The cover glass should be in contact with the objective lens. (1) (2) (3) (4)
Chamot and Mason(1960)	Affix a small diameter cover glass to the objective lens with a drop of immersion oil. (1) (2) (3) (4)
Feuerstein (1967)	Stretch clear, food grade polyethyllene film over the objective lens and secure it with a rubber band. (1) (2) (3) (4)
Hays (1967)	Neutralize acids with a base, dry and degas in a vacuum chamber. (4)
Richardson (1968)	Neutralize hydrofluoric acid from an etchant by immersing sample in a saturated solution of ammonium pentaborate for 1 hour followed by distilled water rinse and dry. (4)

(1) Suitable for chemical microscopy and other types of examinations which use corrosive media in the preparation, e.g., fungus mounts.
(2) May introduce some aberrations in the image.
(3) May not be suitable for objective lenses with short working distances.
(4) Suitable for metallographic examinations.

The xenon arc lamp, is used almost exclusively in the more efficient housing described below. This lamp, as shown in Figure 10-9, consists of two tungsten electrodes sealed in a heavy quartz glass envelope that contains xenon gas under high pressure. It operates at about 20 VDC from a specially designed power supply. These lamps fail gradually due to an increase in the gap between the tungsten electrodes. When the lamp fails to start or has a varying light output, the lamp should be replaced.

Changing of the xenon arc lamp must be accomplished with extreme care. The glass envelope of the lamp must not be subjected to any sharp impacts or bending stresses because these, combined with the high pressure of the contained gas can cause an explosion of the lamp. It should be emphasized that the used lamps are even more likely to explode than new lamps because the glass has been embrittled by the high operating temperature. The replacement of these lamps is usually referred to professional service personnel, however if the operator undertakes the following procedure carefully, the lamp can be replaced safely. NOTE: Do not attempt to remove the failed lamp if the original, heavy plastic shipping cover is unavailable; in this case refer the replacement to professional service personnel!

To change the xenon arc lamp, remove the power from the lamp and allow it to cool to room temperature, if it has been operating. At this point wear an approved safety head shield, a laboratory coat, fully buttoned and thin leather gloves. Open the lamp housing and carefully place the heavy plastic shipping cover around the lamp and fasten it securely. Record the position of the polarity markings on the lamp. Then slightly loosen the lower electrode clamp that holds the lamp rigidly to the housing. If the lamp is threaded into the lower electrode assembly, turn the lower metal portion of the lamp to loosen it 1/8 of a turn. Carefully release the top electrode clamp and remove it from the lamp. Fully loosen the lower electrode clamp, remove the lamp and place it in a packing box for disposal. Remove the new lamp from its packing box, place the correct electrode of the lamp into the lower electrode clamp and partially tighten the clamp. Replace the upper electrode clamp, making sure that the flexible lead places no torsional strain on the lamp. Tighten the lower electrode clamp and carefully remove the protective plastic shipping cover (save this for eventual replacement purposes!). Close the lamp housing. To assure a steady arc, the new xenon burner should be "formed" or operated continuously for 8 hours the first time it is started.

A high pressure mercury vapor lamp may be changed in the same manner, however, this should not be undertaken unless an approved mercury spill kit is available in the event of breakage.

Figure 10-10 illustrates a moderately complex illuminator that provides for alignment of the lamp with the optic axis of the microscope. A complex and more efficient illuminator, as shown in Figure 2-7, also includes a centered and focused concave mirror behind the lamp for increased illumination efficiency.

Some microscopes are equipped with a ground glass plate for alignment of the condenser system. This plate may be a permanent fixture as with some of the Unitron instruments or a temporary accessory that is placed on the dust-glass where the

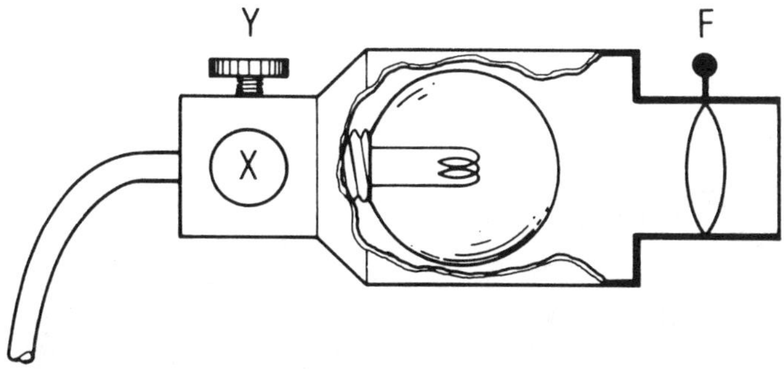

Figure 10-10. A moderately complex illuminator. F is the focus lever for the condenser lens. X and Y are alignment knobs to center the lamp for even illumination of the sample.

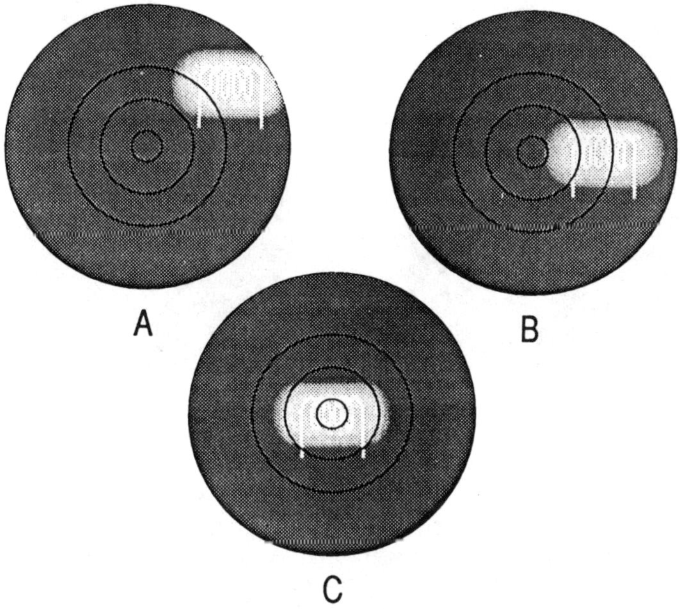

Figure 10-11. Illustration of the alignment process used for a moderately complex illuminator without a collector mirror. The image of the filament is displaced both vertically and horizontally in (A). (B) shows the filament image after vertical adjustment, and (C) shows the filament correctly positioned after further horizontal adjustment. If the microscope does not have an alignment plate, a reasonable, but small image of the filament can be seen on a sheet of white paper placed at the aperture diaphragm. Redrawn courtesy of Wild Leitz USA, Inc.

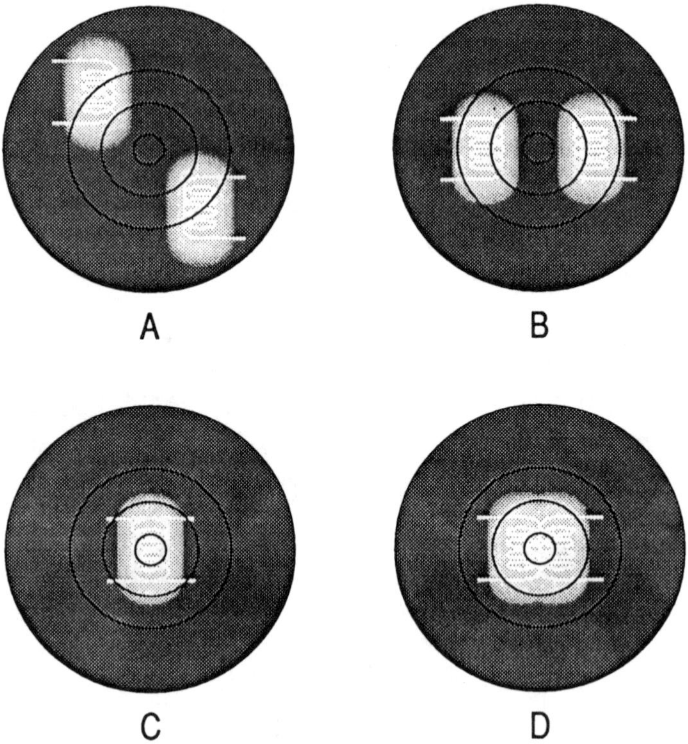

Figure 10-12. Illustration of the alignment process used for a complex illuminator that incorporates a collector mirror, such as shown in Figure 2-7. The images are of the filament and its image as reflected from the collector mirror. (A) shows the filament and its mirror image displaced both vertically and horizontally. (B) shows the images after vertical adjustment, and (C) shows the filament and image correctly positioned after further horizontal adjustment. The images may be move slightly apart, horizontally, in some instruments to provide a larger filament image; this is shown in (D). If the microscope does not have an alignment plate, a reasonable, but small image of the filament can be seen on a sheet of white paper placed at the aperture diaphragm. Redrawn courtesy of Wild Leitz USA, Inc.

illumination exits the base of an upright microscope. If such a device is supplied, by all means use it in conjunction with the manufacturer's alignment procedure.

Since most instruments have no provision for verifying correct alignment of the illuminator, a procedure is described that is based on the use of a reference sample, an eyepiece telescope, see Chapter 7, Section VII, and a neutral density filter, see Chapter, Section VI B. This procedure as applied to the moderately complex and the complex illuminators described above is given in Table 10-9.

TABLE 10-9
PROCEDURE FOR ALIGNMENT OF THE LAMP IN THE ILLUMINATOR

1. Turn on the illuminator lamp and focus on any dust particles in the image plane of a reference brightfield sample* using a 10X objective lens. Open both the field and aperture diaphragms completely.
2. Replace the eyepiece with an eyepiece telescope. Focus the eyepiece telescope on the rear focal plane of the objective lens (the edge of the field of view is sharply defined with a black periphery).
3. At this point, the image(s) of the lamp filament or arc (herein called source) should be visible in the eyepiece telescope. Adjust the brightness using either the intensity adjustment on the power supply or neutral density filters, see Chapter 7, Section VI A, in the light path. Adjust the focus of the source with the condenser focus knob. Note whether one or two images of the source is/are present.
4. If the source is not clearly visible, the lamp is either very much out of alignment or there is a ground glass diffuser, see Chapter 7, Section VI H, in the light path. Check the seating or position of the lamp in the lamp housing and/or remove the diffuser if present. If the diffuser cannot be removed proceed to 5a.
5. If an incandescent filament lamp is being aligned, the lamp should be rotated until the major surface of the filament coil is normal to the illumination axis, see Figure 10-8.
6. If only one image of the source is visible, there is no collector mirror in the illuminator. Adjust the lamp position adjustments to center the source image in the field of view, see Figure 10-11. If two images are present this adjustment should bring both images to the center of the field, as shown in Figure 10-12a. Some manufacturers increase the area of the source by adjusting the two images to be side-by-side as shown in Figure 10-12d.

7. If the two images cannot be adjusted as shown in Figures 10-11 and 10-12, the collector mirror requires adjustment. In this case, temporarily place a paper card between the lamp and the mirror and center the one source image as described in step 6. Remove the card and center the second source image using the mirror adjustments.
8. Alignment of the lamp is complete, replace the eyepiece telescope with the eyepiece and remove any neutral density filters used in the alignment procedure.

5a. Replace the eyepiece telescope with the eyepiece and remove any neutral density filters used in the alignment procedure. Place a sheet of clean white paper on the stage and adjust the lamp until an evenly illuminated circle of light is seen on the paper. This completes the alignment procedure for this type of illuminator.

*This is a sample with no detail such as a clear part of a mounted slide for the transmitted light microscope or a clean first surface mirror for the incident light microscope. A first surface mirror for the alignment of an upright incident light microscope must have front and back surfaces that are parallel, otherwise the mirror must be leveled in the same fashion as a sample, see Chapter 3, Section II A.

VII. PROFESSIONAL MAINTENANCE

Even though the modern optical microscope is a high precision instrument it can reasonably be expected to have a very long life only if correctly used and if properly maintained by a qualified, professional service organization. The key to the successful maintenance of a microscope is dependent on the quality of the professional service. No microscope should be entrusted to a service organization that does not have a staff of experienced service personnel and a reputation in the service community that enlists the microscope manufacturer's support.

A microscope must be serviced as both an precision optical system and a mechanical system. The various elements that make up a comprehensive service schedule is given in Table 10-10.

Following each service an instrument test and image evaluation should be performed to the satisfaction of the microscopist. Other nonroutine procedures which can be performed by the professional service organization are installation of xenon lamps and parfocalization of objective lenses, see Chapter 2, Section III C.

TABLE 10-10
REQUIREMENTS FOR COMPREHENSIVE PROFESSIONAL SERVICE OF MICROSCOPES

REQUIREMENTS	FREQUENCY YEARLY	FREQUENCY 6 MO. INTERVAL
1. DISASSEMBLY:	Each instrument shall be dismantled for access to all critical assemblies.	
2. LUBRICATION:	Moving parts requiring lubrication shall be degreased and relubricated with non-drying lubricants.	Lubricate only exposed items most subject to wear.
3. REFITTING:	Precision tolerances shall be critically adjusted to assure the proper performance of the instrument.	
4. CLEANING:	The instrument stand shall be thoroughly cleaned.	The instrument stand shall be thoroughly cleaned.
5. OPTICS:	Optical elements shall be cleaned using factory approved procedures aligned to factory tolerances. Recommendations for replacement of defective or substandard optics shall be made.	Exposed optical surfaces shall be cleaned using factory approved procedures. Recommendations for replacement of defective or substandard optics shall be made.

REFERENCES

Bennett, H. S., Proc. Royal Micros. Soc., 21/6, 352, (1986).

Chamot, E. M. and C. W. Mason, Handbook of Chemical Microscopy, 2nd Ed., John Wiley and Sons, Inc., N.Y. (1960) p. 43.

Coles, A. C., Watson's Microscope Record, 7, 16, January (1926).

Feuerstein, F. S., private communication, (1967).

Hays, C., private communication, (1967).

Hewlett, P. S., J. Sci. Instrum., 42, 127, (1965).

Richardson, J. H., Metallography, 1, 149, (1968).

Spencer, C. J. D. and W. T. Welford, J. of Sci. Instrum., 38, 328, (1961).

Appendix I—Immersion Fluids

FLUID	REFRACT. INDEX(1)	ABBE #	VISC-OSITY	COMMENTS
Water	1.333	55.6	Low	For special objective lens
Glycerin	1.473	67.7	Med	For special objective lens
Monobromo-naphthalene	1.659	5.9	Low	For special objective lens
Cedarwood Oil	1.515	48.4	Med	Will harden on exposure
Cargille, type A	1.5150	41.6	Lo	Little entrained air
Cargille, type B	1.5150	42.9	Hi	Used for medical work
Cargille, type NVH	1.5150	46.0	VHi	Used to bridge wide gaps between slide and lenses
Cargille, type VH	1.5150	47.7	VVHi	Used for inverted metallographs
Cargille, type LF	1.5150	42.2	low	Very low fluorescence (2)
Cargille, type DF	1.5150	44.3	low	VV low fluorescence
Cargille, type FF	1.4790	57.3	low	Fluorescence free
Cargille, Type 37	1.5150	42.4	Hi	Standardized at 37°C.

(1) Refractive index listed is ND at 23°C unless otherwise specified. Data from R.P. Cargille Laboratories, 55 Commerce Road Cedar Grove, NJ 07009 and "International Critical Tables of Physics, Chemistry and Technology" Vol I, pp 12-13, McGraw Hill Book Co., Inc. N.Y., London (1926).

(2) No Longer available - Contains 34% PCB (Polychlorinated Biphenyls) (Note: This is equivalent to type A produced prior to 1972)

Appendix II—Long Working Distance Lenses (>10X)

MAGNIFICATION	N.A.	WORKING DISTANCE MM	TUBE LENGTH MM	MISC.	MANUFACTURER AND PART DESIGNATION
10X	0.18	14	Inf.	1	Leitz 569050
H20X	0.40	8.3	Inf.	1	Leitz 569001
L20X	0.40	6.6	Inf.	2	Leitz 569003
UM20 (13X)	0.33	15	170	1	Leitz 559004
ELWD 20X	0.40	10.5	210	3,4	Nikon 78775
ELWD 20X	0.40	8.5	210	3,5	Nikon 78861
LWD 20X	0.40		Inf.	3,4	Olympus 1-LM545
ULWD 20X	0.40		Inf.	3,4	Olympus 1-LM546
25X	0.31	12.9		3	B&L 31-13-03
H32X	0.60	5.7	Inf.	1	Leitz 569002
L32X	0.60	4.0	Inf.	3	Leitz 569004
UM32X (19X)	0.30	16	170	1	Leitz 559005
40X	0.45	5.8	170	1	Unitron FF40X
ELWD 40X	0.50	10.1	210	3,4	Nikon 78776
ELWD 40X	0.50	9.8	210	3,5	Nikon 78862
40X	0.52	5.2	250	1,6	Reichert NOIMS
50X	0.45	12.5		3	B&L 31-13-04
50X	0.60	6.0		3	B&L 31-13-05
LWD 50X	0.60		Inf.	3,4	Olympus 1-LM555
ULWD 50X	0.60		Inf.	3,4	Olympus 1-LM556

1. Corrected for use with a quartz window 1.8 mm thick.
2. Corrected for use with or without a cover glass 0.17 mm thick.
3. Corrected for use without a cover glass.
4. Planachromat.
5. Brightfield/darkfield planachromat.

Index

Aberration
 chromatic - 490
 chromatic, axial - 61
 chromatic, lateral - 61, 490
 spherical - 57, 69, 102, 490
 spherical, chromatic difference of - 57
Absorption, light, selective - 184
Accessories, polarizing microscope - 367
Accessory slot - 188, 367
Amplitude contrast - 204
Analyzer, polarizing - 37, 172, 185, 188, 367
 cap, polarizing - 183
Angle of inclination - 134
Angular measurement - 230
Annulus, phase - 153
Anoptral contrast - 205
Antireflection coating - 8
Aperometer - 490, 496
Aperture diaphragm - 51, 98, 126
Aperture, field limiting - 83
Areal analysis - 266
Artifact, optical - 136
Astigmatism - 1, 63, 490
Auxiliary telescope - 375
Axial chromatic aberration (see Aberration, chromatic, axial)

Background for sample - 473
Base, microscope - 31, 40, 331
Beam splitter - 166, 172
Becke line - 278, 417
Bench magnifier - 5
Bertrand lens (see Lens, Bertrand)
Binocular body - 33, 386
Body tube - 33, 40
Borescope - 304
Brightness contrast - 134
Brightness, image - 8, 75
Bunsen-Roscoe reciprocity law - 402, 404
Burker grid - 233

Camera
 flexible bellows - 397
 integral lens - 392
 large format - 397
 photographic - 385
 press - 462
 single lens reflex - 394
 video - 327, 385
 w/o integral lens - 393
Camera Lucida - 325, 385
Cathetometer - 222
CCTV (see closed circuit television)
Centering telescope (see Eyepiece telescope)
Characteristic curve, photographic film - 400
Chromatic aberration - 490
Closed circuit television - 327
Coating, antireflection - 8
"Color" - 194
Color
 absorption - 130
 interference - 130, 189
Coma - 65
Compensator
 Babinet - 369

Compensator (continued)
  Berek tilting - 372
  Brace-Koehler rotatory - 372
  Ehringhaus tilting - 372
  plate, first order red - 369
  plate, quarter wave - 369
  plate, quartz wedge - 369
  Soliel-Babinet - 381
Condenser
  final - 51
  pancratic - 153
Condenser system - 38, 48, 110, 119, 140
  Abbe - 136
Cones, eye - 1
Conjugate foci - 96
Conoscopy - 271, 284
Contrast - 2, 8, 159, 347
  brightness - 147
  color - 130, 147
  evaluation of - 490
  photographic film - 400
Cornea - 1
Counting chamber - 232
Cover glass - 59, 122
  quartz - 210
Cover slip (see Cover glass)
Curvature of field - 65, 414
Cytofluorometry - 207

Darkfield stop - 140
Depth of field - 73
  in photomacrography - 449
Depth of focus - 74, 417
Diaphragm
  aperture - 51, 98, 126
  field - 51, 98, 126
  grain isolating - 188
  iris - 38
Dichroism - 184
Differential interference contrast - 166
  independent reference - 173
  self reference - 169
Diffusing screen - 368
Discussion bridge - 327
Dispersion, optical - 285
Dispersion staining - 285
Distortion - 65, 226, 490
  barrel - 65
  pincushion - 65
Diverting system
  glass prism - 121
  Smith - 119
Doping - 206
Drawtube, graduated - 33, 59

Electronic image enhancement - 215
Emission, fluorescent - 206
Ergonomics - 221, 483, 487
Excitation, fluorescent - 206
Exit pupil - 83
Exposure calibration - 420
Exposure control - 394
Exposure meter - 420
Exposure monitor - 390
Exposure time, determination of - 419
External illuminator alignment - 93
Eye - 1
Eye relief - 86
Eyepiece - 8, 33, 83
  care of - 320, 500
  comparison - 374
  compensating - 86, 90
  cross-wire - 222, 380
  filar - 369
  fixed reticle - 90, 222
  goniometer - 230
  grain size - 375
  Huygens - 87
  image shearing - 374
  integrating - 267
  interchangeable reticle - 222
  noncompensating - 63, 86
  pinhole - 380
  pointer - 375
  protractor - 230, 381
  Ramsden - 90
  screw micrometer - 261, 369
  setting - 380
  simple - 86

Eyepiece (continued)
turret - 374
widefield - 93
Wright - 381
Eyepiece bridge, comparison - 307
Eyepiece telescope - 155, 161, 165, 203, 375
Eyepiece tube - 386

Fading, fluorescent - 209
Faraday effect - 185
Farsightedness - 1
Fiber optics - 304
Field curvature - 414
Field diaphragm - 51, 98, 126
Field finder slide - 353
Field limiting aperture - 83
Field number - 86
Field of view - 86
Filter
absorption - 359
barrier - 207, 365, 469
color - 359
color balancing - 360
color temperature conversion - 360
compensating - 363
excitation - 207, 365
fluorescence microscopy - 365
heat absorbing - 363
infrared - 363
interference - 359
neutral density - 358
optical - 358
polarizing - 365, 473
solar - 303
suppression (see Filter, barrier)
ultraviolet absorbing - 363
Filters for photomacrography - 450
Flatness of field - 414, 490
Fluochrome - 206
Flutter, image - 245
Focus
criteria for - 417
fine, calibration of - 235
Focus drift - 390
Focus mechanism - 33
Focusing
automatic - 325
photomicrographic - 415, 439
powered - 324
Fovea - 2

Gamma, photographic film - 402
Geometric thickness - 233
Glove box - 304
Grain boundaries, etched - 134
Grain size, sample - 258
Graininess, photographic film - 400
Ground glass screen - 368

Hand lens (see Lens, hand)
Hemacytometer - 232
Hot cell, radioactive - 304
Howard cell and disk - 232
Hyperopia - 1

Illumination - 38, 42
coaxial - 134
coherent - 70, 149
correction for color film - 424
critical (See illumination, Nelson)
crossed polarized - 189
darkfield - 140
differential interference contrast - 166
epi - 110
homogenization of - 333
Koehler - 93, 96, 125, 273, 285, 450, 462
mixed - 210
modified Koehler - 273, 285
Nelson - 93, 273, 287
Nomarski DIC - 169, 174
oblique - 134
parallel polarized - 189
plane polarized - 189

Illumination (continued)
  vertical - 110
Illuminator
  alignment of - 95
  calcite crystal - 121
  external - 331, 461, 469
  fiber optic - 333, 469
  glass slip - 119
Illuminator lamp alignment - 95
Image
  computerized analysis of - 215, 268
  conoscopic - 83
  erect - 10
  real - 8, 83
  secondary - 172
  virtual - 3, 8, 56, 83
Image brightness - 75
Image drift - 419
Immersion fluid - 51, 54, 59, 473
Immersion oil, nonfluorescing - 207, 210
Interference fringe - 243
  order of - 245
  spacing of - 249
Interferometer
  two beam - 166
  Michelson/Linnik - 249
  Mirau - 249
  multiple beam - 237, 254
  Nomarski - 245
  polarizing - 245
Interferometry, two beam - 237, 243
  two beam, independent referencing - 245
  two beam, self referencing - 245
Intermediate image distance - 87
Intermediate optics - 80
Interpupillary acuity - 35
Interpupillary spacing - 35
Iris, eye - 1

Kerr effect - 196

Lambda plate - 179
Lamp - 42
  alignment procedure - 509
  carbon arc - 46, 363
  flash - 45, 333, 363, 469
  fluorescent - 469
  incandescent - 42, 502
  installation of - 502
  laser - 45
  mercury vapor - 46, 209, 506
  ring - 469
  sodium vapor - 47, 245
  tungsten halogen - 42
  ultraviolet - 469
  xenon - 43, 123, 363, 486, 505
  zirconium arc - 47
Lamp housing - 47
Lateral chromatic aberration (see Aberration, chromatic, lateral)
Length measurement
  axial - 233
  precision and accuracy - 228
  transverse - 222
Lens
  Barlow - 302
  Bertrand - 37, 81, 159, 161, 165, 188, 203, 253, 271, 331
  condenser (see Condenser system)
  corrector - 80
  eye - 83
  field - 83
  Fresnel - 123
  hand - 5
  objective - 8, 33, 37, 51, 77, 122
  objective, dry - 56
  objective, immersion - 54, 67
  objective, plano - 78
  positive - 3
  relay - 298, 304, 394
  tube - 81
Lenses, Telan - 37, 80
Light, polarized - 181
Lighting, laboratory - 487
Limb - 31, 40
Line width measurement - 228

Lineal analysis - 266
Linen tester - 5
Liquid, refractive index - 184

Macrocamera, flexible
bellows - 439
Magnification - 2, 56
empty - 73
for integral lens camera - 393
linear - 56
minimum useful - 85
photomacrographic - 445
photomicrographic - 413
total - 56, 85
upper limit of - 72
Magnifier, bench - 5
Maintenance, professional - 510
Malassez grid - 233
Material
amorphous - 270
anisotropic - 184, 270
birefringent - 184
ferroelectric - 185
isotropic - 184, 270
radioactive - 304
toxic, microscopy of - 304
Metallograph - 40, 113, 386
Microfluorimetry - 293
Microhardness tester - 348, 355
Microinjector - 342
Micromanipulator - 342
Micrometer, stage - 222, 327, 353
Microphotograph - 385
Microphotometry - 291
Microprojection - 325
Microscope
brightfield, incident light - 110
brightfield, incident light, set-up - 125
brightfield transmitted light - 31
brightfield, transmitted light, set-up - 96
centering - 320
CMO stereo - 18
comparison - 307
confocal scanning laser - 26
darkfield - 140
deep field - 22
differential interference contrast - 166
double view - 311
environment for - 483
fiber optic examining - 298
fluorescence - 206
forensic comparison - 307
Greenough stereo - 11
hole inspection - 311
light scanning - 21
light sectioning - 237
long distance - 298
mask alignment - 310
modulation contrast, - 198
oblique illumination - 135
petrographic - 185, 270, 273
phase contrast - 147
polarized light contrast - 181, 189
portable - 296
projection - 317, 325
remote - 303
selection of - 482
shop - 317
simple - 3
stereo - 10, 439, 477
tandem scanning reflected light - 26
testing of - 488
toolmaker's - 222, 317
travcling - 222
wafer alignment - 310
amplitude contrast - 206
Milk smear slide - 233
Mirror - 38, 47, 249
dichroic - 210, 363, 365
first surface - 38
reference - 254
second surface - 38
Modulation slit - 200
Modulator plate - 200
Myopia - 1

National Institute of Standards and Technology (NIST) - 228, 233
Nearsightedness - 1
Nested cup light trap - 397
Net micrometer - 232
Neubauer grid - 232
Nosepiece, quick change clutch - 35, 185
Nosepiece, revolving - 37, 185
Numerical aperture - 51, 140, 493
  determination of - 493
  effective - 52

Objective lens
  achromatic - 77, 415
  adjustment length of - 67
  adjustment of correction collar - 103
  apochromatic - 77
  browning of - 123, 307
  care of - 320, 500
  centration of - 282
  corrosion protection for - 501
  decementation of - 497
  evaluation of lens centering - 490
  fluorite - 61, 77
  homogeneous immersion - 67
  infinity corrected - 69
  long working distance - 122
  oil immersion - 140
  parfocal - 67
  plano - 65, 78, 123, 390
  radiation resistant - 123, 307
  resolution of - 490, 493
  semi-apochromatic - 77
  strain free - 80, 185
  strain in - 496
  water immersion - 140
Objective system
  catadioptric - 302
  catoptric - 302
  dioptric - 302, (see also Lens, objective)

Ocular (see Eyepiece)
Opacity - 134
Opal screen - 368
Optical activity - 184
Optical comparator - 222
Optical crystallography - 270
Orthoscopy - 271

Pancratic condenser - 153
Parfocality - 37, 67
Particle Atlas, Mc Crone - 290
Particle number counting - 230
Particle spacings - 268
Petrography - 271, 274
Phase annulus - 153
Phase contrast
  negative - 152
  positive - 152
  variable - 153
Phase, light - 134
Phase plate - 152
Phase ring - 153
Phases, material - 134
Photographic enlarger - 438
Photographic media - 399
  black and white (B&W) - 399, 404, 409
  color - 403, 409
  color balance for - 403
  color negative - 403
  color reversal - 403
  instant process - 404
  orthochromatic - 403
  panchromatic - 403
  wet process - 399
Photomacrography - 438
  depth of field in - 449
  exposure determination in - 450
  framing and composition in - 445
  magnification in - 445
  stereoscopic - 476
Photometer - 207
Photometry - 291
Photomicrography - 385
  esthetics in - 410
  cine - 431

Photomicrography (continued)
contrast enhancement for - 426
framing and composition in - 410
slow motion - 431, 434
time-lapse - 431, 433
Photomicroscope - 386
Pinhole cap - 161, 165
Planimeter - 266
Plate phase - 152
Pleochroism - 184
Point counting - 267
Polarization
elliptical - 196
plane - 181
Polarizer - 172, 181, 183, 365
filter - 181
orientation, determination of - 276
plane - 181
prism - 181
reflection - 183
Press, sample leveling - 113
Prism, Wollaston - 169, 174, 245
Profile projector - 320

Quick-change clutch - 185

Ramsden disk - 83
Reading glass - 5
Reciprocity failure - 394, 404
Reflectivity - 134
Reflector
pellicle - 119
plane glass (see reflector, pellicle)
Reflex housing - 397
Refractive index - 134, 184
Resolution
axial - 74
lateral - 69
limit of - 70
photographic film - 400
visual - 2
Resolving power - 70
Reticle - 90, 222, 247
calibration of - 224
cross-wire - 222, 261, 338
micrometer - 91
Reticle cell - 222
Retina - 1
Revolving stage, centration of - 282
RMS thread - 52, 353
Rods, eye - 1

Roscoe-Bunsen reciprocity law - 402

Sample
orientation of - 134, 338
translation of - 334
Schneiderhöhn line - 418
Sector stop - 135
Sedgewick-Rafter cell - 232
Seeing - 3
Shutter, photographic - 394
Slide
field finder - 353
quartz - 210
Slide marker - 357
Slot, accessory - 37, 188
Smith diverting system - 119
Spherical aberration (see Aberration, spherical)
Stage - 37
cold - 347
computer controlled - 336
electroluminescence - 352
environmental - 351
extensometer - 351
height adjustment of - 499
hot - 343
integrating - 267, 338
leveling - 250
mechanical - 38, 222, 262, 34
microelectrophoresis - 350
micrometer - 222, 327, 53, 490
microrefractometer spindle - 339
motor driven - 336
revolving - 37
single axis - 339
spindle - 339
tilting - 477

Stage (continued)
translation - 334
universal - 339
vacuum - 352
Wentworth - 338
Stand - 31, 110, 324
care of - 320, 497
Culpeper - 40, 117
inverted - 38, 113
Le Chatalier - 38
upright - 31, 110
Standard distance of comfortable vision - 56
Star test - 490
Stereo pair photomacrographs - 476
Stereo projector - 481
Stereo viewer - 481
Stereoscope - 477
Stereoscopy - 476
Stop
adjustable, darkfield - 145
central - 140
Structure, linear and periodic - 136

Telan lenses (see Lenses, Telan)
Telescope - 302
Telescopic eyepiece (see Eyepiece telescope)
Test plate - 490
Thoma grid - 233
Three dimensionality - 136
Trinocular body - 35, 386
Tube
binocular - 386
length, mechanical - 33, 67
monocular - 33, 386
Turk grid - 233

Vibration - 390, 486
isolation from - 324
Video enhanced contrast - 215
Viewing and focusing tube for photomicrography - 394
Viewing Screen - 123
Virtual image (see Image, virtual)
Vision, stereoscopic - 2, 476
Volume fraction - 265

Whipple disk - 232
Working distance - 67, 122

Zoom system - 21, 37, 81